D0472131

Includes 2001 FDA Food Code Updates!

ServSafe

Coursebook

Second Edition

Please detach and keep this tab for your records.
To obtain class tracking and score information within **15 days** of your examination date, **record the class number on this answer sheet tab**. You will need this number to retrieve your score at **www.nraef.org/classes**.

Record your class number here.

National Restaurant Association
EDUCATIONAL FOUNDATION
175 West Jackson Boulevard, Suite 1500
Chicago, IL 60604-2814
www.nraef.org

National Restaurant Association
EDUCATIONAL FOUNDATION

DISCLAIMER

While the publisher and author have used their best efforts in preparing this book, they make no representations or warranties with respect to the accuracy or completeness of the contents of this book and specifically disclaim any implied warranties of merchantability or fitness for a particular purpose. No warranty may be created or extended by sales representatives or written sales materials. The advice and strategies contained herein may not be suitable for your situation. You should consult with a professional where appropriate. Neither the publisher nor author shall be liable for any loss of profit or any other commercial damages, including but not limited to special, incidental, consequential, or other damages.

Laws may vary greatly by city, county, or state. This book is not intended to provide legal advice or establish standards of reasonable behavior. Operators who develop food safety-related policies and procedures as part of their commitment to employee and customer safety are urged to use the advice and guidance of legal counsel.

ServSafe is a registered trademark of the National Restaurant Association Educational Foundation.

©2002 National Restaurant Association Educational Foundation

ISBN 1-58280-078-2 (Coursebook without Exam—CB)
ISBN 1-58280-071-5 (Coursebook with Exam—CBX)
ISBN 0-471-23711-6 (Wiley Coursebook without Exam—CB-W)
ISBN 0-471-22517-7 (Wiley Coursebook with Exam—CBX-W)
ISBN 1-58280-080-4 (Texas Cousebook with Exam—CBTX)

All rights reserved. No part of this document may be reproduced, stored in a retrieval system, or transmitted in any form or by any means, electronic, mechanical, photocopying, recording or otherwise, without the prior written consent of the publisher.

Printed in the U.S.A.

10 9 8 7 6

Table of Contents

INTRODUCTION

UNIT I   THE SANITATION CHALLENGE

UNIT II THE FLOW OF FOOD THROUGH THE OPERATION

UNIT III CLEAN AND SANITARY FACILITIES AND EQUIPMENT

UNIT IV SANITATION MANAGEMENT

APPENDIXES

The National Restaurant Association Educational Foundation's

International Food Safety Council®

A Strategic Initiative of the
National Restaurant Association
EDUCATIONAL FOUNDATION

In 1993, the National Restaurant Association Educational Foundation recognized the need for food safety awareness and created the International Food Safety Council® (IFSC). Its mission is to heighten the awareness of the importance of food safety education throughout the restaurant and foodservice industry.

The IFSC envisions a future in which foodborne illness no longer exists. Through its educational programs, publications, and awareness campaigns, the IFSC reaches foodservice instructors, restaurant operators, suppliers, manufacturers, distributors, as well as organizations such as healthcare facilities, colleges and universities, supermarkets, associations, and other stakeholders in the industry.

Active Founding Sponsors

American Egg Board
847.296.7043
www.aeb.org

Cattlemen's Beef Board and National Cattlemen's Beef Association
303.694.0305
www.beeffoodservice.com

ECOLAB, Inc.
800.352.5326
www.ecolab.com

FoodHandler Inc.
800.338.4433
www.foodhandler.com

Heinz North America
800.547.8924
www.heinz.com

San Jamar
800.248.9826
www.sanjamar.com

SYSCO Corporation
281.584.1390
www.sysco.com

Tyson Foods, Inc.
800.424.4253
www.tyson.com

Past Founding Sponsors
Campbell's

Active Campaign Sponsors

Bunzl Distribution, Inc.
314.997.5959
www.bunzldistribution.com

Cargill, Inc.
800.CARGILL
www.cargillfoods.com

Colgate-Palmolive Company
800.432.8226
www.colpalipd.com

Cooper Instrument Corporation/Atkins Temptec
800.835.5011
www.cooperinstrument.com

Farquharson Enterprises, Ltd.
856.863.8883

Jones Dairy Farm
800.635.6637
www.jonesdairyfarm.com

Monsanto Company
314.694.1000
www.monsanto.com

North American Association of Foodservice Equipment Manufacturers (NAFEM)
312.821.0201
www.nafem.org

Orkin Commercial
800.ORKIN.NOW
www.orkin.com

The Procter & Gamble Company
800.332.7787
www.pgbrands.com

Produce Marketing Association
302.738.7100
www.pma.com

UBF Foodsolutions North America
800.272.6296
www.ubffoodsolutions.com

U.S. Foodservice, Inc.
410.312.7100
www.usfoodservice.com

For more information on the National Restaurant Association Educational Foundation's International Food Safety Council, visit our Web site at www.nraef.org or contact us at:
National Restaurant Association
Educational Foundation
International Food Safety Council
175 West Jackson Boulevard, Suite 1500
Chicago, IL 60604-2814
312.715.1010

Acknowledgements

The development of the *ServSafe® Coursebook* and the refinements found in this second edition would not have been possible without the expertise of our many advisors and manuscript reviewers. The National Restaurant Association Educational Foundation is pleased to thank the following people for their time, effort, and dedication.

Marie-Luise Baehr
Sodexho, Inc.

Cheryl Barsness
Alaska Seafood Marketing Institute

Robert Bosselman, Ph.D., RD, FMP
Florida State University

Jeff Carletti
San Jamar/KatchAll

Elaine Cash
Daydots

Larry Clark
Damon's International

Harry D'Ercole
Enrico's Restaurant

Gary DuBois
Taco Bell Corporation

Deborah Fitzgerald
Cooper Instrument Corporation

Kristen Forrestal
Darden Restaurants, Inc.

Peter Good
Peter Good Seminars, Inc.

David Goronkin
Buffets, Inc.

Joe Grosdidier
ECOLAB, Inc.

Steven F. Grover, R.E.H.S.
National Restaurant Association

Margaret Hardin, Ph.D.
National Pork Producers Council

Harold Harlan, Ph.D.
National Pest Control Association

Jean Hayden
Ohio Department of Health

Alice M. Heinze, RD, MBA
American Egg Board

Michael P. Hiza, RD
Manchester Community Technical College

Jane Lindeman
National Cattlemen's Beef Association

Jennifer A. McDuffee
Orkin Commercial

Mark McFarlane
Briazz

Kathy Means
Produce Marketing Association

Bruce Moilan
Chester-Jensen Co., Inc.

Nevin B. Montgomery
National Frozen Food Association, Inc.

Mauro Mordini
Lipton Foodservice

Robert Munnis, CSFP
Atkins Temptec

Anne Munoz-Furlong
Food Allergy and Anaphylaxis Network

Susan Pitman, MA, RD
International Food Information Council

Alain Porte, MBA
GlaxoSmithKline

David J. Poulter
Buffets, Inc.

Mary Sandford
Burger King Corporation

Jill A. Snowdon
American Egg Board

J.R. (Jim) Starnes, FMP
Tyson Foods, Inc.

Lisa Wright
Jack in the Box Inc.

Ron Yudd
Points of Profit Leadership

Michael Zema, CCE, FMP
Elgin Community College

A Message From

THE NATIONAL RESTAURANT ASSOCIATION EDUCATIONAL FOUNDATION

Food safety is nonnegotiable. Serving safe food is not an option. It's our obligation as restaurant and foodservice professionals. Proper training is one of the best ways to create a culture of food safety within our establishments.

By opening this book, you have made a significant commitment to promoting food safety. We applaud you for that commitment.

The ServSafe® program has become the industry standard in food safety training. It is accepted in almost all United States jurisdictions requiring employee certification. The ServSafe program provides accurate, up-to-date information for all levels of employees on all aspects of handling food, from receiving and storing to preparing and serving. It gives you the opportunity to learn science-based information on how to run a safe establishment—information all your employees need in order to be a part of the food safety team. This new edition incorporates the latest information from the 2001 FDA Food Code in a realistic manner.

In many ways, this newly revised edition of the *ServSafe Coursebook* should be attributed to you. Changes found here are based on the feedback we received from readers of the first edition—trainers and trainees alike. It brings you the most current regulations, best practices, and science-based information. With enhanced readability and a more engaging look—including more color photos, charts, and visual aids—your best source for food safety training is now even better. Expanded lists of additional resources are another highlight, featuring dozens of useful Web sites. All in all, the result is an even more valuable training tool.

Your food safety education does not end once you are certified in the ServSafe program. You have the responsibility to take your new knowledge back to the unit and make your coworkers part of the food safety culture.

New, science-based information emerges regularly that significantly impacts food safety. It is our strong recommendation that managers and supervisors become recertified in the ServSafe Food Protection Manager Certification Program every five years.

Whether you are new to the restaurant industry or continuing your career, your participation in ServSafe training will make you more qualified to serve safe food and to spread that knowledge throughout the industry.

Thank you for making the commitment to food safety training and for becoming an active part of the food safety culture within the rapidly growing restaurant, foodservice, and hospitality industry.

Features of the ServSafe® Coursebook

We have designed the *ServSafe Coursebook* to enhance your ability to learn and retain comprehensive food safety knowledge. Here are key features you will find in each chapter of this book.

TEST YOUR FOOD SAFETY KNOWLEDGE

Each chapter begins with five True or False questions designed to test your prior knowledge of some of the concepts presented in the chapter. The answers to these questions appear in Appendix A. If you wish to explore the concepts behind the questions further, see the page reference after each question.

TABLE OF CONTENTS

Chapter content is organized under the major headings identified in this section.

LEARNING OBJECTIVES

Learning objectives identify what you should be able to do after completing each chapter. These objectives are linked to the tasks required to keep your establishment safe.

KEY TERMS

These terms are important for a thorough understanding of the chapter content. They are **highlighted** throughout the chapter, where either they are explicitly defined or their meanings are made clear within the paragraphs in which they appear. Each key term is also defined in the Glossary.

EXHIBITS

Exhibits are placed throughout each chapter to visually reinforce the key concepts presented in the text. They are referenced by the chapter number followed by a letter, and they include charts, photographs, illustrations, and tables.

Key Point

Throughout each chapter, icons appear in the margins of the page. These icons emphasize concepts presented in the text that are important to your understanding of food safety.

ICONS

Icons appear throughout each chapter in the margins of the page. They emphasize concepts presented in the text that are important to your understanding of food safety. While Key Point icons are the most common type, icons related to HACCP, personal hygiene, cross-contamination, and time-temperature abuse are also included.

A CASE IN POINT

These real-world scenarios give you the opportunity to apply some of the food safety concepts you have learned in each chapter.

TRAINING TIPS

These tips are designed to help you teach some of the food safety concepts in each chapter to others back at the establishment. They help you bring the food safety message one level deeper into your organization. Many of the tips are intended for short preshift or postshift training sessions.

DISCUSSION QUESTIONS

These questions are designed to make you think about some of the more important food safety concepts presented in the chapter.

MULTIPLE-CHOICE STUDY QUESTIONS

These questions are designed to test your knowledge of the food safety concepts presented in the chapter. If you have difficulty answering them, you should review the content again. Answers to these questions are provided in Appendix A.

ADDITIONAL RESOURCES

In this section you will find resources—books, articles, and Web sites—that will enable you to further explore the food safety concepts presented in each chapter.

ANSWERS

The answers to the *Test Your Food Safety Knowledge* questions and the *Multiple-Choice Study Questions* are found in Appendix A.

The answers to the *A Case in Point* activities and the *Discussion Questions* are found in the *ServSafe Instructor Guide*.

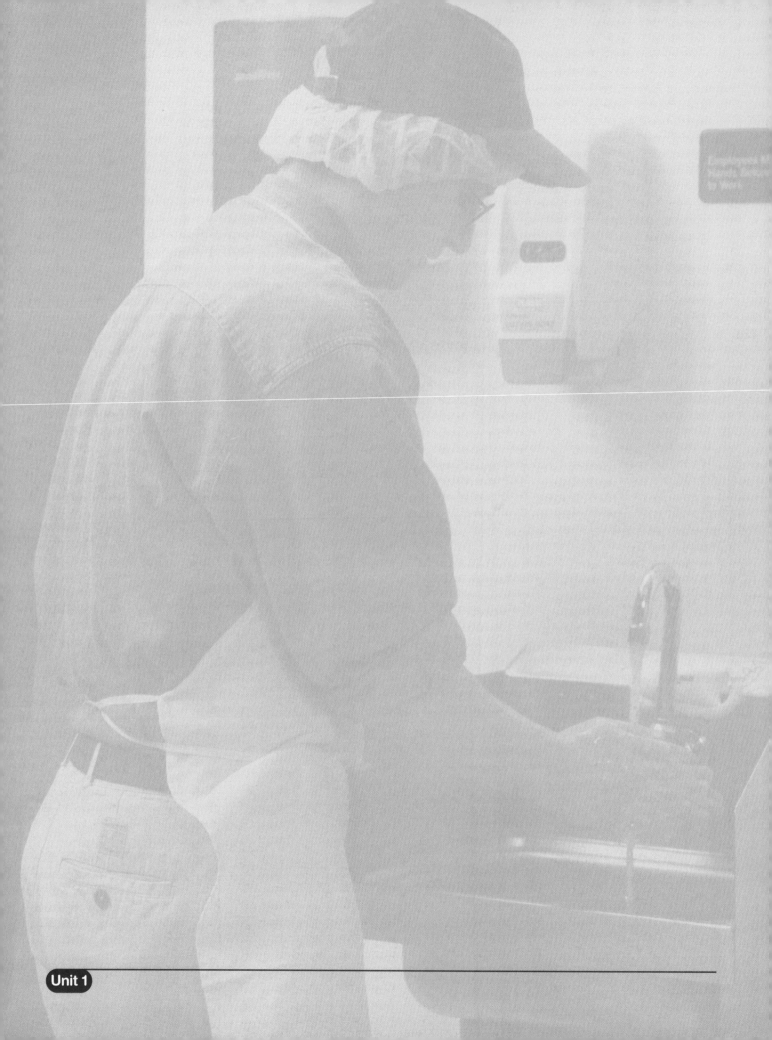

UNIT 1

THE SANITATION CHALLENGE

Chapter 1
Providing Safe Food

Knowledge

TEST YOUR FOOD SAFETY KNOWLEDGE

1. **True or False:** Improperly washed hands can cause a foodborne illness. *(See page 1-9.)*

2. **True or False:** Preschool-age children may be more likely than adults to become ill from contaminated food. *(See page 1-6.)*

3. **True or False:** Food has been time-temperature abused any time it has been allowed to remain too long at temperatures favorable to the growth of microorganisms. *(See page 1-8.)*

4. **True or False:** Cross-contamination occurs when hands that have just handled raw chicken touch a cooked chicken. *(See page 1-9.)*

5. **True or False:** Potentially hazardous food is generally dry, low in protein, and highly acidic. *(See page 1-6.)*

For answers, please turn to Appendix A.

Key Terms

Foodborne illness
Foodborne-illness outbreak
Warranty of sale
Reasonable care defense
Hazard Analysis Critical Control Point (HACCP)
Flow of food
FDA Food Code
Immune system
Potentially hazardous food
Heat-treated

Ready-to-eat food
Contamination
Biological, chemical, and physical hazards
Time-temperature abuse
Cross-contamination
Food-contact surface
Personal hygiene
Clean
Sanitary

Table of Contents

Learning Objectives

After completing this chapter, you should be able to

○ Explain the dangers of foodborne illness.

○ Identify high-risk populations for foodborne illness and explain why they are at risk.

○ Identify the characteristics of potentially hazardous food.

○ Identify three types of contamination associated with food.

○ Describe how foodborne illness occurs.

○ Identify the three key practices that can help ensure food safety.

When diners eat out, they expect safe food, clean surroundings, and well-groomed workers. Overall, the restaurant and foodservice industry does a good job of meeting these demands, but there is still room for improvement.

The risk of foodborne illness impacts the industry. Several factors account for this. They include:

○ The emergence of new foodborne pathogens (disease-causing microorganisms)

○ The importation of food from countries where food safety practices may not be well developed

○ Changes in the composition of food, which may leave fewer natural barriers to the growth of microorganisms

○ Increases in the purchase of take-out and home meal replacement (HMR) food

○ Changing demographics, with an increased number of individuals at high risk for contracting foodborne illness

○ Employee turnover rates that make it difficult to manage an effective food safety system

In the face of these challenges, all establishments must take the necessary steps to help ensure that the food they serve is safe. The first step is to develop a food safety system that includes effective and ongoing employee training.

THE DANGERS OF FOODBORNE ILLNESS

Foodborne illnesses are the greatest danger to food safety. A **foodborne illness** is a disease carried or transmitted to people by food. The Centers for Disease Control and Prevention (CDC) defines a **foodborne-illness outbreak** as an incident in which two or more people experience the same illness after eating the same food. A foodborne illness is confirmed when laboratory analysis shows that a specific food is the source of the illness.

Each year, millions of people are affected by foodborne illness, although the majority of cases are not reported and do not occur at restaurants and foodservice establishments. However, the cases that are reported and investigated help us understand some of the causes of illness, as well as what we, as restaurant and foodservice professionals, can do to control these causes in each of our establishments. The most commonly reported causes of foodborne illnesses are: failure to cool food properly, failure to cook and hold food at the proper temperature, and poor personal hygiene.

Key Point

A foodborne-illness outbreak is an incident in which two or more people experience the same illness after eating the same food.

Fortunately, every restaurant and foodservice establishment, no matter how large or small, can take steps to ensure the safety of the food it prepares and serves to its customers.

The Costs of Foodborne Illness

National Restaurant Association figures show that a foodborne-illness outbreak can cost an establishment thousands of dollars. It can even be the reason an establishment is forced to close.

If your establishment is implicated in a foodborne-illness outbreak, your costs may include increased insurance premiums, as well as lawyer and court fees. You may have to pay for testing food supplies and employees, and may spend time and money retraining employees and cleaning and sanitizing the establishment. Food supplies that may or may not be contaminated will have to be discarded. Other risks could include: lowered employee morale and absenteeism, embarrassment and bad publicity, loss of customers and sales, and loss of prestige and reputation. *(See Exhibit 1a.)*

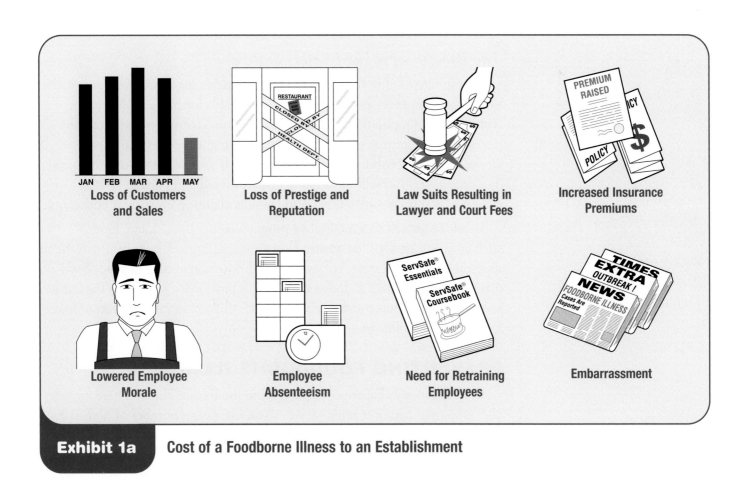

Exhibit 1a   Cost of a Foodborne Illness to an Establishment

Today, customers are very willing to sue to obtain compensation for injuries they feel they have suffered as a result of the food they were served. Under the federal Uniform Commercial Code, a plaintiff bringing about a lawsuit must prove all of the following:

○ The food was unfit to be served

○ The food caused the plaintiff harm

○ In serving the food, the establishment violated the **warranty of sale,** that is, the rules stating how the food must be handled

If the plaintiff wins the lawsuit, he or she can be awarded two types of damages. Compensatory damages can be awarded for lost work, lost wages, and medical bills. Punitive damages can be awarded in addition to normal compensation, punishing the defendant for wanton and willful neglect.

If you have a quality food safety program in place, however, you can use a **reasonable care defense** against a food-related lawsuit. A reasonable care defense requires proof your establishment did everything that could be reasonably expected to ensure that the food served was safe. Evidence of written standards, training practices, procedures such as a HACCP plan and its documentation, and positive inspection results are the keys to this defense.

The Benefits of a Food Safety System

Serving safe food is vital to your establishment's success. A well-designed, food safety program can help protect your establishment's employees, customers, and reputation. Repeat business from customers and increased job satisfaction among employees can lead to higher profits and better service. Your establishment may benefit directly from reduced or minimized insurance costs, and will most likely benefit by reducing health-code violations and becoming less open to lawsuits claiming injury and negligence.

An added benefit to a food safety program is that by handling food safely, you also preserve its quality. Safe foodhandling will help maintain the appearance, flavor, texture, consistency, and nutritional value of food. Food that is stored, prepared, and served properly is more likely to provide the quality your customers deserve and demand. Safe foodhandling can also lead to lower food costs due to less waste.

PREVENTING FOODBORNE ILLNESS

There are many challenges to preventing foodborne illness. These include high employee-turnover rates, service to an increasing number of high-risk customers, and the service of potentially hazardous food. Establishing a comprehensive food safety program, however, greatly reduces the likelihood of causing foodborne illness.

Training Employees in Food Safety

One of the challenges managers typically face is a high employee-turnover rate. Preparing and serving safe food in public establishments is a serious responsibility. All restaurant and foodservice employees need to be trained in the procedures that can protect the public and themselves from foodborne illnesses.

Food Safety Programs

An establishment should have an effective, proactive program based on preventing food safety hazards before they occur. A reactive program that corrects a problem after it has occurred is not an effective system. The **Hazard Analysis Critical Control Point (HACCP)** program, which is discussed throughout this text, is a proactive, comprehensive, science-based, food safety system that allows operators to continuously monitor their establishments and reduce the risk of foodborne illness.

The key to the HACCP system is the emphasis on how food flows through the operation. This **flow of food** is the path food takes from receiving and storage through preparation and cooking, holding, serving, cooling, and reheating. An establishment's HACCP plan identifies the points in the operation at which contamination or growth of microorganisms can occur. Control procedures can then be implemented based on the hazards identified at those points. HACCP will be covered in more detail in Chapter 9.

Both the National Restaurant Association and the Food and Drug Administration's (FDA) Food Code encourage an establishment to develop and use a HACCP-based, food safety system to prevent foodborne illness. The **FDA Food Code** is a science-based reference for retail restaurants and establishments on how to prevent foodborne illness. Local, state, and federal regulators often use the FDA Food Code as a model when developing or updating their own food safety regulations, and to ensure consistency with national regulatory policy.

HACCP Principle

The HACCP system is designed to prevent food safety hazards from occurring.

Populations at High Risk for Foodborne Illness

The demographics of our population show an increase in the percentage of people at high risk of contracting a foodborne illness, sometimes with serious consequences. *(See Exhibit 1b on the next page.)* They include:

○ Infants and preschool-age children

○ Pregnant women

○ Elderly people

○ People taking certain medications, such as antibiotics and immunosuppressants

○ People with weakened immune systems (who have recently had major surgery, are organ-transplant recipients, or who have pre-existing or chronic illnesses)

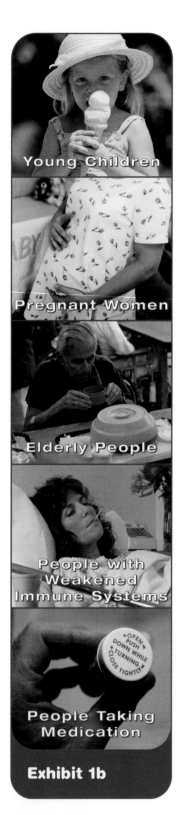

Young Children

Pregnant Women

Elderly People

People with Weakened Immune Systems

People Taking Medication

Exhibit 1b

People at High Risk for Foodborne Illness

Young children are more at risk for contracting foodborne illnesses because they have not yet built up adequate **immune systems** (the body's defense system against illness) to deal with some diseases. This knowledge is especially important for quick-service restaurants since, according to a recent survey, 58 percent of families with children eat at such restaurants at least once a week.

Elderly people are more at risk because their immune systems and resistance may have weakened with age. In addition, as people age, their senses of smell and taste are diminished, so they may be less likely to detect "off" odors or tastes, which indicate that food may be spoiled.

Because these populations are susceptible to foodborne illness, it is of particular concern when they consume potentially hazardous food (or ingredients) that are raw or have not been cooked to the minimum internal temperatures identified in Chapter 7. In all cases, these high-risk consumers should be advised of any potentially hazardous foods (or ingredients) that are raw or not fully cooked. Tell them to consult a physician before regularly consuming this type of food.

Food Most Likely to Become Unsafe

Although any food can become contaminated, most foodborne illnesses are transmitted through food in which microorganisms are able to grow rapidly. Such food is classified as **potentially hazardous food.** This food typically has a history of being involved in foodborne-illness outbreaks, has a natural potential for contamination due to production and processing methods, is often moist and high in protein, and has a neutral or slightly acidic pH.

The FDA Food Code identifies potentially hazardous food *(see Exhibit 1c)* as any food that consists in whole, or in part, of the following:

○ Milk and milk products

○ Shell eggs (except those treated to eliminate *Salmonella* spp.)

○ Meats, poultry, and fish

○ Shellfish and edible crustacea (such as shrimp, lobster, crab)

○ Baked or boiled potatoes

○ Tofu or other soy-protein foods

○ Garlic-in-oil mixtures

○ Plant foods—including fruit and vegetables—that have been **heat-treated** (cooked, partially cooked, or warmed)

○ Raw seeds and sprouts

○ Sliced melons

○ Synthetic ingredients (such as textured soy protein in meat alternatives)

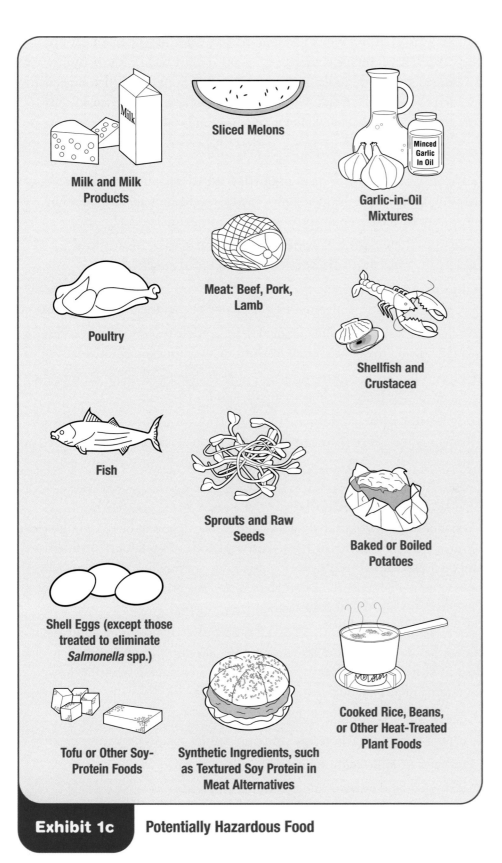

Milk and Milk Products

Sliced Melons

Garlic-in-Oil Mixtures

Minced Garlic In Oil

Poultry

Meat: Beef, Pork, Lamb

Shellfish and Crustacea

Fish

Sprouts and Raw Seeds

Baked or Boiled Potatoes

Shell Eggs (except those treated to eliminate *Salmonella* spp.)

Tofu or Other Soy-Protein Foods

Synthetic Ingredients, such as Textured Soy Protein in Meat Alternatives

Cooked Rice, Beans, or Other Heat-Treated Plant Foods

Key Point

Potentially hazardous food is often moist, high in protein, and has a neutral or slightly acidic pH.

Exhibit 1c   **Potentially Hazardous Food**

Care must be taken when handling **ready-to-eat food,** which may also be considered unsafe because it is intended to be eaten without further washing or cooking. Proper cooking reduces the number of microorganisms on food to safe levels. Food that has been properly cooked is considered ready-to-eat, as is washed whole or cut fruits and vegetables.

Potential Hazards to Food Safety

Unsafe food usually results from **contamination,** which is the presence of harmful substances in the food. Some food safety hazards are introduced by humans or by the environment, and some occur naturally.

Food safety hazards are divided into three categories: **biological hazards, chemical hazards,** and **physical hazards.**

○ **Biological hazards** include certain bacteria, viruses, parasites, and fungi, as well as certain plants, mushrooms, and fish that carry harmful toxins.

○ **Chemical hazards** include pesticides, food additives, preservatives, cleaning supplies, and toxic metals that leach from cookware and equipment.

○ **Physical hazards** consist of foreign objects that accidentally get into the food, such as hair, dirt, metal staples, and broken glass.

By far, biological hazards pose the greatest threat to food safety. Disease-causing microorganisms are responsible for the majority of foodborne-illness outbreaks.

HOW FOOD BECOMES UNSAFE

Foodborne illness is caused by several factors, which can be placed into one of three categories: **time-temperature abuse, cross-contamination,** and poor **personal hygiene.** Reported cases of foodborne illness usually involve more than one factor in each of these categories. A well-designed food safety system will control these factors.

Time-Temperature Abuse: Food has been time-temperature abused any time it has been allowed to remain too long at temperatures favorable to the growth of foodborne microorganisms. Common factors responsible for foodborne illness include:

○ Failing to hold or store food at required temperatures

○ Failing to cook or reheat food to temperatures that kill microorganisms

○ Failing to cool food properly

○ Preparing food a day or more before it is served

Cross-Contamination: Cross-contamination occurs when microorganisms are transferred from one surface or food to another. Common factors responsible for foodborne illness include:

Key Point

Disease-causing microorganisms are responsible for the majority of foodborne-illness outbreaks.

Key Point

A well-designed food safety system will establish controls to prevent time-temperature abuse, cross-contamination, and poor personal hygiene.

Cross-Contamination

Cross-contamination occurs when microorganisms are transferred from one surface or food to another.

○ Adding raw, contaminated ingredients to food that receives no further cooking

○ Food-contact surfaces (such as equipment or utensils) that are not cleaned and sanitized before touching cooked or ready-to-eat food

○ Allowing raw food to touch or drip fluids onto cooked or ready-to-eat food

○ Hands that touch contaminated (usually raw) food and then touch cooked or ready-to-eat food

○ Contaminated cleaning cloths that are not cleaned and sanitized before being used on other food-contact surfaces

Poor Personal Hygiene: Individuals with poor personal hygiene can offend customers, contaminate food or food-contact surfaces, and cause illness. Common factors responsible for foodborne illness include:

○ Employees who fail to wash their hands properly after using the restroom or whenever necessary

○ Employees who cough or sneeze on food

○ Employees who touch or scratch sores, cuts, or boils, and then touch food they are preparing or serving

KEY PRACTICES FOR ENSURING FOOD SAFETY

The key to food safety lies in controlling time and temperature throughout the flow of food, practicing good personal hygiene, and preventing cross-contamination. It is important to establish standard operating procedures that focus on these areas.

Controlling Time and Temperature

Microorganisms pose the largest threat to food safety. Like all living organisms, they cannot survive or reproduce outside certain temperature ranges. As outlined in *Exhibit 1d*, time and temperature must be controlled throughout the flow of food. Each of these steps will be discussed in further detail in later chapters.

Practicing Good Personal Hygiene

Training employees in good personal hygiene practices is the responsibility of every manager. Features of a good personal hygiene program include:

○ **Proper handwashing.** Hands and fingernails should be washed and cleaned thoroughly before and after handling food, between each task, and before using food-preparation equipment.

Receiving: Receive and store food quickly.

Storage: Store food at its recommended temperatures.

Preparation: Minimize time spent in the temperature danger zone of 41°F (5°C) to 140°F (60°C).

Cooking: Cook food to its required minimum internal temperature for the appropriate amount of time.

Holding: Hold hot food at 140°F (60°C) or higher and cold food at 41°F (5°C) or lower.

Cooling: Cool cooked food from 140°F (60°C) to 70°F (21°C) within two hours and from 70°F (21°C) to 41°F (5°C) or lower within an additional four hours, for a total cooling time of six hours.

Reheating: Reheat food to an internal temperature of 165°F (74°C) for fifteen seconds within two hours.

Exhibit 1d

Controlling Time and Temperature Throughout the Flow of Food

○ **Strictly enforced rules regarding eating, drinking, and smoking.** These activities should be prohibited while preparing or serving food, or while in areas used for washing and storing equipment and utensils.

○ **Preventing employees who are ill from working with food.** All illnesses should be reported to the manager.

○ **General cleanliness.** Insist on daily bathing, clean hair, and clean clothing.

Preventing Cross-Contamination

Employees must be carefully trained to recognize and prevent cross-contamination of microorganisms between food and food-contact surfaces. Some ways to prevent cross-contamination include the following:

○ Make sure employees wash their hands frequently when working with raw food. They should never touch raw food and then touch ready-to-eat food without washing their hands.

○ Do not allow raw food to touch or drip fluids onto cooked or ready-to-eat food.

○ Clean and sanitize food-contact surfaces (such as equipment or utensils) that touch contaminated food before they come in contact with cooked or ready-to-eat food.

○ Clean and sanitize cleaning cloths between each use.

Food-contact surfaces may be direct or indirect. A direct food-contact surface includes any equipment or utensil surface that normally touches food, such as tableware, cutting boards, knives, and other utensils used to prepare food, and counters where food is prepared. An indirect food-contact surface is a surface food might drain, drip, or splash onto during preparation, such as the backsplash of a counter. Food that splashes on an indirect food-contact surface may drain down onto a direct food-contact surface and contaminate it.

Food-contact surfaces, cleaning cloths, and sponges must be cleaned and sanitized to prevent cross-contamination. *Clean* simply means free of visible soil. *Sanitary,* on the other hand, means that the number of microorganisms on the surface has been reduced to safe levels.

Hands must also be washed regularly to prevent cross-contamination.

THE FOOD SAFETY RESPONSIBILITIES OF A MANAGER

The manager's basic food safety responsibilities are to serve customers safe food and to train employees in safe foodhandling practices. The manager's positive and supportive attitude toward food safety is critical. This attitude should be based on up-to-date knowledge of the regulations affecting the restaurant and foodservice industry.

Meeting Food Safety Regulations

To stay in operation, your establishment must comply with city, county, and state sanitation codes. The regulatory agency evaluating your establishment shares your commitment to serving safe food, and may have the authority to assess fines and close an establishment that serves unsafe food. Therefore, it is in your best interest to work with local authorities. During evaluations, the regulatory agency may identify deficiencies requiring your attention.

The FDA recommends that local and state health departments hold the person in charge of a restaurant or foodservice establishment responsible for knowing and demonstrating the following information:

○ Diseases carried or transmitted by food and their symptoms

○ Points in the flow of food where hazards can be prevented, eliminated, or reduced, and how procedures meet the requirements of the local code

○ The relationship between personal hygiene and the spread of disease, especially concerning cross-contamination, hand contact with ready-to-eat food, and handwashing

○ How to keep injured or ill employees from contaminating food or food-contact surfaces

○ The need to control the length of time potentially hazardous food remains at temperatures at which disease-causing microorganisms can grow

○ Hazards involved in the consumption of raw or undercooked meat, poultry, eggs, and fish

○ Safe cooking temperatures and times for potentially hazardous food items—such as meat, poultry, eggs, and fish

○ Safe temperatures and times for storing, holding, cooling, and reheating potentially hazardous food

○ Correct procedures for cleaning and sanitizing utensils and food-contact surfaces of equipment

○ Types of toxic materials used in the operation, and how to safely store, dispense, use, and dispose of them

○ The need for equipment that is sufficient in number and capacity, and is properly designed, constructed, located, installed, operated, maintained, and cleaned

○ Approved sources of potable water and the importance of keeping it clean and safe

Key Point

It is critically important for management to have a positive and supportive attitude toward food safety.

○ How the operation complies with the principles of a HACCP-based food safety system, as some local health departments require a HACCP-based system

○ Rights, responsibilities, and authorities the local code assigns to employees, managers, and the local health department

Marketing Food Safety

Make it clear to your employees and customers that your operation takes food safety very seriously.

Show your employees through actions that top management is involved in and supports food safety policies, and that food safety training for managers and all employees is a high priority. Offer training courses, and update and evaluate them regularly. Discuss food safety expectations. Document foodhandling procedures, use them in regular inspections, and update them as necessary. Show employees that safe foodhandling is appreciated—consider awarding certificates for training and give out small rewards for good food safety records. Set a good example by following all food safety rules yourself.

Show your customers that your employees know and follow safety rules. Make sure your employees' appearance reflects your concern for food safety. Consider having employees wear food safety pins or buttons, or use place mats and posters to get your message across. Be sure your employees can answer simple food safety questions from customers.

Marketing your food safety efforts will help assure your customers that you are committed to serving safe food.

RESPONDING TO A FOODBORNE-ILLNESS OUTBREAK

The primary responsibility of restaurant and foodservice employees is to ensure that customers are served quality food that has been prepared safely. By following standard operating procedures and implementing an effective food safety program, all establishments should be able to meet this responsibility.

Appendix G lists some guidelines to follow in the event of a foodborne-illness complaint.

SUMMARY

Foodborne illness is a major concern to the restaurant and foodservice industry. A foodborne illness is a disease carried or transmitted to people by food. A foodborne-illness outbreak involves two or more people experiencing the same illness after eating the same food. Some segments of the population are more susceptible to foodborne illnesses than others. These categories are called high-risk populations.

Some food has a history of involvement in foodborne-illness outbreaks. This is called potentially hazardous food. Typically having a natural potential for contamination due to production and processing methods, potentially hazardous food is often moist, high in protein, and has a neutral or slightly acidic pH.

An incident of foodborne illness can be very expensive for an establishment, including legal liability, damage to reputation, and other related factors. However, a well-designed food safety program protects your customers, your employees, and your reputation. The Hazard Analysis Critical Control Point (HACCP) program helps you monitor your operation and reduce the opportunities for foodborne illness. Because food can become unsafe at any step in the flow of food, the HACCP system focuses on the flow of food in your operation, identifying points at which contamination can occur, and implementing controls for those points.

Key practices for ensuring food safety include controlling time and temperature, practicing strict personal hygiene, and preventing cross-contamination. Also receive, store, prepare, cook, hold, serve, cool, and reheat food using methods that maintain its safety.

People pose a major risk to safe food, especially foodhandlers who do not practice personal hygiene. You must carefully train, monitor, and reinforce food safety principles in your establishment. Establishing a well-designed food safety system can help protect your customers by preventing foodborne-illness outbreaks, and can help the establishment avoid the potentially high costs associated with them.

A CASE IN POINT

Case Study

A restaurant manager received a call from a customer who had purchased a take-out pizza from his restaurant the previous night. The customer told him that her children, ages three and five, suffered from abdominal cramps, diarrhea, and fever, and claimed that the pizza had made them sick. The manager asked her whether anyone else ate the pizza and whether they had the same symptoms. She told him that her husband had eaten it, but felt fine. The manager then asked her what kind of pizza she ordered, and was told that it contained mushrooms, black olives, and extra cheese. The manager told her it couldn't possibly have been the pizza that made the children sick since it contained no meat and her husband was healthy. He suggested that her children had the stomach flu, wished them well, and hung up.

Did the manager handle this situation correctly? What did he do right? What did he do wrong?

TRAINING TIPS

1. Food Safety on Trial: "But your honor, I thought *they were washing their hands.*"

Purpose: *After completing this activity, trainees will be able to identify the legal liability faced by establishments.*

Time: 30 minutes

Directions: Present the group with a case study of a foodborne-illness outbreak. Use one of the Case-In-Point examples from this text or create a detailed fictitious case. Note: *For real-life examples, refer to the Morbidity and Mortality Weekly Report Web site.* (See page 1-24).

Divide the group into three teams: a plaintiff team (representing the persons who claim they became ill), a defense team (representing the establishment that served the food), and a jury. Allow about ten minutes for the plaintiff and defense to plan their presentations, then another ten minutes for each side to present its argument before the judge (the instructor). Questions to be argued should include the following:

○ Was the food unfit to serve?

○ Did the food cause the plaintiff harm?

○ Did the restaurant or establishment violate the "warranty of sale"?

○ Did the restaurant or establishment exhibit "reasonable care"?

After both sides have presented their arguments, give the jury five minutes to deliberate and determine the fate of the restaurant or establishment. No mistrials or appeals!

2. Knowledge and Application of Food Safety

Purpose: *After completing this activity, trainees should be able to identify the information managers are responsible for at the establishment, as recommended by the FDA.*

Time: one hour

Directions: The FDA recommends that local and state regulatory agencies hold the person in charge of a restaurant or foodservice establishment responsible for knowing and applying the information listed below. Assign teams of two or three people. Depending upon the number of teams, assign each team one or more topics from the list. Ask each team to prepare a discussion of current food safety knowledge related to their assigned topics. After the teams have had adequate time to prepare, ask them to present their information to the group. After the presentation, solicit discussion. You might ask group members how well the management teams in their own establishments follow the FDA recommendations.

The FDA recommends that the person in charge of a restaurant or foodservice establishment know and apply the following information:

○ Diseases carried or transmitted by food and their symptoms

○ Points in the flow of food where hazards can be prevented, eliminated, or reduced, and how procedures meet the requirements of the local code

○ The relationship between personal hygiene and the spread of disease, especially concerning cross-contamination, hand contact with ready-to-eat food, and handwashing

○ How to keep injured or ill employees from contaminating food or food-contact surfaces

○ The need to control the length of time potentially hazardous food remains at temperatures at which disease-causing microorganisms can grow

○ Hazards involved in the consumption of raw or undercooked meat, poultry, eggs, and fish

○ Safe cooking temperatures and times for potentially hazardous foods— such as meat, poultry, eggs, and fish

○ Safe temperatures and times for storing, holding, cooling, and reheating potentially hazardous food

○ Correct procedures for cleaning and sanitizing utensils and food-contact surfaces of equipment

○ Types of toxic materials used in the operation, and how to safely store, dispense, use, and dispose of them

○ The need for equipment that is sufficient in number and capacity, and is properly designed, constructed, located, installed, operated, maintained, and cleaned

○ Approved sources of potable water and the importance of keeping it clean and safe

○ How the operation complies with the principles of a HACCP-based food safety system, as some local health departments require a HACCP-based system

○ Rights, responsibilities, and authorities the local code assigns to employees, managers, and the local health department

3. Developing a Food Safety Complaint Form for Your Establishment

Purpose: *To involve your management team in the development of a standard complaint questionnaire form for your establishment. Note:* Involving the management team will help clarify your policies regarding how to handle a possible foodborne-illness complaint.

Time: ½ day

Directions: Every establishment should have a standard complaint questionnaire form on hand to record possible foodborne-illness complaints. These forms are essential in order to ensure that complaints are handled in a professional manner, information is fully documented, and all appropriate questions are included.

Ask management team members to help write and review questions for the form. It is a good idea to look at similar forms from other establishments. Ask your local regulatory agency or State Restaurant Association for input. If possible, you might want to have a lawyer review the form. When the content of the form is approved, print copies and make them accessible to the management team.

Inform your team that it is essential to use these forms whenever a customer has a foodborne-illness complaint. Let the team know how you want a complaint to be handled. They need to know what they should say to the customer and what steps to take once the form has been filled out. It may be a good idea to role-play scenarios regarding a customer complaint. This will allow managers to practice and can expose any weaknesses that may exist in your program.

4. Marketing Your Safe Food

Purpose: *To involve your management team in a brainstorming session to determine different ways to market the food safety program in your establishment to both employees and customers.*

Time: ½ day

Directions: Conduct a brainstorming session with your management team to determine different ways to market your food safety systems and procedures to employees and customers. If you already market your systems and procedures, use the brainstorming session to improve your current marketing practices.

Bring up the following questions in the session:

Employees

○ How do we demonstrate to our employees that we are committed to food safety?

○ What food safety systems are currently in place? Do foodhandlers understand these systems?

○ What training commitment have we made regarding food safety? How is this training made available to employees?

○ What documentation do we keep related to food safety? Do we monitor critical control points? If so, do employees understand how to accurately monitor and document critical control points?

Customers

○ How is our commitment to food safety visible to our customers?

○ What food safety principles do our customers see being put into practice by our staff?

○ What written materials (signs, plaques, posters, decals, statements on packaging, etc.) are in place to inform our customers of our commitment to food safety?

Based on the input received from your management team, come up with a plan to market your food safety systems and procedures to employees and customers, or to improve your current marketing practices. Implement the plan and evaluate it at management meetings. If necessary, modify your plan based on the evaluation.

DISCUSSION QUESTIONS

1. What is the difference between a foodborne illness and a foodborne-illness outbreak?

2. What are the potential costs associated with foodborne-illness outbreaks?

3. Why are the elderly at higher risk for contracting foodborne illnesses?

4. What are the three major types of hazards to food safety?

5. A chef cuts up a salmon on a cutting board, then thoroughly rinses the cutting board and knife in warm water. She then uses the same cutting board and knife to slice fresh parsley. Is this an acceptable foodhandling practice? Why or why not? On what key food safety practice does this example focus?

MULTIPLE-CHOICE STUDY QUESTIONS ???

Multiple Choice

1. Why do elderly people have a higher risk of contracting a foodborne illness?

   A. They are more likely to spend time in a hospital.
   B. Their immune systems are likely to be weaker than those of younger people.
   C. Their allergic reactions to chemicals used in food production might be greater than younger people.
   D. They are likely to have diminished appetites and do not want to cook for themselves.

2. Which type of food would be the most likely to cause a foodborne illness?

   A. Tomato juice
   B. Cooked rice
   C. Stored whole wheat flour
   D. Dry powdered milk

3. Your restaurant is closed Sunday and Monday. Tuesday morning you open the restaurant and notice that the refrigerator is not running. When you check the thermometer inside, it reads 50°F (10°C). What should you do with the fresh ground beef in the refrigerator?

   A. Cook and serve it within two hours.
   B. Freeze it right away.
   C. Discard it.
   D. Rechill it immediately to 41°F (5°C) or lower.

4. Which of the following is *not* a common characteristic of potentially hazardous food items?

   A. They are moist.
   B. They are dry.
   C. They are neutral or slightly acidic.
   D. They are high in protein.

5. When a foodborne illness occurs, it is usually caused by one of three factors. These factors are:

   A. Time-temperature abuse, poor personal hygiene, and physical hazards
   B. Time-temperature abuse, chemical hazards, and physical hazards
   C. Time-temperature abuse, cross-contamination, and chemical hazards
   D. Time-temperature abuse, cross-contamination, and poor personal hygiene

6. In order for a foodborne illness to be considered an "outbreak," how many people must experience the illness after eating the same food?

   A. 1   B. 2   C. 10   D. 20

7. The largest threat to food safety comes from which of the following hazards?

   A. Pesticides   C. Microorganisms
   B. Hair   D. Food additives

8. To prevent cross-contamination, foodhandlers should not

   A. touch raw meat and then touch cooked or ready-to-eat food.
   B. allow food to remain at temperatures above 41°F (5°C).
   C. check food temperatures when receiving food.
   D. hold food at temperatures below 140°F (60°C).

9. Foodhandlers must practice all of the following hygienic practices *except*

   A. proper handwashing.
   B. daily bathing.
   C. wearing clean clothing to work.
   D. getting periodic AIDS tests.

For answers, please turn to Appendix A.

ADDITIONAL RESOURCES

Books and Periodicals

Jay, J. M. 1996. *Modern Food Microbiology.* New York: Chapman & Hall.

McCoy, J. J. 1990. *How safe is our food supply?* New York: F. Watts.

National Research Council. 1998. *Ensuring safe food: From production to consumption.* Washington D.C.: National Academy of Sciences.

Olson, D. G. 1998. Irradiation of food. *Food Technology,* 52 (1):56-65.

Web Sites

American Public Health Association (APHA)

http://www.apha.org

APHA is concerned with a broad set of issues affecting personal and environmental health, including programs and policies related to chronic and infectious diseases, federal and state funding for health programs, and professional education in public health.

Association for Food and Drug Officials (AFDO)

http://www.afdo.org

AFDO is a leader and trusted resource in people developing strategies to resolve and promote public health and consumer protection issues related to the regulation of food, drugs, medical devices, and consumer products.

Centers for Disease Control and Prevention (CDC)

http://www.cdc.gov

The mission of the CDC is to promote health and quality of life by preventing and controlling disease, injury, and disability. To prevent and control foodborne illness, the CDC collects data on outbreaks. This Web site provides general information on foodborne illnesses and their prevention.

FDA-Center for Food Safety and Applied Nutrition (CFSAN)

http://www.cfsan.fda.gov

As the center within the FDA responsible for food safety and nutrition, CFSAN promotes and protects public health by researching and implementing guidelines, policies, and standards to ensure that food is safe, nutritious, wholesome, and properly labeled. This Web site provides a wealth of information on food safety and sanitation, including corresponding guidelines, policies, and standards.

FDA Food Code

http://vm.cfsan.fda.gov/~dms/foodcode.html

As the basis for many local sanitation codes, as well as the basis for information in this textbook, the FDA Food Code, available at this Web address, is a useful resource for information relating to food safety for the restaurant and foodservice industry.

Food Marketing Institute (FMI)

http://www.fmi.org

The FMI is a nonprofit association conducting programs in research, education, industry relations, and public affairs on behalf of its members, which include large multistore chains, small regional firms, and independent supermarkets.

Institute of Food Technologists

http://www.ift.org

The Institute of Food Technologists is a scientific, educational society with an interest in providing a safe and wholesome food supply. Their goal is to provide guidance on relevant issues in the field of food. The Web site provides scientific articles on food and food safety, and lists daily happenings in the food industry.

International Association for Food Protection

http://www.foodprotection.org

Members of the International Association for Food Protection are a diverse group, representing all areas of food protection, industry, government, and academia. Members are kept current on rapidly changing technologies, innovations, and regulations in the area of food safety through two monthly scientific journals, as well as annual meetings.

International Dairy-Deli-Bakery Association (IDDBA)

http://www.iddba.org

IDDBA is a resource for information and services across all food channels for the dairy, deli, and bakery industries. This Web site provides information relevant to all people who work in the dairy, deli, and bakery industries.

International HACCP Alliance

http://haccpalliance.org

The International HACCP Alliance was formed to assist the meat and poultry industry in preparing for mandatory HACCP. The Alliance promotes public health and safety by facilitating uniform development and implementation of HACCP. This Web site houses documents relating to food safety, HACCP, and its implementation.

Morbidity and Mortality Weekly Report (MMWR)

http://www.cdc.gov/mmwr

The MMWR is prepared by the Centers for Disease Control and is available—free-of-charge—on this Web site. The reports contain data based on weekly accounts of reportable diseases from state health departments. Report databases can be searched for information on foodborne-illness outbreaks.

National Association of Convenience Stores (NACS)

http://www.cstorecentral.com

NACS is a proactive organization representing the convenience store segment. This informative Web site provides up-to-date industry happenings and contains reports outlining vital industry statistics.

National Center for Infectious Diseases (NCID)

http://www.cdc.gov/ncidod

NCID is one of the centers of the Centers for Disease Control. Its mission is to prevent illness, disability, and death caused by infectious diseases around the world. NCID accomplishes this mission by conducting surveillance, epidemic investigations, epidemiologic and laboratory research, and training. It also sponsors public education programs to develop, evaluate, and promote prevention and control strategies for infectious diseases. This Web site serves as another great resource for information on foodborne illness.

National Environmental Health Association

http://www.neha.org

The National Environmental Health Association's goals include fostering cooperation and understanding among environmental health professionals while remaining solidly committed to improving the environment in cities, towns, and rural areas throughout the world. This Web site includes many useful resources for restaurant and foodservice inspectors, sanitarians, and the industry.

National Food Processors Association (NFPA)

http://www.nfpa-food.org

The NFPA is the voice of the food processing industry on scientific and public policy issues involving food safety, nutrition, technical and regulatory matters, and consumer affairs. Their Web site offers many resources on food safety and food processing for the entire food industry.

National Restaurant Association

http://www.restaurant.org

The National Restaurant Association is the leading business association for the restaurant industry. Together with the National Restaurant Association Educational Foundation, the Association's mission is to represent, educate, and promote the rapidly growing restaurant and foodservice industry. This Web site should be your starting place for all issues and concerns related to your restaurant. This Web site has it all, from tips for running your establishment to vital data on your customers' spending habits.

U.S. Department of Agriculture–Food Safety and Inspection Service (USDA-FSIS)

http://www.fsis.usda.gov

The Food Safety and Inspection Service (FSIS), the public health agency in the U.S. Department of Agriculture responsible for ensuring that the nation's commercial supply of meat, poultry, and egg products is safe, wholesome, and correctly labeled and packaged. This Web site contains a wealth of information on food safety relating to meat, poultry, and eggs.

Chapter 2
The Microworld

Knowledge

TEST YOUR FOOD SAFETY KNOWLEDGE

1. **True or False:** Proper cooking will destroy all pathogens in food. *(See page 2-4.)*

2. **True or False:** A foodborne intoxication is caused by eating food that contains a pathogen. *(See page 2-20.)*

3. **True or False:** Eating food contaminated by foodborne pathogens is the leading cause of foodborne illness. *(See page 2-2.)*

4. **True or False:** Foodborne bacteria will not grow at refrigeration temperatures. *(See page 2-6.)*

5. **True or False:** Highly acidic food typically does not support the growth of foodborne microorganisms. *(See page 2-5.)*

For answers, please turn to Appendix A.

Table of Contents

Learning Objectives

After completing this chapter, you should be able to

○ Identify the four basic types of microbial contaminants, give examples, and describe preventive actions for each.

○ Differentiate between foodborne infection and foodborne intoxication and identify the major causes of each.

○ Identify microbial risks associated with various types of food.

○ Identify the conditions that allow bacteria to grow.

Key Terms

Microorganisms
Pathogens
Toxins
Bacteria
Virus
Parasite
Fungi
Spoilage microorganism
Spore
pH
Lag phase
Log phase

Vegetative microorganism
Stationary phase
Death phase
FAT TOM
Temperature danger zone
Aerobic
Anaerobic
Facultative
Water activity (a$_w$)

Host
Mold
Yeast
Foodborne infection
Foodborne intoxication
Foodborne toxin-mediated infection
Food irradiation

In the previous chapter, you learned that foodborne microorganisms pose the greatest threat to food safety, and that disease-causing microorganisms are responsible for the majority of foodborne-illness outbreaks. In this chapter, you will learn about the microorganisms that cause foodborne illness, as well as the conditions they require in order to grow. When you understand these conditions, you will begin to see how the growth of foodborne microorganisms can be controlled, a topic that will be covered in greater detail in later chapters.

Microorganisms are small, living beings that can be seen only with a microscope. While not all microorganisms cause disease, some do. These are called **pathogens.** Eating food contaminated with foodborne pathogens or their **toxins** (poisons) is the leading cause of foodborne illness.

MICROBIAL CONTAMINANTS

There are four types of microorganisms that can contaminate food and cause foodborne illness: **bacteria, viruses, parasites,** and **fungi.**

These microorganisms can be arranged into two groups: **spoilage microorganisms** and pathogens. Mold is an example of a spoilage microorganism. While moldy food has an unpleasant appearance, smell, and taste, it seldom causes illness. However, pathogens such as *Salmonella* spp. and the virus that causes hepatitis A cannot be seen, smelled, or tasted, but food contaminated by these pathogens often causes some form of illness when ingested.

BACTERIA

Of all microorganisms, bacteria are of greatest concern to the manager. Knowing what bacteria are and understanding the environment in which they grow is the first step in controlling them.

Basic Characteristics of Bacteria that Cause Foodborne Illness

Bacteria that cause foodborne illness have some basic characteristics:

○ They are living, single-celled organisms
○ They may be carried by a variety of means: food, water, humans, or insects
○ Under favorable conditions, they can reproduce very rapidly
○ Some can survive freezing
○ Some turn into **spores,** a change that protects the bacteria from unfavorable conditions
○ Some can cause food spoilage; others can cause illness
○ Some cause illness by producing toxins as they multiply, die, and break down

Bacterial Growth

To grow and reproduce, bacteria need:

○ Food

○ Appropriate level of acidity **(pH)**

○ Proper temperature

○ Adequate time

○ The necessary level of oxygen

○ Ample moisture

Their growth can be broken down into four progressive stages (phases): lag, log, stationary, and death. *(See Exhibit 2a.)*

When bacteria are first introduced to food, they go through an adjustment period, called the **lag phase.** In this phase, their numbers are stable as they prepare for growth. To control their number and prevent food from becoming unsafe, it is important to prolong the lag phase as long as possible. You can accomplish this by controlling the bacteria's requirements for growth in food: time, temperature, moisture, oxygen, and pH. For example, by refrigerating food, you can keep bacteria in the lag phase. If these conditions are not controlled, bacteria can enter the next phase, the **log phase**, where they will grow remarkably fast.

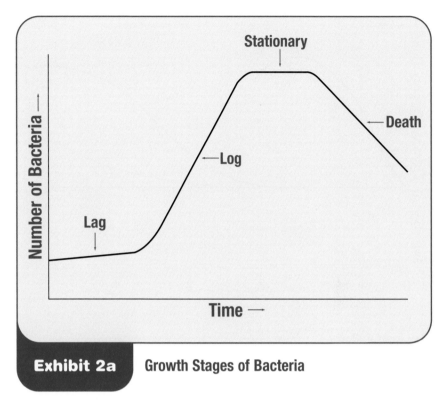

Exhibit 2a Growth Stages of Bacteria

Bacteria reproduce by splitting in two. Those in the process of reproduction are called **vegetative microorganisms.** As long as conditions are favorable, bacteria can grow and multiply very rapidly, doubling their number as often as every twenty minutes. *(See Exhibit 2b.)* This is called exponential growth, and it occurs in the log phase. Food will

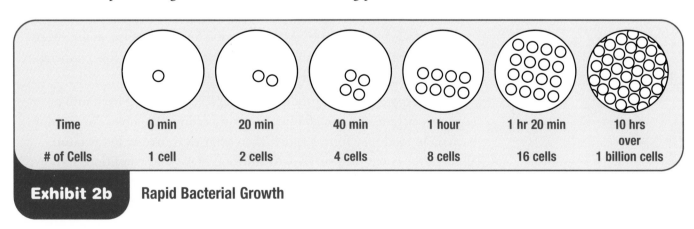

Exhibit 2b Rapid Bacterial Growth

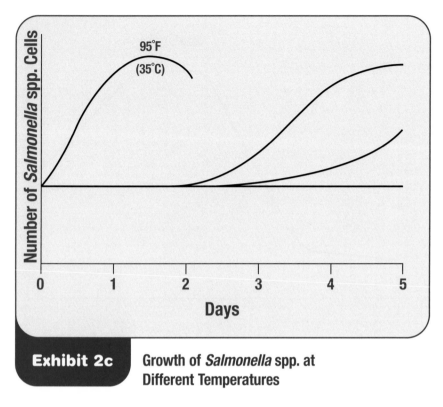

Exhibit 2c   Growth of *Salmonella* spp. at Different Temperatures

rapidly become unsafe during the log phase.

Bacteria can continue to grow until nutrients and moisture become scarce, or conditions become unfavorable. Eventually the population reaches a **stationary phase**, in which just as many bacteria are growing as are dying. When the number of bacteria dying exceeds the number growing, the population declines. This is called the **death phase.**

The time required for bacteria to adapt to a new environment (lag phase) and to begin a rapid rate of growth (log phase) depends on several factors, such as temperature. *Exhibit 2c* shows how different temperatures affect the growth rate of *Salmonella.* As the graph shows, at warmer temperatures (95°F [35°C]), *Salmonella* grows more quickly than at colder temperatures (44°F and 50°F [7°C and 10°C]). At even colder temperatures (42°F [6°C]), *Salmonella* doesn't grow at all—but notice that it doesn't die either (prolonging the lag phase). This is why refrigerating food properly helps keep it safe.

Vegetative Stages and Spore Formation

Although vegetative bacteria may survive low—even freezing—temperatures, they can be killed by high temperatures. For example, pathogenic bacteria can be killed during proper cooking. *(This will be discussed in Chapter 7.)* Some types of bacteria, however, have the ability to change into a different form, called a spore. The spore's thick wall protects the bacteria against unfavorable conditions, such as high or low temperature, low moisture, and high acidity.

While a spore cannot reproduce, it is capable of turning back into a vegetative organism when conditions again become favorable. For example, bacteria in food may form a spore when exposed to freezer temperatures, allowing the bacteria to survive. As the food thaws and conditions improve, the spore can turn back into a vegetative cell and begin to grow in the food.

Since spores are so difficult to destroy, it is important to cook, cool, and reheat food properly.

Conditions that Support the Growth of Microorganisms

All microorganisms, except viruses, are able to grow in food if the conditions are right. To grow, microorganisms need food, water, appropriate temperatures, oxygen, and proper pH. Therefore, you must control these necessities to control their growth of microorganisms.

FAT TOM: What Microorganisms Need to Grow

The conditions that favor the growth of most foodborne microorganisms can be remembered by the acronym **FAT TOM.** *(See Exhibit 2d.)* Each of these conditions for growth will be explained in more detail in the next several paragraphs.

Food

 To grow, foodborne microorganisms need nutrients, specifically proteins and carbohydrates. These substances are commonly found in potentially hazardous food items such as meat, poultry, dairy products, and eggs. *(See page 1-7.)*

Acidity

Foodborne microorganisms typically do not grow in highly acidic or highly alkaline food. They grow best in food with a neutral to slightly acidic pH. The pH of a substance tells how

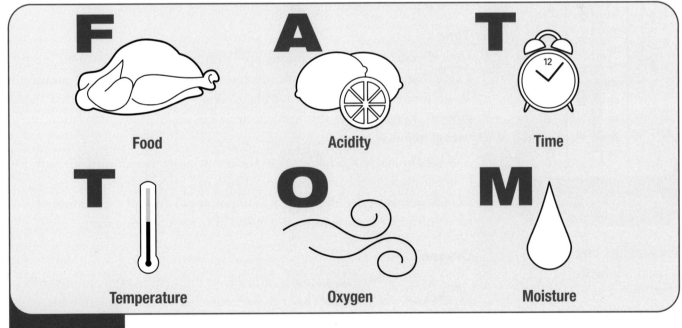

Exhibit 2d	**FAT TOM**

Conditions favoring the growth of most foodborne microorganisms (except viruses) can be remembered by the acronym FAT TOM.

Key Point

Pathogenic bacteria grow well in food with a pH between 4.6 and 7.5.

Key Point

Listeria monocytogenes and *Yersinia enterocolitica* are able to grow at refrigeration temperatures.

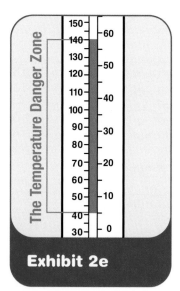

The Temperature Danger Zone

°F	°C
150	
140	60
130	
120	50
110	40
100	
90	30
80	
70	20
60	
50	10
40	
30	0

Exhibit 2e

Temperature and Bacterial Growth
Most foodborne microorganisms grow well at temperatures between 41°F and 140°F (5°C and 60°C).

acidic or alkaline it is. The pH scale ranges from 0 to 14.0. Food with a pH between 0 and 7.0 is acidic, while food with a pH between 7.0 and 14.0 is alkaline. A pH of 7.0 is neutral.

Some microorganisms grow better than others in food with a specific pH. *(See Exhibit 2f.)* Yeasts and molds tend to grow in acidic food (with a pH below 4.6) such as fruit, fruit juice, and jam. Pathogenic bacteria grow well in food with a pH between 4.6 and 7.5 (slightly acidic to neutral), which includes most of the food we eat. A few bacteria have been known to grow in food with a pH below 4.6 and above 7.5. For example, *E. coli* O157:H7 has been found to reproduce in unpasteurized apple juice, which has a pH below 4.0.

Temperature

Most foodborne microorganisms grow well between the temperatures of 41°F and 140°F (5°C and 60°C). *(See Exhibit 2e.)*

This range is known as the **temperature danger zone.** However, exposing microorganisms to temperatures outside the danger zone does not necessarily kill them. Refrigeration temperatures, for example, may only slow their growth. Some bacteria—such as *Listeria monocytogenes* and *Yersinia enterocolitica*—are able to grow at refrigeration temperatures. Bacterial spores can often survive extreme heat and cold. Food must be handled very carefully when it is thawed, cooked, cooled, and reheated since it can be exposed to the temperature danger zone during these times.

Time

Foodborne microorganisms need sufficient time to grow. This means that even under favorable conditions, microorganisms need enough time to move from the lag phase (slow growth) to the log phase (rapid growth). Keep in mind that some bacteria can double their population every twenty minutes.

If contaminated food remains in the temperature danger zone for four hours or more, pathogenic microorganisms can grow to levels high enough to make someone ill. Therefore, it is important to control the amount of time potentially hazardous food remains in the temperature danger zone.

Oxygen

Different microorganisms have different oxygen requirements for growth. They can be categorized as aerobic, anaerobic, or facultative. **Aerobic** microorganisms require oxygen to grow. **Anaerobic** microorganisms can grow only when oxygen is absent. Growth of anaerobic microorganisms has been known to occur in thick, heat-treated

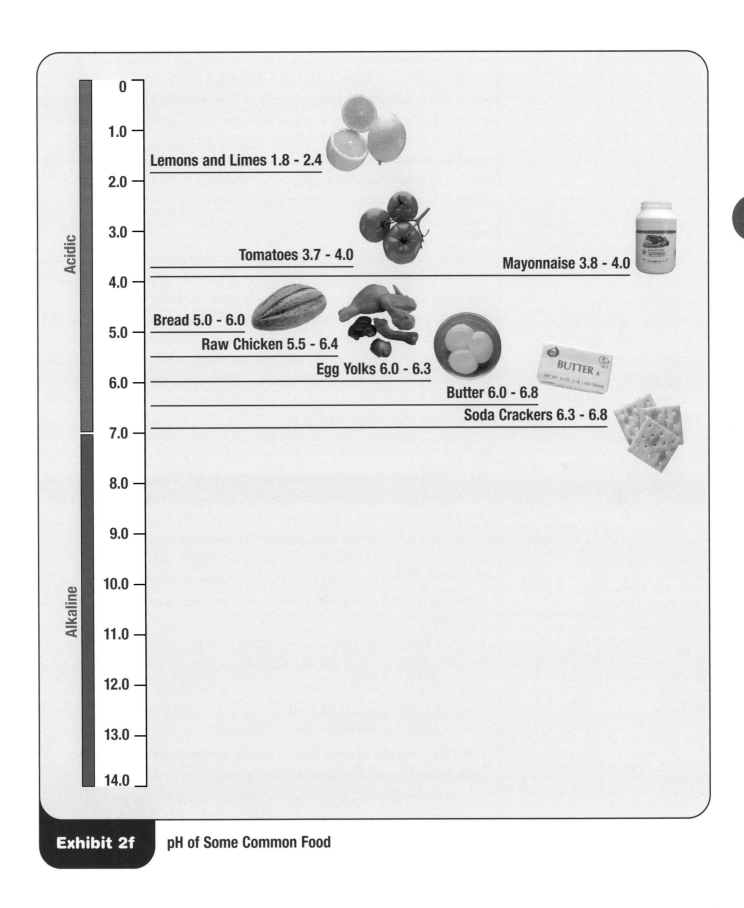

Exhibit 2f   pH of Some Common Food

plant food that has been temperature abused, untreated garlic-in-oil mixtures, and foil-wrapped baked potatoes that have been temperature abused. **Facultative** microorganisms can grow either with or without the presence of oxygen. Most microorganisms that cause foodborne illness are facultative. Manufacturers use different types of packaging to control the oxygen requirements of the most likely type of microorganism found on the food. For example, vacuum packaging is used to control the growth of aerobic microorganisms that can be found on bacon.

Moisture

Because most foodborne microorganisms require water to grow, they grow well in moist food. The amount of moisture available in a food for microorganisms to grow is called its **water activity (a_w).** It is measured on a scale of 0 through 1.0, with water having a water activity of 1.0. Most microorganisms that cause foodborne illness grow best in foods with water activities between .85 and .97, although some can grow in foods with lower water-activity levels. Potentially hazardous food items typically have a water activity of .85 or higher. *Exhibit 2g* presents the water activity levels of some common food items.

Multiple Barriers for Controlling the Growth of Microorganisms

FAT TOM is the key to controlling the growth of microorganisms in food. By putting in place as many barriers as possible, you remove conditions that support bacterial growth. The list below provides some simple steps that, when combined, can create multiple barriers to *control* the growth of pathogens and ensure food safety. To kill or reduce the level of pathogens, food must be cooked to required minimum internal temperatures.

○ **Make food more acidic.** Add vinegar, lemon juice, lactic acid, or citric acid.

○ **Raise or lower food temperature**. Move food out of the temperature danger zone by cooking it to the proper temperature, by refrigerating it to 41°F (5°C) or lower, or by freezing it.

○ **Lower the water activity of food.** Dry food by adding sugar, salt, alcohol, or acid. Food can also be air-dried or freeze-dried to remove water.

○ **Limit the amount of time food is in the temperature danger zone.** Prepare food in small batches, as close to service as possible.

Information about major foodborne illnesses caused by bacteria, as well as ways to prevent them, is presented in *Exhibit 2h beginning on page 2-10. (Biological toxins are discussed in Chapter 3.)*

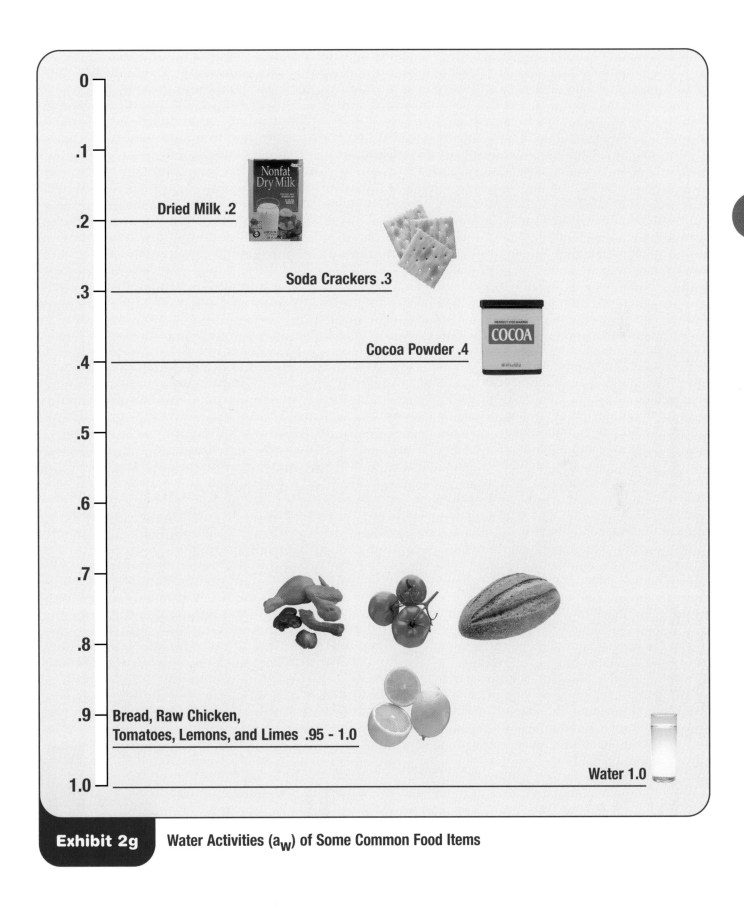

0

.1

Dried Milk .2

.2

Soda Crackers .3

.3

Cocoa Powder .4

.4

.5

.6

.7

.8

.9   Bread, Raw Chicken,
Tomatoes, Lemons, and Limes .95 - 1.0

Water 1.0

1.0

Exhibit 2g   Water Activities (a$_w$) of Some Common Food Items

Exhibit 2h: Major Foodborne Illnesses Caused by Bacteria

Foodborne Illness	Salmonellosis (nontyphoid)	Shigellosis (bacillary dysentery)	Listeriosis
Bacteria	*Salmonella* spp.	*Shigella* spp.	*Listeria monocytogenes*
Characteristics of Bacteria	Does not form spores; facultative; some can survive pHs below 4.5	Does not form spores; facultative; some produce shiga toxin	Does not form spores; facultative; resists freezing, drying, and heat; can grow at refrigeration temperatures
Type of Illness	Infection (possibly toxin-mediated)	Toxin-mediated infection	Infection
Symptoms	Abdominal cramps, headache, nausea, fever, diarrhea, and vomiting; may cause severe dehydration in infants and elderly; may cause arthritic symptoms 3 to 4 weeks later	Diarrhea (may be bloody), abdominal pain, fever, nausea, cramps, vomiting, chills, fatigue, dehydration	Nausea, vomiting, diarrhea, headache, persistent fever, chills, backache, meningitis, encephalitis, septicemia, and intrauterine or cervical infections in pregnant women, which may result in spontaneous abortion or stillbirth; most often affects fetuses and infants
Incubation Period	6 to 48 hours; usually 12 to 36 hours	12 to 50 hours; usually 1 to 3 days	3 to 70 days; usually about 3 weeks
Duration of Illness	1 to 2 days (sometimes longer)	Usually 4 to 7 days; can be indefinite (depends on treatment)	Indefinite (depends on treatment); high fatality rates in immunocompromised people
Source	Water, soil, insects, domestic and wild animals, and the human intestinal tract, especially as carriers; raw meat, poultry, seafood	Human intestinal tract; flies; frequently found in water polluted by feces	Soil, water, and damp environments; humans; domestic and wild animals; fowl
Food Involved in Outbreaks	Poultry and poultry salads; meat and meat products; fish; shrimp; milk; shell eggs and egg products, such as improperly cooked custards, sauces, and pastry creams; tofu and other protein foods; sliced melons, sliced tomatoes, raw sprouts, and other fresh produce	Salads (potato, tuna, shrimp, chicken, and macaroni); lettuce; raw vegetables; milk and dairy products; poultry	Unpasteurized milk and cheese; ice cream; frozen yogurt; raw vegetables; poultry and meat; seafood; prepared and chilled ready-to-eat food (e.g., soft cheese, deli foods, pâté)
Preventive Measures	Avoid cross-contamination; refrigerate food; thoroughly cook poultry to at least 165°F (74°C) for at least 15 seconds and cook other food to minimum internal temperatures; properly cool cooked meat and meat products; properly handle and cook eggs; ensure that employees avoid cross-contaminating food and food-contact surfaces by practicing good personal hygiene	Avoid cross-contamination; ensure that employees practice good personal hygiene; use sanitary food and water sources; control flies; properly cool food	Use only pasteurized milk and dairy products; cook food to proper internal temperatures; avoid cross-contamination; clean and sanitize surfaces; thoroughly wash raw vegetables

(Continued on next page.)

Exhibit 2h

Exhibit 2h: Major Foodborne Illnesses Caused by Bacteria (Continued)

Foodborne Illness	Staphylococcal Gastroenteritis	Clostridium perfringens Gastroenteritis	Bacillus cereus Gastroenteritis
Bacteria	*Staphylococcus aureus*	*Clostridium perfringens*	*Bacillus cereus*
Characteristics of Bacteria	Does not form spores; facultative; can survive high acidity (to pH 2.6); resists drying and freezing	Forms spores; anaerobic	Forms spores; facultative
Type of Illness	Intoxication	Toxin-mediated infection	Intoxication (emetic); toxin-mediated infection (diarrheal)
Symptoms	Nausea, retching, abdominal cramps, diarrhea; in severe cases—headache, muscle cramping, changes in blood pressure and pulse rate	Abdominal pain, diarrhea, nausea, dehydration (fever, headache, and vomiting usually absent)	Nausea and vomiting, sometimes abdominal cramps or diarrhea (emetic); watery diarrhea, abdominal cramps, pain, nausea (diarrheal)
Incubation Period	1 to 7 hours; usually 2 to 4 hours	8 to 22 hours; usually 10 to 12 hours	30 minutes to 6 hours (emetic); 6 to 15 hours (diarrheal)
Duration of Illness	2 to 3 days	Usually 24 hours; may last 1 to 2 weeks	Less than 24 hours (emetic); 24 hours (diarrheal)
Sources	Humans: skin, hair, nose, throat, and infected sores; animals	Humans and animals (intestinal tracts), soil; soil contaminated with feces	Soil and dust; cereal crops
Food Involved in Outbreaks	Reheated food, meat and meat products, poultry, egg products and other protein food, sandwiches, milk and dairy products, cream-filled pastries, salads (egg, tuna, chicken, potato, and macaroni)	Cooked meat, meat products, poultry, stew, gravy and beans that have been temperature abused	Rice products; starchy food (potatoes, pasta, and cheese products); food mixtures, such as sauces, puddings, soups, casseroles, pastries, salads (emetic), meats, milk, vegetables, and fish (diarrheal)
Preventive Measures	Avoid contamination from unwashed bare hands, practice good personal hygiene, exclude employees with skin infections from foodhandling and preparation tasks, properly refrigerate food, rapidly cool prepared food	Use careful time and temperature control in cooling and reheating cooked food	Careful time and temperature control, adequate cooking of food, proper cooling of food

(Continued on next page.)

Exhibit 2h: Major Foodborne Illnesses Caused by Bacteria (Continued)

Foodborne Illness	Botulism	Campylobacteriosis	Hemorrhagic colitis
Bacteria	*Clostridium botulinum*	*Campylobacter jejuni*	Shiga toxin-producing *Escherichia coli*, including O157:H7 and O157:NM
Characteristics of Bacteria	Forms spores; anaerobic	Does not form spores	Does not form spores; facultative; has survived freezing and high acidity (pHs below 4.0); can grow at refrigerator temperatures
Type of Illness	Intoxication	Infection	Toxin-mediated infection
Symptoms	Fatigue, weakness, vertigo followed by blurred or double vision, difficulty speaking and swallowing, dry mouth; eventually leading to paralysis and death	Diarrhea (watery or bloody); fever, nausea, and vomiting; abdominal pain, headache, and muscle pain	Diarrhea (watery, may become bloody); severe abdominal cramps and pain, vomiting, mild or no fever; may cause kidney failure in some people; symptoms more severe in the very young
Incubation Period	4 hours to 8 days; usually 18 to 36 hours	Usually 2 to 5 days, with a range of 1 to 10 days	2 to 8 days; usually 3 to 4 days
Duration of Illness	Several days to 1 year	7 to 10 days (relapses common); self-limiting	2 to 8 days
Sources	Present on almost all food of either animal or vegetable origin, soil, water	Domestic animals (sheep, pigs, cattle, poultry, and pets); raw milk	Animals; particularly found in the intestinal tracts of cattle and humans
Food Involved in Outbreaks	Food that was underprocessed or temperature abused in storage, canned low-acid food, untreated garlic-in-oil products, temperature-abused sautéed onions in butter sauce, leftover baked potatoes, stews, meat/poultry loaves; potential risks for MAP (modified atmosphere packaging) and *sous vide* products	Unpasteurized milk and dairy products, raw poultry, nonchlorinated or fecal-contaminated water	Raw and undercooked ground beef, imported cheeses, unpasteurized milk and apple cider/juice, roast beef, improperly cured dry salami, lettuce, nonchlorinated water, alfalfa sprouts
Preventive Measures	Do not use home-canned products; use careful time and temperature control for *sous vide* items and all large, bulky foods; purchase only acidified garlic-in-oil mixtures and keep refrigerated; sauté onions to order; rapidly cool leftovers	Thoroughly cook food, especially poultry, to minimum safe internal temperatures; use pasteurized milk and treated water, avoid cross-contamination	Thoroughly cook ground beef to at least 155°F (68°C) for 15 seconds; avoid cross-contamination, practice good personal hygiene, use only pasteurized milk, dairy products, and juices

(Continued on next page.)

Exhibit 2h: Major Foodborne Illnesses Caused by Bacteria (Continued)

Foodborne Illness	Vibrio spp. (noncholerae) Gastroenteritis/Septicemia	Yersiniosis
Bacteria	Vibrio parahaemolyticus and Vibrio vulnificus	Yersinia enterocolitica
Characteristics of Bacteria	Does not form spores; more common in warmer months	Does not form spores; facultative; can survive at pH below 4.5; can grow at low temperatures
Type of Illness	Infection	Infection
Symptoms	Fever, chills, diarrhea, abdominal cramps, nausea, vomiting, headache for both; severe cases of V. vulnificus include decreased blood pressure and septicemia; fatality from V. vulnificus is high in immunocompromised people	Fever, diarrhea, severe abdominal pain, vomiting
Incubation Period	V. parahaemolyticus: 4 to 96 hours, usually 12 to 24 hours; V. vulnificus: 12 hours to several days, usually 38 hours	1 to 11 days; usually 24 to 48 hours
Duration of Illness	V. parahaemolyticus: 1 to 8 days; V. vulnificus: days to weeks; death within a few days for immunocompromised people	Days to weeks; chronic in some individuals
Sources	Oysters and other shellfish, (shrimp, lobsters, clams, and mussels), especially from the Gulf of Mexico	Domestic pigs are primary reservoir, soil, water, wild animals, rodents
Food Involved in Outbreaks	Raw or partially cooked oysters, raw or undercooked shellfish (clams and mussels)	Meat (pork, beef, lamb), oysters, fish, raw milk, contaminated pasteurized milk and raw unpasteurized milk, tofu, nonchlorinated water
Preventive Measures	Avoid eating raw or undercooked seafood (particularly oysters), avoid cross-contamination, freezing does not completely destroy these bacteria, purchase from approved sources	Minimize cross-contamination from pork, thoroughly cook food to minimum safe internal temperatures, ensure that facilities and equipment are properly sanitized, follow proper storage procedures, use only sanitary, chlorinated water supplies

Key Point

Viruses cannot reproduce outside a living cell.

Key Point

Viruses can be transmitted from person to person, from people to food, and from people to food-contact surfaces.

Health Alert

Practicing good personal hygiene and minimizing bare-hand contact with ready-to-eat food are important ways to prevent the contamination of food by foodborne viruses.

VIRUSES

Viruses are the smallest of the microbial contaminants. They consist of genetic material wrapped with an outer layer of protein. While a virus cannot reproduce outside a living cell, once inside a human cell, it will reproduce more viruses. Viruses are responsible for several foodborne illnesses, such as hepatitis A and infections caused by Norwalk virus and rotavirus. Rotavirus infections are the leading cause of severe gastroenteritis among infants and young children worldwide.

Basic Characteristics of Viruses

Viruses share some basic characteristics.

○ Unlike bacteria, they rely on a living cell to reproduce.

○ They are not complete cells.

○ Unlike bacteria, they do not reproduce in food.

○ Some may survive freezing and cooking.

○ They can be transmitted from person to person, from people to food, and from people to food-contact surfaces.

○ They usually contaminate food through a foodhandler's improper personal hygiene.

○ They can contaminate both food and water supplies.

Practicing good personal hygiene is an important way to prevent the contamination of food by foodborne viruses. It is especially important to minimize bare-hand contact with ready-to-eat food. Information about major foodborne illnesses caused by viruses, as well as ways to prevent them, is presented in *Exhibit 2i.*

Exhibit 2i: Major Foodborne Illnesses Caused by Viruses

Foodborne Illness	Hepatitis A	Norwalk Virus Gastroenteritis	Rotavirus Gastroenteritis
Virus	*Hepatovirus* or hepatitis A virus	Norwalk and Norwalk-like viral agents	Rotavirus
Type of Illness	Infection	Infection	Infection
Symptoms	Mild or no illness, then sudden onset of fever, general discomfort, fatigue, headache, nausea, loss of appetite, vomiting, abdominal pain, and jaundice after several days	Nausea, vomiting, diarrhea, abdominal cramps, headache, and mild fever	Vomiting and diarrhea, abdominal pain, and mild fever (illness more common in children than adults)
Incubation Period	10 to 50 days, average 30 days	Usually 1 to 2 days	Usually 1 to 3 days
Duration of Illness	1 to 2 weeks; severe cases may last several months	Usually 1 to 3 days, usually self-limiting	Usually 4 to 8 days
Sources	Human intestinal tracts; feces-contaminated water	Human intestinal tract and feces-contaminated water	Human intestinal tract; feces-contaminated water
Food Involved in Outbreaks	Water and ice, shellfish, salads, deli meats and sandwiches, fruits and fruit juices, milk and milk products, vegetables, any food that will not receive a further heat treatment	Water, shellfish (especially raw or insufficiently steamed clams and oysters), raw vegetables, fresh fruit and salads, contaminated water	Water and ice, raw and ready-to-eat food (salads, fruits, and hors d'oeuvres), contaminated water
Preventive Measures	Obtain shellfish from approved sources, prevent cross-contamination from hands, ensure foodhandlers practice good personal hygiene, clean and sanitize food-contact surfaces, use sanitary water sources	Obtain shellfish from approved sources, prevent cross-contamination from hands, ensure foodhandlers practice good personal hygiene, thoroughly cook food to required minimum internal temperatures; use sanitary, chlorinated water	Ensure foodhandlers practice good personal hygiene, thoroughly cook food to required minimum internal temperatures; use sanitary, chlorinated water

Exhibit 2i

PARASITES

Parasites are organisms that need to live in or on a host organism to survive. A **host** is a person, animal, or plant on which another organism lives and takes nourishment. Parasites can live inside many animals that humans use for food, such as cattle, poultry, pigs, and fish. Foodborne parasites include protozoa, roundworms, and flatworms. Parasites require many different hosts to carry out their life cycles; however, they are typically passed to humans through the meat of an animal host. To prevent foodborne illness caused by parasites, make sure food has been properly frozen, use proper cooking techniques, avoid cross-contamination, use sanitary water supplies, and follow proper handwashing procedures, especially after using the restroom.

Basic Characteristics of Foodborne Parasites

Parasites share some basic characteristics.

○ They are living organisms that need a host to survive.
○ They grow naturally in many animals—such as pigs, cats, and rodents—and can be transmitted to humans.
○ Most are very small, often microscopic, but larger than bacteria.
○ They may be killed by proper cooking or freezing.
○ They pose hazards to both food and water.

Information about major foodborne illnesses caused by parasites, as well as ways to prevent them, is presented in *Exhibit 2j.*

FUNGI

Fungi range in size from microscopic, single-celled organisms to very large, multicellular organisms. They are found naturally in air, soil, plants, animals, water, and some food. **Molds, yeasts,** and mushrooms are examples of fungi. The fungi of concern to restaurants and foodservice establishments are molds and yeasts. *(Mushroom toxins are discussed in Chapter 3.)*

Molds

Individual mold cells can usually be seen only with a microscope. However, fuzzy or slimy mold colonies, consisting of a large number of cells, are often visible to the naked eye. Bread mold is an example. The spores produced by molds are not the same as the spores produced by bacteria. Molds use spores for reproduction.

Molds are responsible for the spoilage of food. This spoilage results in discoloration and the formation of odors and off-flavors. Molds are able to

Key Point

Parasites are typically passed to humans through an animal host.

Exhibit 2j: Major Foodborne Illnesses Caused by Parasites

Foodborne Illness	Trichinosis	Anisakiasis	Giardiasis
Parasite	*Trichinella spiralis*	*Anisakis simplex*	*Giardia duodenalis* (formerly *G. lamblia*)
Characteristic of Parasite	Roundworm	Roundworm	Protozoan
Type of Illness	Infection	Infection	Infection
Symptoms	Nausea, diarrhea, abdominal pain, occasionally vomiting, swelling around the eyes, and fever; later: muscle soreness, thirst, extreme sweating, chills, hemorrhaging (bleeding), and fatigue	Severe abdominal pain, cramping, vomiting, nausea, tingling or tickling sensation in throat, vomiting or coughing up worms	Fatigue, nausea, intestinal gas, weakness, weight loss, and abdominal cramps
Incubation Period	2 to 28 days; depends on number of larvae ingested	A few hours to 1 week	3 to 25 days; usually about 7 to 10 days
Duration of Illness	Several days to more than 30 days; depends on treatment and health status of individual	Up to 3 weeks	Usually 1 to 2 weeks; illness may last months
Sources	Domestic pigs; wild game, such as bears and walruses	Marine fish, but not freshwater species; usually herring	Wild animals, especially beavers and bears; domestic animals (dogs and cats); intestinal tract of humans
Food Involved in Outbreaks	Undercooked pork or wild game, pork and nonpork sausages (ground meat may be contaminated by meat grinders)	Raw, undercooked, or improperly frozen seafood, especially cod, haddock, fluke, Pacific salmon, herring, flounder, monkfish, and fish used for sashimi and ceviche	Contaminated water and ice, salads and (possibly) other raw vegetables
Preventive Measures	Cook pork and other meat to minimum internal cooking temperatures; wash, rinse, and sanitize equipment, such as sausage grinders and utensils used in the preparation of raw pork and other meat	Obtain seafood only from certified sources; when serving raw or undercooked seafood, only use sashimi-grade fish that has been properly frozen; avoid eating raw or partly cooked fish and shellfish unless it has been properly treated to eliminate parasites; fish intended to be eaten raw should be frozen at -4°F (-20°C) or lower for 7 days in a freezer, or at -31°F (-35°C) or lower for 15 hours in a blast chiller	Use sanitary, chlorinated water supplies; ensure that foodhandlers practice good personal hygiene; wash raw produce carefully.

(Continued on next page.)

Exhibit 2j

Exhibit 2j: Major Foodborne Illnesses Caused by Parasites (Continued)

Foodborne Illness	Toxoplasmosis	Intestinal Cryptosporidiosis	Cyclosporiasis
Parasite	*Toxoplasma gondii*	*Cryptosporidium parvum*	*Cyclospora cayetanensis*
Characteristic of Parasite	Protozoan; not directly passed from person to person	Protozoan; sporocysts are resistant to most chemical disinfectants	Protozoan
Type of Illness	Infection	Infection	Infection
Symptoms	Often, there are no symptoms. When symptoms do occur, they include enlarged lymph nodes in head and neck, severe headaches, severe muscle pain, and rash	Severe, watery diarrhea; may have no symptoms	Watery diarrhea, loss of appetite, weight loss, bloating, gas, abdominal cramps, nausea, vomiting; muscle aches, mild or no fever; fatigue
Incubation Period	5 to 20 days; depends on immune system of individual	6 to 14 days; average is 7 days	Days to weeks; usually about 1 week
Duration of Illness	Depends on treatment (relapses common)	2 days to 4 weeks, usually self-limiting; indefinite and life-threatening in immunocompromised people	A few days to a month, or longer; usually several weeks (relapses spanning 1 to 2 months are common)
Sources	Animal feces (especially felines), mammals, birds	The intestinal tract of humans, cattle, and other domestic animals; drinking water contaminated with runoff from farms or slaughterhouses	The intestinal tract of humans; contaminated water supplies
Food Involved in Outbreaks	Raw or undercooked meat contaminated with this parasite, especially pork, lamb, venison, and hamburger meat	Water, salads and raw vegetables; milk; raw food, such as apple cider; ready-to-eat food	Water, marine fish, raw milk, raw produce
Preventive Measures	Avoid raw or undercooked meat; thoroughly cook meat to the minimum internal temperature, so there is no pink inside; properly wash hands that come in contact with soil, raw meat, cat feces, or raw vegetables	Ensure that foodhandlers practice good personal hygiene; thoroughly wash produce; use sanitary water	Ensure that foodhandlers practice good personal hygiene, thoroughly wash produce, use sanitary water

grow on almost any food at almost any storage temperature. They can also grow in environments that are moist or dry, have a high or low pH, and are salty or sweet. They typically prefer to grow in and on sweet, acidic food with low water activity. Molds often spoil fruit, vegetables, meat, cheese, and bread because of their water activity and pH. *(See Exhibit 2k.)*

Some molds produce toxins that can cause allergic reactions, nervous system disorders, and kidney and liver damage. For example, aflatoxin, produced by the molds *Aspergillus flavus* and *Aspergillus parasticus,* can cause liver disease.

Food such as corn and corn products, peanuts and peanut products, cottonseed, milk, and tree nuts (such as Brazil nuts, pecans, pistachio nuts, and walnuts) have been associated with aflatoxins.

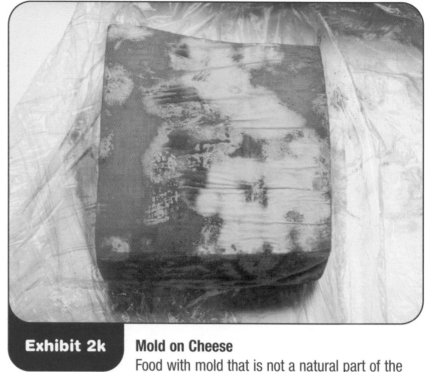

Exhibit 2k **Mold on Cheese**
Food with mold that is not a natural part of the product should always be discarded.

Basic Characteristics of Foodborne Molds

Molds share some basic characteristics.

○ They spoil food and sometimes cause illness.

○ They grow under almost any condition, but grow well in sweet, acidic foods with low water activity.

○ Freezing temperatures prevent or reduce the growth of molds, but do not destroy them.

○ Some molds produce toxins called aflatoxins.

○ Although the FDA recommends cutting away any moldy areas in cheese—at least one inch (2.5 centimeters) around them—to avoid illnesses caused by mold toxins, throw out all moldy food, unless the mold is a natural part of the food (e.g., cheeses such as Gorgonzola, Bleu, Brie, and Camembert).

Although mold cells and spores can be killed by heating them, toxins that may be present are not destroyed by normal cooking methods. Food with molds that are not a natural part of the product should always be discarded.

Exhibit 2l

Yeast on Jelly
Food that has been spoiled by yeast should be discarded.

Yeasts

Some yeasts are known for their ability to spoil food rapidly. Carbon dioxide and alcohol are produced as yeast slowly consumes food. Yeast spoilage may, therefore, produce a smell or taste of alcohol. Yeast may appear as a pink discoloration or slime and may bubble.

Yeasts are similar to molds in that they grow well in sweet, acidic food with low water activity, such as jellies, jams, syrup, honey, and fruit juice. *(See Exhibit 2l.)* Food that has been spoiled by yeast should be discarded.

FOODBORNE INFECTION vs. FOODBORNE INTOXICATION

Foodborne diseases are classified as infections, intoxications, or toxin-mediated infections. Each occurs in a different way.

- ○ **Foodborne infections** result when a person eats food containing pathogens, which then grow in the intestines and cause illness. Typically, symptoms of a foodborne infection do not appear immediately.
- ○ **Foodborne intoxications** result when a person eats food containing toxins that cause illness. The toxin may have been produced by pathogens found on the food or may be the result of a chemical contamination. The toxin could also come from a plant or animal that was eaten. Typically, symptoms of a foodborne intoxication appear quickly, within a few hours.
- ○ Foodborne **toxin-mediated infections** result when a person eats food containing pathogens, which then produce illness-causing toxins in the intestines.

EMERGING PATHOGENS AND ISSUES

Progress has been made in preventing foodborne diseases. For example, typhoid fever (caused by *Salmonella typhii)* was very common in the early Twentieth Century. Typhoid is now almost forgotten in the United States due to improved methods of treating drinking water and sewage, milk sanitation and pasteurization, and shellfish sanitation. Similarly, cholera and trichinosis, once very common, are now also rare in the United States.

In the past, foods involved in foodborne-illness outbreaks were often raw or inadequately cooked meat, poultry, seafood, or unpasteurized milk. Today, some foods previously considered safe have become vehicles for foodborne pathogens. For example, it was assumed for centuries that the internal contents of an egg were sterile and safe to eat raw. Now, however, there are clear indications that, in rare cases, eggs can be contaminated internally with *Salmonella* enteritidis. A number of outbreaks of salmonellosis have been traced to undercooked, contaminated eggs included in traditional recipes, such as eggnog, Hollandaise sauce, Caesar salad, lightly cooked omelets, French toast, and even meringue.

Produce and highly acidic food, once considered safe, are now considered potential vehicles for foodborne pathogens. As detection techniques and foodborne-illness investigations improve, new pathogens causing foodborne illness are being recognized.

These trends have emerged for a variety of reasons. As the population continues to grow, more food is needed. Food is produced in greater quantities and at centralized sources. More food is shipped from greater distances, across state and national borders. It remains important for establishments to purchase food from reputable suppliers and to keep food safe by using proper storage, preparation, and serving practices.

HAZARDS ASSOCIATED WITH FRESHER AND HEALTHIER FOOD

Many changes are occurring that can greatly affect food safety. Natural or organic food without additives or preservatives is becoming more available, along with low-fat, low-calorie versions of everyday food. Also, food from all over the world is becoming more accessible. In response to the public's health consciousness, many products commonly seen on grocery store shelves are now being altered.

Altering the makeup of food is not without risk to its safety. If the stability of a product was traditionally due to the water activity or pH (i.e., the product was too dry or acidic to allow microorganisms to grow), changing its makeup may eliminate that barrier, causing it to become unsafe.

Organic products might also have food safety risks. Organic food is grown and processed without the use of pesticides, additives, and preservatives. Lacking these traditional barriers to microbial growth, organic products might have reduced shelf lives and allow spoilage microorganisms to grow.

TECHNOLOGICAL ADVANCEMENTS IN FOOD SAFETY

Several new technologies are emerging to reduce pathogenic and spoilage microorganisms in food. These include food irradiation, electron pasteurization, high-pressure processing, bacteriocins, and various light treatments, including laser light, ultraviolet light, and pulsed light.

Food irradiation, often called cold pasteurization, involves exposing food to an electron beam or gamma rays, similar to the way microwaves are used to heat food. The FDA has approved the use of irradiation on certain foods in the United States. To date, irradiation is allowed for pork (to control the parasite *Trichinella spiralis),* poultry (to control bacterial contaminants), fruit and vegetables (to prevent premature maturation and to control insect pests), and on herbs, spices, teas, and other dried vegetable substances (to control various microbial contaminants).

Benefits of food irradiation include:

○ Reduction or elimination of pathogens and spoilage microorganisms

○ Replacement of chemical treatment of food

○ Extended shelf life of food

Irradiated foods have only recently become widely available in the United States, despite regulatory approval and repeated endorsements by professional health and nutrition organizations. No evidence exists of harmful effects resulting from the amount or types of radiation used, and nutritional value, appearance, and taste are virtually not affected. However, some consumers remain apprehensive about food exposed to radiation, due to their limited knowledge of the irradiation process. Studies show that once the process and benefits are explained to them, consumers are more receptive to the idea of irradiation and would purchase irradiated products.

It is important to note that food irradiation does not replace proper food storage and handling practices, since food could still become unsafe if cross-contaminated. Therefore, when handling irradiated food, you should follow the same food safety practices you would follow with other food.

SUMMARY

Microbial contaminants are responsible for the majority of foodborne illnesses. Because bacteria, viruses, parasites, and fungi can be introduced at any point in the flow of food, they must be monitored and controlled throughout. Understanding how these microorganisms grow, reproduce, contaminate food, and infect humans is critical to understanding how to prevent the foodborne illnesses they cause.

Of all foodborne microorganisms, bacteria are of greatest concern to the manager. Bacteria are living, single-celled organisms that can be carried by a variety of means. Some cause food spoilage; others cause illness. Some bacteria cause illness by producing toxins as they multiply, die, and break down. Under favorable conditions, bacteria can reproduce very rapidly. Although vegetative bacteria may be resistant to low—even freezing—temperatures, they can be killed by high temperatures, such as those reached during cooking. Some types of bacteria, however, have the ability to form spores, which protect the bacteria from unfavorable conditions. Since spores are so difficult to destroy, it is important to thaw, cook, cool, and reheat food properly to keep bacteria from growing to harmful levels.

The acronym FAT TOM—which stands for Food, Acidity, Time, Temperature, Oxygen, and Moisture—might help you remember the conditions that promote the growth of foodborne microorganisms. Needing nutrients to grow, specifically proteins and carbohydrates, microorganisms grow best in food with a slightly acidic-to-neutral pH. Most foodborne microorganisms grow well between the temperatures of 41°F and 140°F (5°C and 60°C). They also need sufficient time within these temperatures to grow. If contaminated food remains in the temperature danger zone for four hours or more, pathogenic microorganisms can grow to levels high enough to make someone ill. Most microorganisms that cause foodborne illness can grow either with or without the presence of oxygen, and require the moisture in food to grow. FAT TOM is the key to controlling the growth of microorganisms. Multiple barriers denying the microorganisms as many growth-supporting conditions as possible need to be put in place. These barriers include making food more acidic, raising or lowering its temperature, and lowering water activity.

Viruses are the smallest of the microbial contaminants. While a virus cannot reproduce in food, once ingested it will cause illness. Viruses can be transmitted from person to person, from people to food, and from people to food-contact surfaces. They can contaminate both food and water supplies. Some may survive freezing and cooking. Practicing good personal hygiene and minimizing bare-hand contact with ready-to-eat food is an important defense against foodborne illness from viruses.

Parasites are organisms that need to live in, or on, a host organism to survive. They can live inside many animals humans eat, such as cattle, poultry, pigs, and fish. They can be killed by proper cooking and freezing.

Molds and yeasts are examples of fungi, another concern of restaurant and foodservice establishments. Fungi are mostly responsible for the spoilage of many kinds of food, while some molds can produce harmful toxins. Molds are able to grow in a variety of environments, but they typically prefer to grow in and on sweet, acidic food with low water activity. Yeasts are known for their ability to spoil food rapidly. They are similar to molds in that they grow well in sweet, acidic food with low water activity.

Foodborne diseases are classified as infections, intoxications, or toxin-mediated infections. Each occurs in a different way. Foodborne infections result when a person eats food containing pathogens, which then grow in the intestines and cause illness. Typically, symptoms do not appear immediately. Foodborne intoxications result when a person eats food containing illness-causing toxins produced by pathogens found on the food or the result of a chemical contamination. The toxin might also come from a plant or animal that was eaten. Typically, symptoms of foodborne intoxication appear quickly, within a few hours. Foodborne toxin-mediated infections result when a person eats food that contains pathogens, which then produce illness-causing toxins in the intestines.

Our ever-changing world is opening new doors for microorganisms. Many new pathogens are emerging. New product formulations are creating niches for microorganisms that may not have been able previously to grow in the product. However, many new technologies are available and others are being developed to prevent the growth of pathogenic microorganisms.

A CASE IN POINT

Case Study

A day-care center is serving stir-fried rice for lunch. The rice was cooked to the proper temperature for the proper amount of time at 1:00 P.M. The covered rice was then placed on the countertop and allowed to cool to room temperature. At 6:00 P.M., the cook placed it in the refrigerator. At 9:00 A.M. the following day, the rice was combined with the other ingredients for stir-fried rice and cooked to 165°F (74°C) for at least fifteen seconds. The cook covered the stir-fried rice and left it on the range until she gently reheated it at noon. Within an hour of eating the stir-fried rice, however, several of the children began vomiting, and a few had diarrhea. Samples from some of the children revealed the rice as the probable cause of the outbreak.

Based on the information given, was the illness caused by bacteria, a virus, a parasite, or a fungi? What is the name of the microorganism most likely to have caused the outbreak? Is this illness an infection or an intoxication?

TRAINING TIPS

1. Pathogen Scramble

Purpose: *After completing this activity, trainees should be able to identify the major foodborne illnesses caused by bacteria, viruses, and parasites.*

Time: 20 minutes

Directions: Create a word scramble game using the different types of bacteria, viruses, and parasites discussed in Chapter 2. Scramble the name of each pathogen and write it on a piece of paper (one pathogen per page).

Example: suerec sullicab *(Bacillus cereus)*

Then divide your trainees into four or five groups and give each group a bell. Pass out one sheet with a pathogen on it and ask each group to ring the bell when they have unscrambled the pathogen. If they are correct, they get twenty points. (If you choose, you can assign a certain number of points for each pathogen, based on its perceived difficulty. Otherwise, to keep it simple, you can just give twenty points for each one.)

To enhance the learning process, you could require the group that correctly unscrambles the word to give two or three characteristics of the pathogen, and give bonus points for each correct answer. As usual, the team with the most points wins!

2. Twenty Questions

Purpose: *After completing this activity, trainees should be able to identify the major foodborne illnesses caused by bacteria, viruses, and parasites, and be able to identify characteristics, symptoms, sources, food involved, and preventive measures for each of these pathogens.*

Time: 40 minutes

Directions: After discussing the information in Chapter 2, divide trainees into several groups. Assign each group a pathogenic bacteria, virus, or parasite discussed in the chapter. Have each group study its pathogen for ten minutes. Ask them to focus on:

○ Characteristics of the pathogen

○ Symptoms of illness caused by the pathogen

○ Source of the pathogen

○ Food involved in transmitting the pathogen

○ Preventive measures taken against the pathogen

Then, one at a time, ask each group to stand up and solicit up to a total of twenty questions from the other groups. Each group in the audience asks one question and guesses what pathogen the group presenting represents. When a group correctly identifies the pathogen, award points based on the number of unasked questions remaining. For example, if four questions were asked before a correct guess, award the group making the correct guess sixteen points (20 - 4 = 16).

In the unlikely event no group can figure out what pathogen a particular group represents, the group presenting receives twenty points. Play until every group has presented their pathogen. The group with the most points wins!

Note: *If time is an issue, select the six or so most common pathogens and assign larger groups.*

DISCUSSION QUESTIONS

1. Why should potentially hazardous food not be left in the temperature danger zone for more than four hours? Refer to some of the phases of bacterial growth in your explanation.

2. Jams and jellies have water activity levels of about 0.75. Based on this information alone, do they provide an environment favoring bacterial growth? Explain your thinking.

3. In what ways are viruses and bacteria similar? How are they different?

4. What is the difference between a foodborne infection and foodborne intoxication?

5. Why is it important to use multiple barriers to prevent bacterial growth? What are some examples that illustrate your reasoning?

MULTIPLE-CHOICE STUDY QUESTIONS

1. To what does the acronym FAT TOM refer?

   A. The major types of microorganisms that can cause foodborne illness
   B. The conditions supporting the growth of microorganisms
   C. Human health hazards associated with food high in fat and sodium
   D. The federal agency responsible for monitoring food safety

2. All of the following conditions typically support the growth of microorganisms *except*

   A. moisture.
   B. a protein or carbohydrate food source.
   C. high acidity.
   D. temperatures between 41°F and 140°F (5°C and 60°C).

3. A person who has a foodborne infection most likely has eaten a food containing

   A. a ciguatoxin.   C. a plant toxin.
   B. histamine.   D. a live pathogen.

4. Which food is more likely to transmit parasites to humans?

   A. Improperly cooked eggs
   B. Improperly frozen sashimi
   C. Improperly refrigerated milk
   D. Unpasteurized apple juice

5. Which of the following is *not* a basic characteristic of foodborne mold?

   A. It grows well in sweet, acidic food with low water activity.
   B. Freezing temperatures prevent or slow its growth, but do not destroy it.
   C. Its cells and spores may be killed by heating, but the toxins it produces may not be destroyed.
   D. It needs a host to survive.

6. Which of the following statements regarding foodborne intoxication is true?

   A. Symptoms of intoxication often appear days after exposure.
   B. Medical treatment for intoxication can be painful.
   C. Foodborne intoxication is more common than foodborne infection.
   D. Symptoms of intoxication appear quickly, within a few hours.

7. Which of the following microorganisms is most likely to cause a foodborne infection?

   A. *Listeria monocytogenes*
   B. *Staphylococcus aureus*
   C. *E. coli* O157: H7
   D. *Clostridium botulinum*

8. Which of the following microorganisms is most likely to cause a foodborne intoxication?

   A. *Staphylococcus aureus*
   B. *Listeria monocytogenes*
   C. *Campylobacter jejuni*
   D. *Salmonella* spp.

9. Which of the following foods is commonly associated with an outbreak of *Bacillus cereus?*

   A. Cooked rice
   B. Juices
   C. Raw chicken
   D. Shell eggs

10. Which of the following foods is commonly associated with an outbreak of *Vibrio* spp.?

    A. Raw oysters
    B. Unpasteurized milk
    C. Apple cider
    D. Sliced melons

For answers, please turn to Appendix A.

ADDITIONAL RESOURCES

Books and Periodicals

American Public Health Association. 1995. *Control of communicable diseases in man.* Washington D.C.: American Public Health Association.

Buchanan, R. L. and M. P. Doyle. 1997. Foodborne disease significance of *Escherichia coli* O157:H7 and other enterohemorrhagic *E. coli. Food Technology.* 51 (10):69-76.

Bush, R. K. 1995. Seafood allergy and allergens: A review. *Food Technology.* 49 (10):103-116.

Carrera, N. 1997. Avoiding cross-contamination. *Pizza Today.* 16 (5):84.

Cliver, D. O. 1997. Virus transmission via food. *Food Technology.* 51 (4):71-78.

Cliver, D. O., ed. 1990. *Foodborne diseases.* San Diego, CA: Academic Press.

Dulen, J. 1998. Under the microscope. *Restaurants & Institutions.* 108 (21):80.

Durocher, J. 1997. Germ warfare. *Restaurant Business.* 96 (22):123.

Farkas, D. 1996. Creating awareness. *Restaurant Hospitality.* 80 (9):112.

Featsent, A. W. 1998. Food fright! Consumers' perceptions of food safety versus reality. *Restaurants USA.* 18 (6):30.

Grossbauer, S. 1998. Food safety on the Web. *FoodService Director.* 11 (2):140.

Hernandez, J. 1997. Eliminating the *E. coli* threat. *Food Management.* 32 (12):67.

How to prevent foodborne illness. 1995. *FoodService Director.* 8 (2):150.

International Commission on Microbiological Specifications for
Food (ICMSF). 1988. *Microorganisms in foods 1: Their
significance and methods of enumeration,* rev. ed. Toronto:
University of Toronto Press.

ICMSF. 1997. *Microorganisms in foods 3: Vol. 1. Factors affecting life
and death of microorganisms: Vol. 2. Food commodities: Vol. 3.
Microbial ecology of foods.* New York: Academic Press.

ICMSF. 1996. *Microorganisms in foods 5: Characteristics of microbial
pathogens.* London: Blackie Academic & Professional.

Jay, J. M. 1996. *Modern food microbiology.* New York: Chapman & Hall.

Longree, K. and G. Armbruster. 1996. *Quantity food sanitation,* 5th ed.
New York: Wiley.

Marriott, N. G. 1994. *Principles of food sanitation.* New York:
Chapman & Hall.

National Restaurant Association. 1996. *Sanitation survival kit.*
Washington, D.C.: National Restaurant Association.

Neumann, R. 1998. The eight most frequent causes of foodborne
illness. *Food Management.* 33 (6):28.

Puzo, D. P. 1997. USDA estimates financial costs of foodborne illness.
Restaurants & Institutions. 107 (16):80.

Rehe, S. 1990. *Preventing foodborne illness: A guide to safe foodhandling.*
Washington D.C.: United States Department of Agriculture
Food Safety & Inspection Service.

Rubin, K.W. 1995. What to look for, how to prevent it: Food
poisoning. *FoodService Director.* 8 (11):72.

Salmonella and food safety. 1995. *National Culinary Review.*
19 (11):26.

Spake, A. 1997. O is for outbreak. *United States News & World
Report.* 123 (20):70.

Study considers seven foodborne illnesses. 1997. *FoodReview.*
20 (3):37.

Web Sites

American Council on Science and Health

http://www.acsh.org/food

The American Council on Science and Health, Inc., (ACSH) is a consumer education consortium concerned with issues related to food, nutrition, chemicals, pharmaceuticals, lifestyle, the environment, and health. ACSH is an independent, nonprofit, tax-exempt organization. This Web site provides practical, matter-of-fact information on foodborne-illness stories you might hear about in the news.

American Society of Microbiology

http://www.asmusa.org

The mission of the American Society for Microbiology (ASM) is to promote the microbiological sciences and their applications for the common good. To this end, ASM publishes journals and books, and conducts and supports education, training, and public information programs to facilitate the dissemination and application of new microbiological knowledge affecting public interest. This Web site contains excellent information on foodborne illness and teaching microbiology.

Centers for Disease Control and Prevention (CDC)

http://www.cdc.gov

The mission of the CDC is to promote health and quality of life by preventing and controlling disease, injury, and disability. To prevent and control foodborne illness, the CDC collects data on outbreaks. This Web site provides general information on foodborne illnesses and their prevention.

Compendium of Fish and Fishery Product Processes, Hazards, and Controls

http://seafood.ucdavis.edu/haccp/compendium/compend.htm

Developed by the U.S. Food and Drug Administration, the compendium available at this Web site provides useful information on biological, chemical, and physical hazards associated with fish and fish products.

Council for Biotechnology Information

http://www.whybiotech.com

The Council for Biotechnology Information was founded by leading biotechnology companies to create a comprehensive communication campaign about biotechnology. The Council is committed to providing objective, balanced information to help the public better understand and appreciate the benefits of biotechnology.

FDA Bad Bug Book

http://vm.cfsan.fda.gov/~mow/intro.html

Produced by the FDA's Center for Food Safety and Nutrition (FDA CFSAN), this online handbook provides basic facts regarding pathogenic foodborne microorganisms and natural toxins. It brings together information from the FDA, the Centers for Disease Control, the USDA Food Safety Inspection Service, and the National Institutes of Health.

FDA-Center for Food Safety and Applied Nutrition (CFSAN)

http://www.cfsan.fda.gov

As the center within the FDA responsible for food safety and nutrition, CFSAN promotes and protects the public health by researching and implementing guidelines, policies, and standards to ensure that food is safe, nutritious, wholesome, and properly labeled. This Web site provides a wealth of information on food safety and sanitation, including corresponding guidelines, policies, and standards.

FDA Food Code

http://vm.cfsan.fda.gov/~dms/foodcode.html

As the basis for many local sanitation codes, as well as the basis for information in this textbook, the FDA Food Code, available at this Web address, is a useful resource for information relating to food safety information for the restaurant and foodservice industry.

Hepatitis Control Report

http://www.hepatitiscontrolreport.com

This site is an online quarterly newsletter devoted to news on the public-health control of viral hepatitis. Its mission is to provide accurate and balanced reporting of developments in hepatitis epidemiology, control programs, and public policy.

International Atomic Energy Agency

http://www.iaea.org

As a specialized agency within the United Nations system, the International Atomic Energy Agency (IAEA) serves as the world's central intergovernmental forum for scientific and technical cooperation in the nuclear field. This Web site houses up-to-date information on all issues concerning atomic energy, including irradiation.

International Food Information Council (IFIC) Foundation

http://ific.org

IFIC collects and disseminates scientific information on food safety, nutrition, and health. They work with an extensive roster of scientific experts to help translate research into understandable and useful information for opinion leaders and, ultimately, consumers. This Web site provides easy-to-understand information on foodborne illness and health-related stories circulating in the news.

Journal of Emerging Infectious Diseases

http://www.cdc.gov/ncidod/eid

This journal provides information on emerging infections and diseases. It is published by the National Center for Infectious Diseases and is available—free-of-charge—on this site in several languages. The journal databases can be searched for foodborne pathogen-related information.

Monsanto Company

http://www.monsanto.com

The Monsanto Company is a leading provider of agricultural solutions to growers worldwide. They provide top-quality, cost-effective, and integrated approaches to help farmers improve their productivity and produce better quality food. This Web site contains information on agricultural biotechnology.

Morbidity and Mortality Weekly Report (MMWR)

http://www.cdc.gov/mmwr

The MMWR is prepared by the Centers for Disease Control and Prevention and is available—free-of-charge—on this Web site. The reports contain data based on weekly accounts of reportable diseases from state health departments. Report databases can be searched for information on foodborne-illness outbreaks.

National Center for Infectious Diseases (NCID)

http://www.cdc.gov/ncidod

NCID is one of the centers of the Centers for Disease Control. Its mission is to prevent illness, disability, and death caused by infectious diseases around the world. NCID accomplishes this mission by conducting surveillance, epidemic investigations, epidemiologic and laboratory research, and training. It also sponsors public education programs to develop, evaluate, and promote prevention and control strategies for infectious diseases. This Web site serves as another great resource for information on foodborne illness.

National Restaurant Association

http://www.restaurant.org

The National Restaurant Association is the leading business association for the restaurant industry. Together with the National Restaurant Association Educational Foundation, the Association's mission is to represent, educate, and promote the rapidly growing restaurant and foodservice industry. This Web site should be your starting place for all issues and concerns related to your restaurant. This Web site has it all, from tips for running your establishment to vital data on your customers' spending habits.

Chapter 3

Contamination, Food Allergies, and Foodborne Illness

Learning Objectives

After completing this chapter, you should be able to

○ Identify the three types of contamination (biological, chemical, and physical) and give examples of each.

○ Identify ways in which food can become contaminated.

○ Identify methods to prevent biological, chemical, and physical contamination.

Knowledge

TEST YOUR FOOD SAFETY KNOWLEDGE

1. **True or False:** Fish that has been properly cooked will be safe to eat. *(See page 3-4.)*

2. **True or False:** Cooking can destroy the toxins in toxic wild mushrooms. *(See page 3-5.)*

3. **True or False:** Copper utensils and equipment can cause an illness when used to prepare acidic food. *(See page 3-5.)*

4. **True or False:** Cleaning products may be stored with packages of food. *(See page 3-7.)*

5. **True or False:** Most biological toxins found in seafood, plants, and mushrooms occur naturally and are not caused by the presence of microorganisms. *(See page 3-2.)*

For answers, please turn to Appendix A.

Key Terms

Biological contaminant
Biological toxins
Chemical toxins
Ciguatera poisoning
Scombroid poisoning
Histamine
Chemical contaminant
Toxic metal poisoning
Physical contaminant
Food allergy

Food is considered contaminated when it contains hazardous substances. These substances may be biological, chemical, or physical. The most common food contaminants are **biological contaminants** that belong to the microworld—bacteria, parasites, viruses, and fungi. Most foodborne illnesses result from these contaminants, but **biological toxins** and **chemical toxins** are also responsible for many foodborne illnesses. While biological and chemical contamination pose a significant threat to food, the danger from physical hazards should also be recognized.

TYPES OF FOODBORNE CONTAMINATION

Ensuring the safety of food is the manager's most important job. A thorough understanding of the causes and prevention of various types of contamination can help you keep food safe.

Biological Contamination

As you learned in Chapter 2, a foodborne intoxication occurs when a person eats food containing toxins. The toxin may have been produced by pathogens found on the food or may be the result of a chemical contamination. The toxin could also come from a plant or animal that was eaten. Toxins in seafood, plants, and mushrooms are responsible for many cases of foodborne illness in the United States each year. Most of these biological toxins occur naturally and are not caused by the presence of microorganisms. Some occur in animals as a result of their diet.

Seafood Toxins

The ciguatera toxin occurs in certain predatory tropical reef fish, such as amberjack, barracuda, grouper, and snapper. *(See Exhibit 3a.)* Ciguatera accumulates in the tissue of these large, predatory fish after they eat smaller fish that have fed upon certain species of toxic algae. When a person eats fish containing this toxin, an illness may result, requiring weeks or months of

Exhibit 3a **Ciguatera Toxin**
Because the ciguatera toxin cannot be smelled or tasted and is not destroyed by cooking, predatory tropical reef fish, such as snapper, should be purchased only from approved suppliers.

recovery. Symptoms of **ciguatera poisoning** include vomiting, severe itching, nausea, dizziness, hot and cold flashes, temporary blindness, and, sometimes, hallucinations. Because the ciguatera toxin cannot be smelled or tasted and it is not destroyed by cooking, it is very important to purchase predatory tropical reef fish only from approved suppliers.

Shellfish may contain toxins that occur because of the algae upon which they feed. Illnesses caused by shellfish poisoning vary and are specific to the type of toxin consumed. Paralytic shellfish poisoning (PSP), the most serious type, may lead to respiratory failure and even death if respiratory support is not provided. PSP is generally associated with mussels, clams, cockles, and scallops. Since cooking may not destroy shellfish toxins, it is important to purchase shellfish from approved suppliers who can certify that the shellfish have been harvested from safe waters. *(See Exhibit 3b.)*

Exhibit 3b

Shellfish Toxins
Purchase shellfish from approved suppliers who can certify the shellfish have been harvested from safe waters.

Scombroid poisoning is one of the more common forms of illness caused by fish toxins in the United States. It occurs when scombroid species of fish—such as tuna, mackerel, bluefish, skipjack, swordfish, and bonito—are time-temperature abused. Under these conditions, bacteria associated with the fish produce the toxin **histamine.** This odorless, tasteless chemical causes scombroid intoxication when consumed. Symptoms of the illness include flushing and sweating, a burning, peppery taste in the mouth, dizziness, nausea, and headache. Sometimes a facial rash, hives, edema, diarrhea, and abdominal cramps will follow. Scombroid poisoning, also known as histamine poisoning, has been associated with other fish species as well, such as mahi-mahi, marlin, and sardines. Histamine is not destroyed by cooking or freezing. Since time-temperature abuse during the harvesting process may cause scombroid fish to become unsafe, it is important to purchase these fish from reputable suppliers who practice strict time-temperature controls.

Some fish toxins are systemic: that is, they occur as a natural part of the fish. An example of a potentially toxic fish is pufferfish, which contains tetrodotoxin in its liver, skin, and other organs. Consuming the toxin in this fish can produce rapid and violent death. Few cases of this type of poisoning

Key Point

Purchase scombroid fish, such as tuna and swordfish, from reputable suppliers who practice strict time-temperature controls.

have been reported in the United States. Cooking may not destroy systemic fish toxins. Pufferfish should be eaten only if they are handled and prepared by a trained and licensed chef.

The following general procedures should be practiced to guard against a seafood-specific foodborne illness.

- Purchase fish from reputable suppliers who maintain strict time-temperature controls.
- Refuse fish that has been thawed and refrozen.
- Check temperature. The temperature of fresh fish must be 41°F (5°C) or lower upon arrival. However, this temperature standard will not prevent a scombroid poisoning if the fish has been temperature-abused during the harvesting process or prior to being received. This is why it is critical to purchase fish from reputable suppliers.
- Thaw frozen fish at refrigerator temperatures of 41°F (5°C) or lower.

Remember that toxins are not living organisms, so cooking or freezing may not destroy them. Therefore, most of the preventive measures taken against biological toxins in seafood must be implemented when purchasing and receiving. Chapter 5 will discuss proper receiving practices in greater detail.

Plant Toxins

Plant toxins are another form of biological contamination. Most poisonings caused by plants result when toxic plants have been used in medicinal home remedies. Foodborne illnesses have occurred after people have consumed rhubarb leaves, jimsonweed, and the root of water hemlock. Some illnesses have occurred after animals have eaten toxic plants and people have consumed the byproducts of those animals. For example, people have become ill after consuming milk from cows that had eaten snakeroot. People have also become ill from consuming honey produced by bees that had gathered nectar from mountain laurel or rhododendrons.

Some plants may be toxic in their raw state but safe when properly cooked. For example, fava beans and red kidney beans may be unsafe if eaten when raw or undercooked. In general, toxic plant species and products prepared with them should be avoided. Only commercially processed honey and properly cooked beans should be used.

Mushroom Toxins

Foodborne-illness outbreaks associated with mushrooms are almost always caused by the consumption of wild mushrooms collected by amateur mushroom hunters. Most cases occur when toxic mushroom species are confused with edible species. The symptoms of intoxication vary depending

Key Point

Toxic plant species and products prepared with them should not be used in a foodservice establishment.

upon the species consumed. Some mushroom toxins will destroy internal organs; others cause convulsions, hallucinations, and coma; still others produce nausea, vomiting, abdominal cramping, and diarrhea.

Cooking or freezing will not destroy the toxins produced by toxic wild mushrooms. Establishments should not use wild mushrooms or products made with them. All mushrooms should be purchased from approved suppliers. *(See Exhibit 3c.)*

Exhibit 3d on the next page summarizes information about common biological toxins, the sources of these toxins, the food with which they are often associated, and preventive measures that can be taken to keep these toxins from causing foodborne illness.

Exhibit 3c **Mushroom Toxins**
Restaurant and foodservice establishments should not use wild mushrooms or products made with them.

Chemical Contamination

Chemical contaminants are responsible for many cases of foodborne illness. Contamination can come from a variety of substances normally found in restaurant and foodservice establishments. These include toxic metals, pesticides, and chemicals.

Toxic Metals

Utensils and equipment that contain toxic metals—such as lead, copper, brass, zinc, antimony, and cadmium—can cause a **toxic metal poisoning**. If acidic food is stored in or prepared with this type of equipment, it can leach these metals from the item and become contaminated. For example, storing tomato sauce in a copper pot or lemonade in a pewter pitcher could lead to a foodborne illness. Only food-grade utensils and equipment should be used to prepare and store food.

Improperly installed carbonated-beverage dispensers can also create a hazard. If carbonated water is allowed to flow back into the copper supply lines, it could leach copper from the line and contaminate the beverage. Beverage-dispensing systems should be installed and maintained only by professionals who will ensure that a proper backflow-prevention device is installed.

Key Point

Only food-grade utensils and equipment should be used to prepare and store food.

Exhibit 3d: Biological Toxins (Biological Contaminants)

Biological Toxin	Source of Contamination	Associated Food	Preventive Measures
Seafood Toxins			
Ciguatera Toxin	Fish that have eaten algae containing the toxin	Predatory tropical reef fish, such as amberjack, barracuda, grouper, and snapper	Cooking does not destroy these toxins; purchase predatory tropical reef fish only from approved suppliers
Scombroid Toxin (histamine)	Histamine produced by bacteria in some fish when they are time-temperature abused	Primarily occurs in tuna, bluefish, mackerel, swordfish, skipjack, roundfish, and bonito; other fish, such as mahi-mahi, marlin, and sardines, have also been implicated in histamine poisoning	Cooking does not destroy histamine; because time-temperature abuse during the harvesting process may cause the fish to become unsafe, it is important to purchase from reputable suppliers
Shellfish Toxins	Shellfish that have eaten a type of algae containing the toxin	Shellfish, especially mollusks, such as mussels, clams, cockles, and scallops	Cooking may not destroy these toxins; purchase these shellfish from approved suppliers who can certify they are harvested from safe waters
Systemic Fish Toxins	Toxins that are a natural part of some fish	Pufferfish, moray eels, and freshwater minnows	Cooking may not destroy systemic fish toxins; pufferfish should be handled and prepared by properly trained chefs
Plant Toxins	Toxins that are a natural part of some plants	Fava beans, rhubarb leaves, jimsonweed, water hemlock, and apricot kernels; honey from bees that have gathered nectar from mountain laurel; milk from cows that have eaten snakeroot, jimsonweed, and other toxic plants	Cooking may not destroy these toxins; avoid these plant species and products prepared with them
Fungal Toxins	Toxins that are a natural part of some varieties of fungi	Poisonous varieties of mushrooms and other fungi	Cooking does not destroy these toxins; do not use wild mushrooms; purchase mushrooms only from approved suppliers

Exhibit 3d

Chemicals and Pesticides

Chemicals such as cleaning products, polishes, lubricants, and sanitizers can contaminate food if they are improperly used or stored. Follow the directions supplied by the manufacturer when using these chemicals. Exercise caution when using chemicals during operating hours to prevent contamination of food and food-preparation areas. Store chemicals away from food, utensils, and equipment used for food. Keep them in a locked storage area in their original containers. If chemicals must be transferred to smaller containers or spray bottles, label each container appropriately.

Pesticides are often used in kitchens and in food-preparation and storage areas to control pests, such as rodents and insects. If used, pesticides should be applied only by a licensed pest control operator (PCO). All food should be wrapped or stored prior to application of the pesticide. Pesticides should be stored with the same care as other chemicals used in the establishment.

Exhibit 3f on the next page summarizes information about some common chemical contaminants, their sources, the food with which they are often associated, and measures that can be taken to prevent them from causing foodborne illness.

Physical Contamination

Physical contamination results from the accidental introduction of foreign objects into food. Common **physical contaminants** include metal shavings from cans *(see Exhibit 3e)*, staples from cartons, glass from broken light bulbs, blades from plastic or rubber scrapers, fingernails, hair, bandages, and dirt. The contaminant may even be a natural part of the food itself, such as the bones in chicken or fish. Closely inspect the food you receive and take steps to ensure that it will not be physically contaminated during the flow of food in your operation.

Key Point Chemicals should not be stored with food.

Key Point Pesticides should only be applied by a trained pest control operator (PCO).

Exhibit 3e **Physical Contamination** Metal shavings in an opened can might contaminate the food inside.

Exhibit 3f: Chemical Contaminants

Chemical Toxin	Source of Contamination	Associated Food	Preventive Measures
Toxic Metals	Utensils and equipment containing potentially toxic metals, such as lead, copper, brass, zinc, antimony, and cadmium	Any foods, but especially high-acid food, such as sauerkraut, tomatoes, and citrus products; the acidity of this food can cause metal ions to leach into its liquids Carbonated beverages; the carbonated water used to make the beverage might leach copper ions from copper water supply lines	Use only food-grade storage containers Use metal and plastic containers only for their intended use Use only food-grade brushes on food; do not use paintbrushes or wire brushes Do not use enamelware, which may chip and expose the underlying metal Do not use equipment or utensils made of materials that contain lead (such as pewter) for food preparation Do not use zinc-coated (galvanized) equipment or utensils for food preparation Use a backflow-prevention device to prevent carbonated water in soft drink-dispensing systems from flowing back into the copper water-supply lines
Chemicals	Cleaning products, polishes, lubricants, and sanitizers		Follow manufacturers' directions for storage and use; use only recommended amounts Store away from food, utensils, and equipment used for food Store in a dry cabinet in original, labeled containers, apart from other chemicals that might react with them Tools used for dispensing chemicals should never be used on food If chemicals must be transferred to smaller containers or spray bottles, label each container appropriately Use only food-grade lubricants or oils on kitchen equipment or utensils
Pesticides	Used in kitchens and food-preparation and storage areas to control pests, such as rodents and insects		Pesticides should only be applied by a licensed professional; wrap or store all food before pesticides are applied

Exhibit 3f

FOOD ALLERGIES

Six to seven million Americans have food allergies. A **food allergy** is the body's negative reaction to a particular food protein. Depending on the person, allergic reactions may occur immediately after the food is eaten or several hours later. The reaction could include some or all of the following symptoms:

○ Itching in and around the mouth, face, or scalp

○ Tightening in the throat

○ Wheezing or shortness of breath

○ Hives

○ Swelling of the face, eyes, hands, or feet

○ Gastrointestinal symptoms, including abdominal cramps, vomiting, or diarrhea

○ Loss of consciousness

○ Death

Employees should be aware of the most common food allergens, including milk and dairy products, eggs and egg products, fish, shellfish, wheat, soy and soy products, peanuts, and tree nuts.

Your employees should be able to inform customers of menu items that contain these potential allergens. Designate one person per shift to answer customers' questions regarding menu items. To help customers with allergies enjoy a safe meal at your establishment, keep the following points in mind.

○ Be able to fully describe each of your menu items when asked. Tell customers how the item is prepared and identify any "secret" ingredients used.

○ If you don't know if an item is free of an allergen, tell the customer. Urge the customer to order something else.

○ When preparing food for a customer with allergies, ensure that the food makes no contact with the ingredient the customer is allergic to. Make sure all cookware, utensils, and tableware are allergen-free to prevent food contamination.

○ Serve menu items as simply as possible to customers with allergies. Sauces and garnishes are often the source of allergic reactions. Serve these items on the side.

SUMMARY

Biological and chemical toxins are responsible for many foodborne-illness outbreaks. Most occur naturally and are not caused by the presence of microorganisms. Some occur in the animal as a result of its diet. Since toxins are not living organisms, cooking or freezing typically will not destroy them. Most measures taken to prevent foodborne intoxication center on proper purchasing and receiving.

There are several things a manager can do to help prevent seafood-specific toxins from causing foodborne illness. Purchase seafood from reputable suppliers who maintain strict time-temperature controls and can certify the seafood has been harvested from safe waters. Make sure fish is received at 41°F (5°C) or lower, and refuse fish that has been thawed and refrozen. Thaw fish at refrigeration temperatures.

Toxins are a natural part of certain plants. Some plants may be toxic in their raw state but safe when properly cooked. In general, avoid toxic plant species and products prepared with them. Foodborne-illness outbreaks associated with mushrooms are almost always caused by the consumption of misidentified wild mushrooms. Establishments should not use wild mushrooms or products made with them. All mushrooms should be purchased from approved suppliers.

Chemical contaminants are responsible for many cases of foodborne illness. Contamination can come from a variety of substances normally found in restaurant and foodservice establishments, such as toxic metals, pesticides, and cleaning chemicals. Use only food-grade utensils and equipment to prepare and store food. Cleaning products, polishes, lubricants, and sanitizers should be used as directed. Be cautious when using these chemicals during operating hours and store them properly. If used, pesticides should be applied only by a licensed pest control operator (PCO).

Physical contamination results from the accidental introduction of foreign objects into food. Closely inspect the food you receive and take steps to ensure food will not become physically contaminated during its flow through your operation.

Many people have food allergies. Employees should be aware of the most common food allergens, which include milk and dairy products, eggs and egg products, fish, shellfish, wheat, soy and soy products, peanuts, and tree nuts. You should be able to inform customers of these and other potential food allergens that may be included in food served at your establishment.

A CASE IN POINT I

Case Study

Roberto receives a shipment of frozen mahi-mahi steaks. The steaks are frozen solid at the time of delivery and the packages are sealed and contain a lot of ice crystals, indicating they have been time-temperature abused. Roberto accepts the mahi-mahi steaks and thaws them in a refrigerator at a temperature of 38°F (3°C). The thawed fish steaks are then held at this temperature during the evening shift and are cooked to order. The cooks follow all appropriate guidelines for preparing, cooking, and serving the fish, monitoring time and temperature throughout the process. Unfortunately, these fish steaks are implicated in an outbreak of scombroid intoxication.

Explain why this may have happened.

A CASE IN POINT II

Case Study

A busboy, Mark, has been asked to spray for ants in the dry-storage room. He sprays along the ants' trail and then places the can of spray on the top shelf in the dry-storage room, along with the other pesticides and foodservice chemicals. Mark then returns to busing tables.

What has Mark done correctly? What unsafe practices are evident?

TRAINING TIPS

1. Pick Your Poison

Purpose: *After completing this activity, trainees will be able to identify foods often associated with biological toxins and determine preventive measures that can be taken to keep these foods from causing an illness.*

Time: 20 minutes

Directions: Using *Exhibit 3d* as a reference, create a restaurant menu that includes (but is not limited to) the list of foods often associated with biological toxins. Divide trainees into teams of two. Give each team a copy of the menu and ask them to do the following:

1. Circle menu items often associated with a biological toxin

2. Determine preventive measures that can be taken to keep these foods from causing an illness

Ask the teams to present their ideas to the rest of the group. Discuss these ideas and create a list of preventive measures for each menu item.

2. "You Found What In Your Soup?"

Purpose: *After completing this activity, trainees will be able to identify common physical and chemical contaminants and determine methods to prevent them from occurring.*

Time: 30 minutes

Directions: Ask trainees to list as many examples of physical and chemical contamination as they can think of in exactly two minutes.

When time has expired, randomly solicit examples of physical contamination. After each example is given, ask trainees if they have ever experienced this type of physical contamination, as either a foodhandler or customer. If so, ask the following questions:

○ How do you think the contamination might have happened?

○ Did it cause or could it have caused illness or injury?

○ If not, what other type of "damage" did it cause?

○ How would you have responded to such an incident?

○ How could it have been prevented from happening?

After the examples of physical contamination have been exhausted, repeat the process with examples of chemical contamination. Create a checklist of safe-foodhandling practices and policies that will prevent both physical and chemical contamination from occurring in a restaurant or foodservice establishment.

3. Perform a Physical or Chemical Hazard Inspection

Purpose: *To determine how food might become physically or chemically contaminated in your facility and to identify corrective actions that must be taken.*

Time: several days

Directions: Using the information identified in the sections on physical and chemical contamination, conduct an inspection of your facility to identify any possible problems. Solicit help from your foodservice team to do this. Look at the layout of the facility itself and foodhandling procedures followed during receiving and storage, preparation and cooking, and holding and serving. Identify places where food can become physically or chemically contaminated. Look for possible worst-case scenarios.

Next, identify preventive measures that can be taken. Some corrective actions can be made immediately (simple changes in procedures), while others may require more time (equipment changes).

Keep your foodservice team involved in this process, especially if changes will affect their work habits or current procedures. Everyone must be notified of changes in processes or procedures, and the reasons for these changes must be clearly explained.

4. Develop a "Customer Complaint" Form

Purpose: *To develop a form that documents customer complaints regarding menu items. Information on this form can then be used to identify corrective actions that may be needed in the foodservice establishment's processes, procedures, or facility.*

Time: several weeks

Directions: Create a customer complaint form with the following headings:

Date   **Item**   **Food Complaint**   **Corrective Action**

Require employees to document incidents that resulted in the customer sending food back to the kitchen. Analyze customer complaints to determine if the complaint was the result of a hazard requiring a change in current processes or procedures.

Share customer complaints with employees and get their input regarding corrective actions and preventive measures. Explain to employees the importance of documenting customer complaints. Make sure they are not penalized for telling the truth, especially since the complaint may be the result of a mistake they made.

DISCUSSION QUESTIONS

1. How does the scombroid toxin differ from the other seafood toxins listed in *Exhibit 3d?*

2. What preventive measures should be taken to guard against a seafood-specific foodborne illness?

3. A restaurant is serving broiled red snapper along with side dishes of asparagus and rice with wild mushrooms. Which food could contain biological toxins, and why?

4. Why is serving orange juice from an enamelware pitcher not recommended?

5. Mike is serving drinks and cannot find the ice scoop. He uses a clean, sanitary glass tumbler as a temporary ice scoop. Why is this not a safe practice?

MULTIPLE-CHOICE STUDY QUESTIONS   ???

1. You have ordered frozen tuna steaks for your restaurant. When the delivery arrives, you notice there is excessive frost and ice in the package, which indicates they have been time-temperature abused. You refuse the delivery. Why?

   A. You suspect the steaks may have been contaminated with a cleaning compound.
   B. You suspect the steaks may contain ciguatera toxins.
   C. You suspect the steaks may cause scombroid poisoning if you serve them.
   D. You believe the supply company may have treated the steaks with an unauthorized preservative.

2. A customer becomes ill after eating grouper. It is discovered that the shipment of grouper contained the ciguatera toxin. This is an example of

   A. chemical contamination.   C. biological contamination.
   B. physical contamination.

3. Which of the following is *not* a common food allergen?

   A. Eggs
   B. Dairy products
   C. Peanuts
   D. Beef

4. You find a piece of glass at the bottom of your ice storage bin. This is an example of

   A. chemical contamination.
   B. physical contamination.
   C. biological contamination.

5. An establishment should do all of the following to guard against a seafood-specific foodborne illness *except*

   A. thaw fish at temperatures higher than 41°F (5°C).
   B. purchase seafood from suppliers who practice time-temperature control.
   C. refuse fish that has been thawed and refrozen.
   D. accept fish that has been received at 41°F (5°C) or lower.

6. Which of the following fish is associated with ciguatera?

   A. Mahi-mahi
   B. Pufferfish
   C. Snapper
   D. Fresh water minnows

7. An establishment should do all of the following to prevent contamination *except*

   A. store food away from chemicals.
   B. keep high-acid food separate from other types of food.
   C. purchase food products from approved suppliers.
   D. use food-grade storage containers.

8. Which of the following is an example of a *physical* contaminant?

   A. A virus present in shellfish
   B. Dirt on a head of lettuce
   C. Sanitizer residue left on a cutting board
   D. Parasites present in a raw fish fillet

9. All of the following can lead to the chemical contamination of food *except*

   A. cooking tomato sauce in a copper pot.
   B. storing orange juice in a pewter pitcher.
   C. using a backflow-prevention device on a carbonated beverage dispenser.
   D. serving fruit punch in a galvanized tub.

10. Which of the following statements is true about fish containing ciguatera or scombroid (histamine) toxins?

    A. Freezing will destroy these toxins.
    B. Cooking will not destroy these toxins.
    C. You can see and smell these toxins.
    D. Cooking will destroy these toxins.

For answers, please turn to Appendix A.

ADDITIONAL RESOURCES

Resources

Books and Periodicals

American Academy of Allergy, Asthma and Immunology
& The Food Allergy Network. 1992. *What you need to know about food allergies.* Washington D.C.: National
Restaurant Association.

Benenson, A., ed. 1995. *Control of communicable diseases in man,*
16th ed. Washington, D.C.: American Public Health Association.

Bush, R. K. 1995. Seafood allergy and allergens: A review.
Food Technology. 49 (10):103-116.

Dulen, J. 1998. Under the microscope. *Restaurants & Institutions.*
108 (21):80.

Grossbauer, S. 1998. Food safety on the Web. *FoodService Director.*
11 (2):140.

How to prevent foodborne illness. 1995. *FoodService Director.* 8 (2):150.

International Commission on Microbiological Specifications
for Food. 1996. *Microorganisms in foods 5: Characteristics
of microbial pathogens.* London: Blackie Academic
& Professional.

National Restaurant Association. 1998. *Sanitation survival kit.*
Washington, D.C.: National Restaurant Association.

Neumann, R. 1998. The eight most frequent causes of foodborne
illness. *Food Management.* 33 (6):28.

Puzo, D. P. 1997. USDA estimates financial costs of foodborne
illness. *Restaurants & Institutions.* 107 (16):80.

Rehe, S. 1990. *Preventing foodborne illness: A guide to safe foodhandling.*
Washington D.C.: United States Department of Agriculture
Food Safety & Inspection Service.

Rubin, K. W. 1995. What to look for, how to prevent it: Food poisoning. *FoodService Director.* 8 (11):72.

Spake, A. 1997. O is for outbreak. *U.S. News & World Report.* 123 (20):70.

Web Sites

Alaska Seafood Marketing Institute

http://www.alaskaseafood.org

The Alaska Seafood Marketing Institute enhances the appeal and popularity of Alaska seafood to the restaurant and foodservice industry and to the general public. This useful Web site contains information on how to market seafood to your customers, how to purchase Alaskan seafood, detailed information on types of Alaskan seafood, and how to handle seafood to maintain safety and quality.

Centers for Disease Control and Prevention (CDC)

http://www.cdc.gov

The mission of the CDC is to promote health and quality of life by preventing and controlling disease, injury, and disability. To prevent and control foodborne illness, the CDC collects data on foodborne-illness outbreaks. This Web site provides general information on foodborne illnesses and their prevention.

Compendium of Fish and Fishery Product Processes, Hazards, and Controls

http://seafood.ucdavis.edu/haccp/compendium/compend.htm

Developed by the FDA, the compendium available at this Web site provides useful information on biological, chemical, and physical hazards associated with fish and fish products.

ECOLAB, INC.

http://www.ecolab.com

An excellent source of information on housekeeping and sanitation supplies for the restaurant and foodservice industry from the world's leading sanitation product supplier.

The Food Allergy and Anaphylaxis Network (FAAN)

http://www.foodallergy.org

FAAN was established in 1991 and works to build public awareness of food allergies through media, education, publishing, advocacy, and research efforts. Its membership includes families, dietitians, nurses, physicians, school staff, representatives from government agencies, and the food and pharmaceutical industries. This Web site contains a wealth of information related to food allergies, including scientific studies, product recall reports, and recipes.

FDA Bad Bug Book

http://vm.cfsan.fda.gov/~mow/intro.html

Produced by the FDA's Center for Food Safety and Nutrition (FDA CFSAN), this online handbook provides basic facts regarding pathogenic foodborne microorganisms and natural toxins. It brings together information from the FDA, the Centers for Disease Control, the USDA Food Safety Inspection Service, and the National Institutes of Health.

FDA-Center for Food Safety and Applied Nutrition (CFSAN)

http://vm.cfsan.fda.gov

As the center within FDA responsible for food safety and nutrition, CFSAN promotes and protects the public health by researching and implementing guidelines, policies, and standards to ensure that food is safe, nutritious, wholesome, and properly labeled. This Web site provides a wealth of information on food safety and sanitation, including corresponding guidelines, policies, and standards.

FDA Food Code

http://vm.cfsan.fda.gov/~dms/foodcode.html

As the basis for many local sanitation codes, as well as the basis for information in this textbook, the FDA Food Code, available at this Web address, is a useful resource for information relating to food safety for the restaurant and foodservice industry.

FDA Seafood Information and Resources

http://vm.cfsan.fda.gov/seafood1.html

The FDA operates an oversight compliance program for fishery products under which responsibility for products safety, wholesomeness, identity, and economic integrity rests with the processor or importer, who must comply with regulations under the Federal Food, Drug and Cosmetic (FD&C) Act. This Web site houses information on the seafood program, foodborne pathogens and contaminants associated with seafood, and HACCP compliance.

International Food Information Council (IFIC) Foundation

http://ific.org

IFIC collects and disseminates scientific information on food safety, nutrition, and health. They work with an extensive roster of scientific experts to help translate research into understandable and useful information for opinion leaders and, ultimately, consumers. This Web site provides easy-to-understand information on foodborne illness and health-related stories circulating in the news.

Morbidity and Mortality Weekly Report

http://www.cdc.gov/mmwr

The MMWR is prepared by the Centers for Disease Control and is available—free-of-charge—on this Web site. The reports contain data based on weekly accounts of reportable diseases from state health departments. Report databases can be searched for information on foodborne-illness outbreaks.

The Mushroom Council

http://www.mushroomcouncil.com

This interesting and useful Web site from The Mushroom Council contains recipes and information about how to handle and store mushrooms to ensure safety and quality.

National Center for Infectious Diseases (NCID)

http://www.cdc.gov/ncidod

NCID is one of the centers of the Centers for Disease Control. Its mission is to prevent illness, disability, and death caused by infectious diseases around the world. NCID accomplishes this mission by conducting surveillance, epidemic investigations, epidemiologic and laboratory research, and training. It also sponsors public education programs to develop, evaluate, and promote prevention and control strategies for infectious diseases. This Web site serves as another great resource for information on foodborne illness.

National Shellfish Sanitation Program Manual of Operations

http://vm.cfsan.fda.gov/~ear/nsspman.html

The National Shellfish Sanitation Program (NSSP) is a voluntary cooperative program between the federal and state governments and industry. The program relies on regulatory controls by the State Shellfish Authority (SSA) to ensure safe mollusks and shellfish. The NSSP Manual of Operations, published by the FDA and available at this Web site, is a guide for establishing state shellfish laws and regulations.

Orkin Commercial

http://www.orkin.com

From the leader in pest control, Orkin's Web site contains useful information on how to prevent and control pests in your foodservice establishment. It also contains a bug guide providing all the information you'll ever want to know about ants, rodents, flies, and birds.

Chapter 4
The Safe Foodhandler

Knowledge

TEST YOUR FOOD SAFETY KNOWLEDGE

1. **True or False:** A foodhandler who doesn't look or act sick could still spread a foodborne illness. *(See page 4-2.)*

2. **True or False:** Wearing gloves is an acceptable substitute for handwashing. *(See page 4-7.)*

3. **True or False:** Foodhandlers must always wash their hands after smoking. *(See page 4-6.)*

4. **True or False:** A foodhandler diagnosed with salmonellosis cannot continue to work at an establishment while he has the illness. *(See page 4-11.)*

5. **True or False:** A foodhandler should not be allowed to work with food if she has a cut on her hand. *(See page 4-11.)*

For answers, please turn to Appendix A.

Key Terms

Gastrointestinal illness
Infected lesion
Carriers
Single-use paper towel
Hand sanitizers
Finger cot
Hair restraint
Jaundice
Hepatitis A

Table of Contents

Learning Objectives

After completing this chapter, you should be able to

○ Identify personal behaviors that can contaminate food.

○ Identify proper handwashing procedures.

○ Respond properly to cuts, wounds, and sores to ensure food safety.

○ Identify procedures that must be followed when using gloves.

○ Identify and respond to employee health problems that pose a threat to food safety.

○ List examples of clothing, accessories, and other items worn by employees that pose a threat to food safety.

○ Identify policies that should be implemented at the establishment regarding eating, drinking, and smoking while working with food.

At every step in the flow of food through the operation—from receiving through final service—foodhandlers can contaminate food and cause customers to become ill. Good personal hygiene is a critical protective measure against foodborne illness, and customers expect it.

You can minimize the risk of foodborne illness by establishing a personal hygiene program that spells out your specific hygiene policies, provides your employees with training on those policies, and enforces established policies. When employees have the proper knowledge, skills, and attitudes toward personal hygiene, you are one step closer to operating a safe food system.

HOW FOODHANDLERS CAN CONTAMINATE FOOD

In previous chapters you learned that foodhandlers can cause an illness when they transfer microorganisms to food they touch. Many times these microorganisms come from the foodhandlers themselves. Foodhandlers can contaminate food when:

- They have a foodborne illness
- They show symptoms of **gastrointestinal illness** (an illness relating to the stomach or intestine)
- They have **infected lesions** (infected wounds or injuries)
- They live with or are exposed to a person who is ill
- They touch anything that may contaminate their hands

Even an apparently healthy person may be hosting foodborne pathogens. With some illnesses, such as hepatitis A, an individual is at the most infectious stage of the disease for several weeks before symptoms appear. With other illnesses, the pathogens may remain in a person's system for months after all signs of infection have ceased. Some people are called **carriers,** because they might carry pathogens and infect others, yet never become ill themselves.

The next three paragraphs will help illustrate the routes by which employees can contaminate food.

A deli foodhandler who was diagnosed with salmonellosis failed to inform his manager that he was ill for fear of losing wages. It was later determined that he was the cause of an outbreak that involved more than two hundred customers through twelve different products.

A foodhandler suffering from diarrhea, a symptom of gastrointestinal illness, didn't wash his hands and made approximately five thousand people ill when he mixed a vat of buttercream frosting with his bare hands and arms. Another large foodborne-illness outbreak was caused by a foodhandler who scratched an infected facial lesion and then handled a large amount of sliced pepperoni.

A foodborne-illness outbreak was traced to a woman who prepared food for a dinner party. The investigation revealed that the woman was caring for her infant son, who had diarrhea. The woman could not recall washing her hands after changing the infant's diaper. As a result, twelve of her dinner guests became violently ill with symptoms that included diarrhea and vomiting.

Simple acts such as nose picking, rubbing an ear, scratching the scalp, touching a pimple or an open sore, or running fingers through the hair can contaminate food. Thirty to fifty percent of healthy adults carry *Staphylococcus aureus* in their noses, and about twenty to thirty-five percent carry it on their skin. If these microorganisms contaminate a foodhandler's hands and they touch food, the consequences can be severe. Because of these factors, foodhandlers must pay close attention to what they do with their hands and maintain good personal hygiene.

DISEASES NOT TRANSMITTED THROUGH FOOD

In recent years, the public has expressed growing concern over communicable diseases spread through intimate contact or by direct exchange of bodily fluids. Diseases such as AIDS (Acquired Immune Deficiency Syndrome), hepatitis B and C, and tuberculosis are not spread through food.

Although these diseases are not transmitted through food, as a manager you should be aware of the following laws concerning employees who are HIV-positive (Human Immunodeficiency Virus), have hepatitis B or C, or have tuberculosis.

○ The Americans with Disabilities Act (ADA) provides civil rights protection to individuals who are HIV-positive or have hepatitis B, and thus prohibits employers from firing people or transferring them out of foodhandling duties simply because they have these diseases.

○ Employers must maintain the confidentiality of employees who have any nonfoodborne illness.

Key Point

Simple acts such as rubbing an ear, scratching the scalp, or touching a pimple or sore can contaminate food.

Key Point

Diseases such as AIDS, hepatitis B or C, and tuberculosis are not spread through food.

Key Point

Managers must train foodhandlers to wash their hands properly and when necessary.

Key Point

There are products on the market that help teach proper handwashing techniques. These products are applied to the hands, which are then washed normally. The employees' hands are then illuminated to reveal how well they were washed.

Before

After

Picture provided courtesy of Brevis Corp.
3310 S. 2700 East, Salt Lake City, UT 84109
800.383.3377

COMPONENTS OF A GOOD PERSONAL HYGIENE PROGRAM

Good personal hygiene is key to the prevention of foodborne illness. Good personal hygiene includes:

○ Following hygienic hand practices
○ Maintaining personal cleanliness
○ Wearing clean and appropriate uniforms and following dress codes
○ Avoiding unsanitary habits and actions
○ Maintaining good health
○ Reporting illnesses

Hygienic Hand Practices

Handwashing

While it may appear fundamental, many foodhandlers fail to wash their hands properly and as often as needed. As a manager, it is your responsibility to train your foodhandlers and then monitor them. Never take this simple action for granted.

To ensure proper handwashing in your establishment, train your foodhandlers to follow these steps. *(See Exhibit 4a.)*

Step 1: Wet your hands with running water as hot as you can comfortably stand (at least 100°F [38°C]).

Step 2: Apply soap. Apply enough soap to build up a good lather.

Step 3: Vigorously scrub hands and arms for at least twenty seconds. Lather well beyond the wrists, including the exposed portions of the arms.

Step 4: Clean under fingernails and between fingers. A nail brush might be helpful.

Step 5: Rinse thoroughly under running water. Turn off the faucet using a **single-use paper towel** if available.

Step 6: Dry hands and arms. Use single-use paper towels or a warm-air hand dryer. Never use aprons or wiping cloths to dry hands after washing.

1. Wet your hands with running water as hot as you can comfortably stand (at least 100°F [38°C])

2. Apply soap

3. Vigorously scrub hands and arms for at least twenty seconds

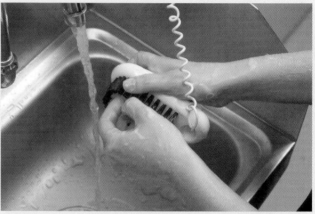

4. Clean under fingernails and between fingers

5. Rinse thoroughly under running water

6. Dry hands and arms with a single-use paper towel or warm-air hand dryer

Exhibit 4a   Proper Handwashing Procedure

Key Point

Hand sanitizers should never be used in place of proper handwashing.

Key Point

Food-handlers must keep fingernails short and clean, and should not wear false nails or nail polish.

Key Point

Hand cuts or sores should be covered with a clean bandage and a glove or finger cot.

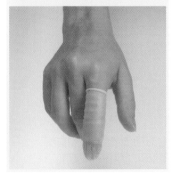

Hand sanitizers (a liquid used to lower the number of microorganisms on the surface of the skin) or hand dips may be used after washing, but should never be used in place of proper handwashing. If hand sanitizers are used, foodhandlers should never touch food or food-preparation equipment until the hand sanitizer has dried.

Foodhandlers must wash their hands after the following activities:

○ Using the restroom

○ Handling raw food (before *and* after)

○ Touching the hair, face, or body

○ Sneezing, coughing, or using a handkerchief or tissue

○ Smoking, eating, drinking, or chewing gum or tobacco

○ Handling chemicals that might affect the safety of food

○ Taking out garbage or trash

○ Clearing tables or busing dirty dishes

○ Touching clothing or aprons

○ Touching anything else that may contaminate hands, such as unsanitized equipment, work surfaces, or wash cloths

Hand Maintenance

In addition to proper washing, hands need other regular care to ensure that they will not transfer microorganisms to food. Food, filth, and harmful substances can get caught under both long and short fingernails. Fingernails should be kept short and clean. Nail polish can disguise dirt under nails and may flake off into food. Nail polish should not be worn if the employee will be handling food. Long fingernails, false fingernails, and acrylic nails may be difficult to keep clean and can break off into food. Therefore, they should not be worn while handling food, unless your jurisdiction allows single-use gloves to be worn over polished and false nails.

Cuts and sores on hands, including hangnails, should be treated and kept covered with clean bandages. If hands are bandaged, clean gloves or **finger cots,** a protective covering, should be worn at all times to protect the bandage and to prevent it from falling off into food. You may need to move the foodhandler to another job, where he or she will not handle food or touch food-contact surfaces, until an injury heals. To ensure food safety, good hand maintenance must be learned and practiced without fail.

Use of Gloves

Gloves can protect hands from cuts and the effects of detergents and chemicals. They can also help keep food safe by creating a barrier between hands and food. *(See Exhibit 4b.)* Some jurisdictions require the use of gloves for foodhandlers who work with ready-to-eat food. However, other jurisdictions will allow an exception to the use of gloves if the establishment has a verifiable written policy on handwashing procedures. Check with your local regulatory agency for requirements in your jurisdiction.

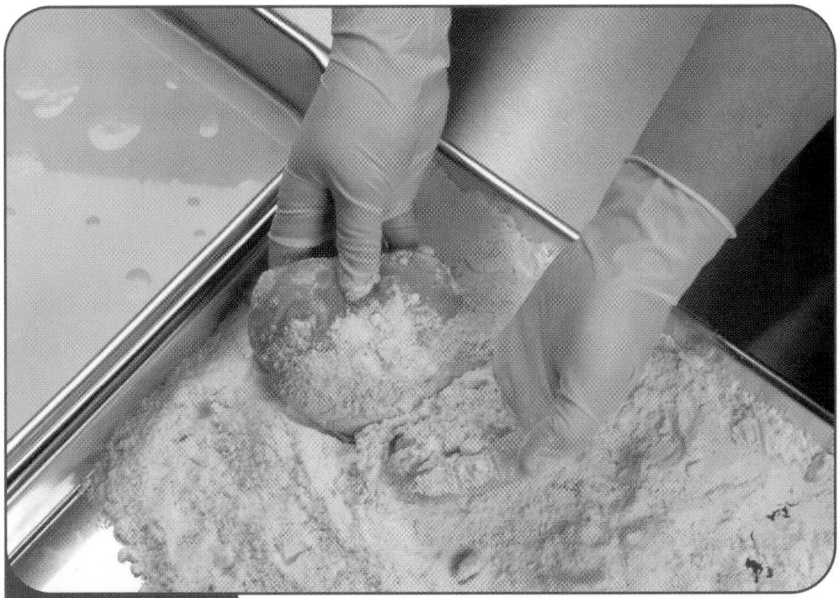

Gloves must never be used in place of handwashing. Hands must be washed before putting on gloves and when changing to a fresh pair. Gloves used to handle food are for single use only and should never be washed and reused. Foodhandlers should change their gloves when necessary. Gloves should be changed:

Exhibit 4b **Gloves**
Gloves can help keep food safe by creating a barrier between hands and food.

○ As soon as they become soiled or torn

○ Before beginning a different task

○ At least every four hours during continual use, and more often when necessary

○ After handling raw meat and before handling cooked or ready-to-eat food

Often, foodhandlers consider gloves more sanitary than bare hands. Because of this false sense of security, they might not change gloves as often as necessary. Managers must reinforce the habit of proper hand sanitation with foodhandlers. An effective handwashing education and compliance program will establish a habit that will help prevent foodborne illness.

Other Good Personal Hygiene Practices

Personal hygiene can be a sensitive subject for some people, but because personal cleanliness is vital to food safety, as a manager, you must address the subject with every foodhandler.

Key Point

Gloves must never be used in place of handwashing.

Key Point

Food-
handlers
must keep their hair
clean, since oily hair
can harbor pathogens.

Key Point

Food-
handlers
must always wear clean
clothes, since dirty
clothes can harbor
pathogens.

Key Point

Food-
handlers
must remove jewelry
prior to preparing or
serving food since it
can harbor pathogens.

General Personal Cleanliness

In addition to following proper hand-hygiene practices, your foodhandlers must maintain personal cleanliness. Foodhandlers should bathe or shower before work. They must also keep their hair clean. Oily, dirty hair can harbor pathogens, and dandruff may fall into food or onto food-contact surfaces.

Proper Work Attire

A foodhandler's attire plays an important role in the prevention of foodborne illness. Dirty clothes may harbor pathogens and give customers a bad impression of your establishment. Therefore, managers should make sure foodhandlers observe strict dress standards. *(See Exhibit 4c.)*

Foodhandlers should:

○ **Wear a clean hat or other hair restraint.** A **hair restraint** will keep hair away from food and keep the foodhandler from touching it. Foodhandlers with facial hair should also wear beard restraints.

○ **Wear clean clothing daily.** The type of clothing chosen should minimize contact with food and equipment, and should reduce the need for adjustments. If possible, foodhandlers should put on work clothes at the establishment.

○ **Remove aprons when leaving food-preparation areas.** For example, aprons should be removed and properly stored prior to taking out garbage or using the restroom.

○ **Wear appropriate shoes.** Wear clean, closed-toed shoes with a sensible, nonslip sole.

○ **Remove jewelry prior to preparing or serving food or while around food-preparation areas.** Jewelry can harbor microorganisms, often tempts foodhandlers to touch it, and may pose a safety hazard around equipment. Remove rings (except for a plain band), bracelets (including medical information jewelry), watches, earrings, necklaces, and facial jewelry (such as nose rings, etc.).

Check with your local regulatory agency regarding requirements. These requirements should be reflected in written policies that are consistently monitored and enforced. All potential employees should be made aware of these policies prior to employment.

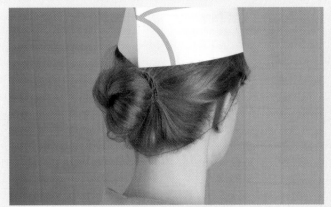

Hair properly restrained

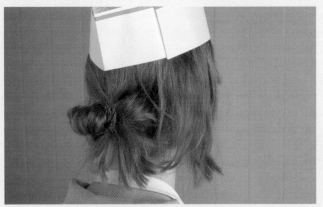

Hair improperly restrained

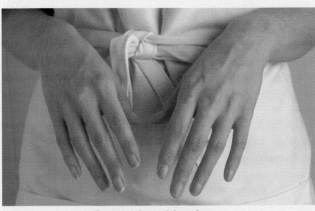

Proper hand hygiene:
clean, short fingernails; no jewelry or nail polish

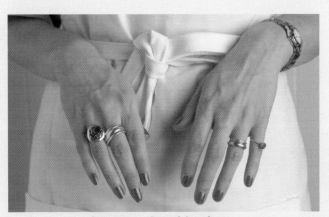

Improper hand hygiene:
long fingernails, jewelry, nail polish

Proper apron: clean

Improper apron: dirty and stained

Exhibit 4c   Proper and Improper Attire

Exhibit 4d Safe and Unsafe Food-Tasting Methods

Policies Regarding Eating, Drinking, Chewing Gum, and Tobacco

Small droplets of saliva can contain thousands of disease-causing microorganisms. In the process of eating, drinking, chewing gum, or smoking, this saliva can be transferred to the foodhandler's hands or directly to the food they are handling. For this reason, managers must implement the following policies in their establishments:

○ Foodhandlers must not smoke or chew gum or tobacco while preparing or serving food, while in food-preparation areas, or while in areas used for equipment and utensil washing.

○ Foodhandlers must not eat or drink while in food-preparation areas or in areas used to clean utensils and equipment (with the exception of chefs properly tasting food). Some jurisdictions allow employees to drink from a covered container with a straw. Check with your local regulatory agency.

○ Foodhandlers should eat, drink, chew gum, or use tobacco products only in designated areas, such as an employee break room.

○ Foodhandlers should never spit in the establishment.

If foods must be tasted during preparation, they must be placed in a separate dish and tasted with a clean utensil. The dish and utensil should then be removed from the food-preparation area for cleaning and sanitizing. *(See Exhibit 4d.)*

Key Point

Foodhandlers must not eat, drink, smoke, or chew gum or tobacco while preparing or serving food.

Policies for Reporting Illness and Injury

Foodhandlers must report health problems to the manager of the establishment before working with food. If they become ill while working, they must immediately report their condition. If the foodhandler's condition could possibly contaminate food or equipment, he or she must stop working and see a doctor. If the foodhandler must refrigerate personal medication while working, it must be stored inside a covered, leakproof container that is clearly labeled.

According to the FDA Food Code, managers must *exclude* from the establishment foodhandlers who have been diagnosed with a foodborne illness, and they must notify the local regulatory agency. Managers must work with the agency to determine when the foodhandler can safely return to work.

Managers must *restrict* foodhandlers from working with or around food if they have the following symptoms:

○ Fever

○ Diarrhea

○ Vomiting

○ Sore throat

○ **Jaundice** (a yellowing of the skin and eyes that could indicate a person is ill with hepatitis A)

Any cuts, burns, boils, sores, skin infections, or infected wounds should be covered with a bandage when the foodhandler is working with or around food or food-contact surfaces. Bandages should be clean, dry, and must prevent leakage from the wound. As previously mentioned, waterproof disposable gloves or finger cots should be worn over bandages on hands. Foodhandlers wearing bandages may need to be temporarily reassigned to duties not involving contact with food or food-contact surfaces.

Vaccination for Hepatitis A

Hepatitis A is a disease causing inflammation of the liver. It is transmitted to food by poor personal hygiene or contact with contaminated water. It infects many people each year, resulting in community-wide outbreaks. Of all foodborne illnesses facing the foodservice industry, hepatitis A is the only one that can be prevented by vaccine. While effective handwashing is a critical practice to prevent contamination, vaccinating your foodhandlers for hepatitis A can provide an additional barrier. This may be especially recommended in areas where hepatitis A outbreaks are highly prevalent.

Health Alert

Managers must not allow foodhandlers to work in the establishment if they have been diagnosed with a foodborne illness.

Health Alert

Foodhandlers must not work with or around food if they have symptoms which include fever, diarrhea, vomiting, a sore throat, or jaundice.

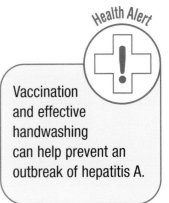

Health Alert

Vaccination and effective handwashing can help prevent an outbreak of hepatitis A.

Key Point

Management should model proper personal hygiene practices at all times.

Key Point

Job tasks and assignments should be planned to prevent the risk of cross-contamination.

MANAGEMENT'S ROLE IN A PERSONAL HYGIENE PROGRAM

Management plays a critical role in the effectiveness of a personal hygiene program. Responsibilities include:

○ Training foodhandlers on personal hygiene policies

○ Modeling proper behavior for foodhandlers at all times

○ Establishing proper personal hygiene policies

○ Revising policies when laws and regulations change, as well as when changes are recognized in the science of food safety, and retraining foodhandlers as necessary

○ Supervising sanitary practices continuously, retraining foodhandlers as necessary

Job Assignments

When job descriptions are developed and responsibilities assigned, consider the risk of cross-contamination and plan tasks to prevent it. The risk may be higher if foodhandlers are required to perform several different duties than if specific foodhandlers are assigned to a single duty. For example, an employee expected to prepare and wrap food, clear off tables, and then return to food-preparation duties could more easily contaminate food than an employee who is assigned to just one of these tasks. By planning tasks to prevent cross-contamination, you will minimize the amount of time needed for supervision and enable employees to follow sanitation rules more easily.

SUMMARY

Foodhandlers can contaminate food at every step in its flow through the establishment. Good personal hygiene is a critical protective measure against contamination and foodborne illness. A successful personal hygiene program depends on trained foodhandlers who possess the knowledge, skills, and attitude necessary to maintain a safe food system.

Foodhandlers have the potential to contaminate food when they have been diagnosed with a foodborne illness, when they show symptoms of a gastrointestinal illness, when they have infected lesions, or when they touch anything that might contaminate their hands. Foodhandlers must pay close attention to what they do with their hands since simple acts such as nose picking or running fingers through the hair can contaminate food. Proper handwashing must also be practiced. This is especially important after using the restroom, before and after handling raw food, after sneezing and coughing, and after smoking, eating, or drinking. It is up to the manager to monitor handwashing to make sure it is thorough and frequent. In addition, hands need other care to ensure they will not transfer contaminants to food. Fingernails should be kept short and clean. Cuts and sores should be covered with clean bandages. Hand cuts should also be covered with gloves or finger cots.

Gloves can create a barrier between hands and food; however, they should never be used in place of handwashing. Hands must be washed before putting on gloves and when changing to a fresh pair. Gloves used to handle food are for single use and should never be washed or reused. They must be changed whenever contamination occurs.

Personal hygiene can be a sensitive subject for some people, but, because it is vital to food safety, it must be addressed with every employee. All employees must maintain personal cleanliness. They should bathe or shower before work and keep their hair clean.

Prior to handling food, foodhandlers must put on a clean hair restraint, put on clean clothing, remove jewelry, and put on appropriate shoes. Aprons should always be removed and properly stored when the employee leaves food-preparation areas.

Establishments should implement strict policies regarding eating, drinking, smoking, and chewing gum and tobacco. These activities should not be allowed when the foodhandler is preparing or serving food or working in food-preparation areas.

Foodhandlers must be encouraged to report health problems to management before working with food. If their condition could contaminate food or equipment, they must stop working and see a doctor. Managers must not allow foodhandlers diagnosed with a foodborne illness to work, and they must notify the local regulatory agency. Managers must restrict foodhandlers from working with or around food if they have symptoms that include fever, diarrhea, vomiting, sore throat, or jaundice.

Management plays a critical role in the effectiveness of a personal hygiene program. By establishing a program that includes specific policies, and by training and enforcing those policies, managers can minimize the risk of causing a foodborne illness.

A CASE IN POINT I

Case Study

Chris works at a quick-service restaurant. She is suffering from seasonal allergies, so she carries a small pack of tissues with her. Her assigned responsibility is to make salads. She washes her hands properly and puts on single-use gloves before she starts her shift. When Chris needs to sneeze, she steps away from the food-preparation area, pulls a clean tissue out of her pocket, sneezes into it, then discards it. Because her medication gives her a dry mouth, Chris keeps a glass of water at her station. Her clearly labeled bottle of allergy medication has to be kept cold, so she stores it in the walk-in refrigerator.

Does this situation represent a threat to food safety? Explain why or why not, and tell what Chris is doing right and wrong in this scenario.

A CASE IN POINT II

Case Study

Marty works for a catering company. A few days ago, he was serving hot food from chafing dishes at an outdoor music festival. He did not wear gloves because he used spoons and tongs to serve the food. His manager noticed that Marty made multiple trips to the bathroom during his four-hour shift. These trips did not interrupt service to customers because there were plenty of staff members on hand and Marty hurried to and from the restroom.

The nearest restroom had soap, separate hot and cold water faucets, and a working hot-air dryer, but no paper towels. Each time Marty used the restroom, he washed his hands quickly and then dried them on his apron. Throughout the following week, the manager of the catering company received several telephone calls from people who had attended the music festival and had eaten their food. They each complained of diarrhea, fever, and chills. One call was from a mother of a young boy who was hospitalized for dehydration. The doctor reported that the boy had shigellosis.

Explain how Marty might have caused an outbreak of shigellosis. What measures should have been taken to prevent it?

TRAINING TIPS

1. Glo Germ™ Activity

Purpose: *After completing this activity, trainees should recognize the importance of proper handwashing.*

Time: 20 minutes

Directions: Ask for a volunteer from the group to participate in an experiment. Apply some iridescent "germs" (harmless orange liquid) from a *Glo Germ* kit to the hands of the volunteer and ask him to rub it in. Have him wash his hands.

When the volunteer returns, turn off the lights, and hold a black light over his hands to see how effectively he washed them. Often, the hands will glow around the fingernails and around watches or jewelry. Walk the volunteer around and use the black light to show the rest of the group how many germs have survived the handwashing process.

Discuss the importance of proper handwashing with the group.

2. Habits with Hands Role-Play

Purpose: *After completing this activity, trainees should be able to identify how hands can cross-contaminate food and the proper times to wash hands.*

Time: 30 minutes

Directions: Ask for two volunteers from the group. Assign one volunteer to role-play a cook, the other a server.

Have the volunteers design a three-minute role-play in which they portray their character performing various tasks. While doing so, they should also depict different ways hands can cross-contaminate food.

For example, the person playing the cook might handle raw poultry and then prepare sandwiches without washing her hands, or the server may sneeze into his hands, wipe them on his apron, and then serve dinner rolls to a guest using his bare hands. The goal is to design scenes with as many violations as possible.

Ask the group to note every incident of cross-contamination they see during the role-play. Afterwards, discuss the incidents noted and develop a list identifying when hands should be washed. Then discuss the proper procedure for handwashing.

3. Creating the Safe Foodhandler

Purpose: *To make foodhandlers aware of personal hygiene standards necessary to keep food safe in your establishment.*

Time: several days

Directions: Develop a colorful wall poster with a drawing of "The Safe Foodhandler." Make the drawing life-size if possible. Note on the drawing all of the aspects of good hygiene from head to toe, from clean and properly restrained hair to proper closed-toe shoes. For added effect, you might want to use photographs of actual employees in your establishment. Make a checklist of basic hygiene standards, such as coming to work healthy, taking a shower or bath before work, etc. Attach the checklist to the poster.

Place your poster in a conspicuous place for all foodhandlers to see.

4. Yuck

Purpose: *After completing this activity, trainees should recognize the importance of proper handwashing.*

Time: several days

Directions: Purchase some petri dishes and petri-agar mix from a science supply store. Follow the directions for mixing the agar for each dish. Issue one petri dish per trainee. Ask each trainee to lightly press a finger or thumb into the agar. Cover the petri dishes and place them in a warm location. Allow the dishes to sit for at least forty-eight hours. As a variation of this activity, you might ask trainees to wash their hands and make another impression in the agar. Separate these impressions from the first set.

After a couple of days, a growth should appear where the trainee left his or her finger or thumb impression in the agar. Compare the impressions left by the unwashed hands to the impression left by the washed hands.

This simple activity will show trainees that microorganisms, while invisible to the unaided eye, can be present on the body. It should also emphasize how important it is to properly wash hands.

5. Gotcha

Purpose: *This activity may help make handwashing a habit at your establishment.*

Time: several weeks

Directions: Create a list of activities after which handwashing must be performed. Share this list with all of your foodhandlers and designate a "gotcha" day. The "gotcha" day will be a handwashing monitoring day. Foodhandlers will monitor one another. Whenever a foodhandler is caught not washing his or her hands after one of the listed activities, fellow foodhandlers should shout "gotcha."

By the end of the day it will be evident to employees that handwashing is easy to forget when they are busy. Ask the staff to continue the "gotcha" activity until handwashing becomes a habit. If necessary, offer incentives to keep the staff involved in this activity.

DISCUSSION QUESTIONS

1. What are the steps for proper handwashing?

2. What personal behaviors can contaminate food?

3. How can sneezing and coughing transmit pathogens to food?

4. Which procedures must foodhandlers follow when using gloves?

5. What employee health problems are a possible threat to food safety? What are the appropriate actions that should be taken?

MULTIPLE-CHOICE STUDY QUESTIONS

1. Which of the following personal behaviors can contaminate food?

   A. Touching a pimple   C. Nose picking

   B. Touching hair   D. All of the above

2. After you've washed your hands, which of the following items should be used to dry them?

   A. Your apron   C. A common cloth towel

   B. Single-use paper towels   D. A wiping cloth

3. Which of the following items can contaminate food?

   A. Rings   C. Earrings

   B. A watch   D. All of the above

4. Which of the following practices is most likely to ensure the safety of the food being prepared?

   A. A foodhandler uses gloved hands to place sandwich meat onto a bun.

   B. A foodhandler sneezes into a tissue before handling food.

   C. A foodhandler uses hand lotion after washing her hands.

   D. A foodhandler drinks from a plastic cup while preparing food.

5. Which of the following is the proper procedure for washing your hands?

   A. Run hot water (at least 100°F [38°C]), moisten hands and apply soap, vigorously scrub hands and arms, apply sanitizer, dry hands.
   B. Run hot water (at least 100°F [38°C]), moisten hands and apply soap, vigorously scrub hands and arms, rinse hands, dry hands.
   C. Run cold water (at least 41°F [5°C]), moisten hands and apply soap, vigorously scrub hands and arms, rinse hands, dry hands.
   D. Run cold water (at least 41°F [5°C]), moisten hands and apply soap, vigorously scrub hands and arms, apply sanitizer, dry hands.

6. Which foodhandler is least likely to contaminate the food she will handle?

   A. A foodhandler who keeps her fingernails long
   B. A foodhandler who keeps her fingernails short
   C. A foodhandler who wears false fingernails
   D. A foodhandler who wears nail polish

7. Kim wore disposable gloves while she formed raw ground beef into patties. When she was finished, she continued to wear the gloves while she sliced hamburger buns. What mistake did Kim make?

   A. She failed to wash her hands and put on new gloves after handling raw meat and before handling the ready-to-eat buns.
   B. She failed to wash her hands before wearing the same gloves to slice the buns.
   C. She failed to wash and sanitize her gloves before handling the buns.
   D. She failed to wear reusable gloves.

8. A foodhandler who has been diagnosed with shigellosis should be

   A. told to stay home.
   B. told to wear gloves while working with food.
   C. told to wash his hands every fifteen minutes.
   D. assigned to a nonfoodhandling position until he is feeling better.

9. Which of the following diseases can be prevented by a vaccine?

   A. Salmonellosis   C. Hepatitis A
   B. Listeriosis   D. Hemorrhagic colitis

10. Management can play a key role in promoting proper personal hygiene by
    A. providing hand lotion in handwashing stations.
    B. providing reusable gloves for foodhandlers.
    C. modeling proper behavior at all times.
    D. permitting smoking in food-preparation areas.

11. Foodhandlers should be restricted from working with or around food if they are experiencing which of the following symptoms?
    A. Soreness, itching, fatigue
    B. Fever, vomiting, diarrhea
    C. Headache, irritability, thirst
    D. Muscle cramps, insomnia, sweating

12. Which of the following policies should be implemented at establishments?
    A. Employees must not smoke while preparing or serving food.
    B. Employees must not eat while in food-preparation areas.
    C. Employees must not chew gum or tobacco while preparing or serving food.
    D. All of the above

13. Stephanie has a small cut on her finger and is about to prepare chicken salad. How should Stephanie's manager respond to the situation?
    A. Send Stephanie home immediately
    B. Cover the hand with a glove or finger cot
    C. Cover the cut with a clean bandage and a glove or finger cot
    D. Cover the cut with a clean bandage

14. Hands should be washed after which of the following activities?
    A. Touching your hair   C. Using a handkerchief
    B. Eating   D. All of the above

For answers, please turn to Appendix A.

ADDITIONAL RESOURCES

Books and Periodicals

Cliver, D. O. 1997. Virus transmission via food. *Food Technology.* 51 (4):71-78.

Frable, F., Jr. 1997. 10 common errors in kitchen planning and ways to avoid them. *Nation's Restaurant News.* 31 (6):22.

Hernandez, J. 1998. Food safety: Preparation and cooking. *Food Management.* 33 (5):90.

Longree, K. and G. Armbruster. 1996. *Quantity food sanitation,* 5th ed. New York: Wiley.

More than an ounce of prevention. 1997. *Best Practices.* 1 (3):6-9.

Web Sites

Activ USA, Inc.

http://www.activusa.com

Activ USA, Inc. is a biopharmaceutical company championing food safety by incorporating new technologies in product- and education-based solutions to known problems in the areas of cross-contamination and first aid treatments.

Centers for Disease Control and Prevention (CDC)

http://www.cdc.gov

The mission of the CDC is to promote health and quality of life by preventing and controlling disease, injury, and disability. To prevent and control foodborne illness, the CDC collects data on outbreaks. This Web site provides general information on foodborne illnesses and their prevention.

Colgate-Palmolive Company

http://www.colpalipd.com

A manufacturer of powerful, performance-tested cleaners and sanitizers, Colgate-Palmolive provides a Web site that contains useful information on its foodservice products and on food safety issues.

Daydots International

http://www.daydots.com

Daydots International is a maker and distributor of many sanitation and food safety tools for restaurants, most of which can be ordered online. Their site also contains articles and information related to food safety.

FDA-Center for Food Safety and Applied Nutrition (CFSAN)

http://www.cfsan.fda.gov

As the center within the the FDA responsible for food safety and nutrition, CFSAN promotes and protects the public health by researching and implementing guidelines, policies, and standards to ensure that food is safe, nutritious, wholesome, and properly labeled. This Web Site provides a wealth of information on food safety and sanitation, including corresponding guidelines, policies, and standards.

FDA Food Code

http://vm.cfsan.fda.gov/~dms/foodcode.html

As the basis for many local sanitation codes, as well as the basis for information in this textbook, the FDA Food Code, available at this Web address, is a useful resource for information relating to food safety for the restaurant and foodservice industry.

FoodHandler Inc.

http://www.foodhandler.com

FoodHandler Inc. is a manufacturer of foodservice safety equipment, such as gloves, aprons, and food-storage systems. The site contains information on products, as well as issues related to food safety.

Hepatitis Control Report

http://www.hepatitiscontrolreport.com

This site in an online quarterly newsletter devoted to news on the public-health control of viral hepatitis. The mission of the site is to provide accurate and balanced reporting of developments in hepatitis epidemiology, control programs, and public policy.

National Center for Infectious Diseases (NCID)

http://www.cdc.gov/ncidod

NCID is one of the centers of the Centers for Disease Control. Its mission is to prevent illness, disability, and death caused by infectious diseases around the world. NCID accomplishes this mission by conducting surveillance, epidemic investigations, epidemiologic and laboratory research, and training. It also sponsors public education programs to develop, evaluate, and promote prevention and control strategies for infectious diseases. This Web site serves as another great resource for information on foodborne illness.

KEEP IN MIND...

The safety of the food you serve at your establishment will depend largely on your understanding of food safety concepts throughout the flow of food, as outlined in Chapters 5 through 8. Food safety also depends on your ability to develop a system that prioritizes, monitors, and verifies the most important food safety practices, which will be discussed in Chapter 9: Principles of a HACCP System.

Hazard Analysis Critical Control Point (HACCP) is a dynamic system combining proper foodhandling procedures, hazard and risk analysis, monitoring techniques, and record keeping. The HACCP system enables you to serve safe food consistently by identifying and controlling possible hazards throughout the flow of food. Since HACCP is dynamic, it allows you to update and improve your food safety system continuously. Because this is a preventive rather than reactive system, the National Restaurant Association and the Food and Drug Administration recommend that restaurants and foodservice establishments develop and use a HACCP-based food safety system. In fact, recognizing that the complete flow of food begins on the farm and continues through service, other government agencies and foodservice industries have accepted HACCP as the best food safety system available. Industry segments such as canning plants and meat, poultry, and seafood producers and processors are currently required by law to use HACCP.

Foodservice operators or managers who have chosen to develop a HACCP plan have found additional benefits, including improved quality, minimized waste, better product consistency, increased customer satisfaction, improved inspection scores, and improved relations with the regulatory community.

Since your application of HACCP begins even before food arrives at your establishment, our presentation of the flow of food begins with purchasing and receiving (Chapter 5) and continues through storage (Chapter 6), preparation (Chapter 7), and service (Chapter 8). In Chapter 9, we pull concepts from these chapters together to explain how the seven principles of HACCP are applied.

As you read through this unit, it is important to keep in mind that the food safety concepts discussed here tell you **what** to do to keep food safe, while HACCP tells you **how** to consistently keep food safe. If you know what to do but do not know how to do it, you will not have an effective and complete food safety system.

UNIT 2

THE FLOW OF FOOD THROUGH THE OPERATION

Chapter 5
Purchasing and Receiving Safe Food

TEST YOUR FOOD SAFETY KNOWLEDGE

1. **True or False:** Upon arrival, a delivery of fresh fish should be received at an internal temperature of 41°F (5°C) or lower. *(See page 5-12.)*

2. **True or False:** Refrigerated *sous vide* products should be received at 41°F (5°C) or lower unless otherwise specified by the manufacturer. *(See page 5-20.)*

3. **True or False:** You should reject a delivery of frozen steaks covered in large ice crystals since the steaks have probably been thawed and refrozen. *(See page 5-19.)*

4. **True or False:** If a sack of flour is dry upon delivery, the contents may still be contaminated. *(See page 5-21.)*

5. **True or False:** Hot fried chicken should be delivered at 140°F (60°C) or higher. *(See page 5-23.)*

For answers, please turn to Appendix A.

Learning Objectives

After completing this chapter, you should be able to

○ Choose a safety-conscious food supplier.

○ Receive deliveries properly.

○ Calibrate thermometers and use them properly.

○ Check the temperature of different types of food during receiving.

○ Determine when to accept or reject different types of food during receiving.

○ Identify some of the microbial risks for different food that can be minimized through proper receiving.

Key Terms

Bimetallic stemmed thermometer
Time-temperature indicator (TTI)
Calibration
Shellstock tags
Modified atmosphere packaging (MAP)
Vacuum-packed food
Sous vide food
Ultra-high temperature (UHT) food

Food safety starts long before you prepare or serve meals in your operation. To be sure the food you serve is safe, you must first control the quality and safety of food that comes in your back door.

Many things can happen to a product at different points within the flow of food. A frozen product that leaves the processor's plant in good condition, for example, may thaw on its way to the distributor's warehouse, which will affect both product quality and safety.

The final responsibility for the safety of food entering your establishment rests with you. You can avoid many potential food safety hazards by making sure products are properly received. Using approved suppliers and inspecting products when they are delivered are the first steps in the process.

CHOOSING A SUPPLIER

A number of factors go into selecting the right suppliers. While choosing a supplier who can deliver safe food is the ultimate goal, the level of service also needs to be considered. Before you accept any deliveries from a supplier, it is your responsibility to be sure food you purchase comes from approved sources. Also check suppliers to see if they meet or exceed the food safety standards you follow in your establishment.

Quality Standards

○ **Make sure suppliers are getting their products from licensed, reputable sources.** Check with your regulatory agency to find out if your suppliers have had any food safety problems or health code violations. Ask other operators what their experiences with a particular supplier have been.

○ **Inspect your supplier's warehouse or plant from time to time, if possible.** See if it is clean and well-run.

○ **Ask your suppliers if they have a HACCP program in place.** (If they supply fresh produce, ask whether they have a Good Agricultural Practices Plan.) If not, ask what precautions or procedures they take to ensure product safety.

○ **Find out if your supplier's employees are trained in food safety.**

○ **Check the condition of the supplier's delivery trucks.** Are they clean and well-maintained? Do they hold refrigerated or frozen products at the proper temperatures? Are raw products separated from processed food and fresh produce?

○ **Check your supplier's shipments for consistent product quality.** Inspect deliveries for unsafe packaging. Broken boxes, leaky packages, or dented cans are signs of careless handling.

○ **Request that suppliers deliver products when your staff has time to receive them properly.**

Rejecting Shipments

Remember, you have the right to refuse any delivery. You should have a company policy about returns, and your suppliers should be aware of it and agree to it. When food doesn't meet acceptable standards, your employees should know what to do. To reject a product or shipment:

○ **Set the rejected product aside.** Keep it separate from other food and supplies.

○ **Tell the delivery person exactly what is wrong with the rejected product.** Use your purchase agreement and company standards to back up your decision to reject the product.

○ **Get a signed adjustment or credit slip from the delivery person before throwing the product away or letting the delivery person remove it.**

○ **Log the incident on the invoice or receiving document.** Note the food involved, including lot number and expiration date if appropriate, the standard that was not met, and the corrective action you took.

Once you have established a relationship with a supplier, continue to be a smart customer. Always inspect deliveries. Randomly check weights and product temperatures, break down cases, and take counts. Don't take anything for granted. You are buying more than products: you're buying service and food safety.

INSPECTION PROCEDURES

If you establish procedures for inspecting products, you can reduce hazards before they enter your establishment.

Here are some general guidelines that can help you improve the way you receive deliveries.

○ **Train employees to inspect deliveries properly.** Ideally, you should assign the responsibility for inspecting and receiving deliveries to specific employees. These employees should be trained to judge product quality, check products for proper temperatures, identify code dates, identify food that has been thawed and refrozen, spot damage or insect infestation, and so on. They also should be authorized to accept, reject, and sign for deliveries.

○ **Plan ahead for shipments.** Have clean hand trucks, carts, dollies, and containers available in the receiving area. Make sure enough space is available in walk-ins and storerooms prior to receiving a shipment. Some operations use a refrigerator and freezer in the receiving area for temporary storage. If products need to be washed or broken down and rewrapped, make work

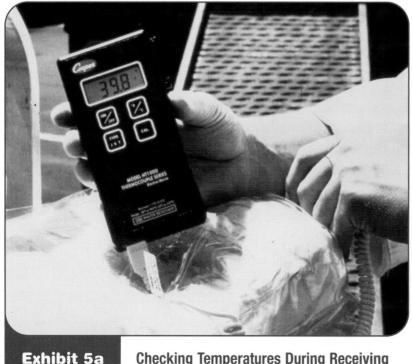

Courtesy of Cooper Instrument Corporation, Middlefield, CT

Exhibit 5a

Checking Temperatures During Receiving
Take sample temperatures of refrigerated and
frozen foods.

space available as close to the receiving area as possible. This will prevent dirt and pests from being brought into storage areas or the kitchen.

○ **Schedule deliveries during off-peak hours.** Arrange it so products are delivered during times when employees have adequate time to inspect them.

○ **Plan a backup menu in case you have to return some food items.** If food is not safe or does not meet your standards, you may have to take an item off the menu, substitute another menu item, or try to arrange delivery from another supplier.

○ **If possible, receive only one delivery at a time.** Inspect and store each delivery before accepting another one to avoid potential confusion and product abuse in the receiving area.

○ **Have the right information available.** Receivers should have a purchase order or order sheet ready to check against a supplier's invoice. The sheet should list quantities, quality specifications, and agreed-upon prices. The sheet should also have room to record the date and time of delivery, product temperatures, and other notes.

○ **Inspect deliveries immediately.** Do a thorough visual inspection to count quantities, check for damaged products, and look for items that might have been repacked or mishandled. Spot-check weights and take sample temperatures of all refrigerated and frozen food. *(See Exhibit 5a.)* There is always a possibility that food—even government-inspected products—may have been mishandled during shipment.

○ **Correct mistakes immediately.** If any products are damaged, not at the correct temperature, or have not been delivered to specifications, do not accept them.

○ **Label all items before storage with the delivery date or the use-by date to ensure proper stock rotation.** Put products away as quickly as possible, especially products requiring refrigeration.

○ **Keep the receiving area clean and well-lighted to discourage pests.**

MONITORING TIME AND TEMPERATURE

Time and temperature play critical roles in the process of maintaining a safe food supply. They affect food quality and safety from the moment food arrives at the back door to the time a prepared meal is presented to the customer.

If a potentially hazardous food is kept in the temperature danger zone longer than four hours, it must be discarded. This four-hour time period begins when the food is taken off the delivery truck and continues through storage, preparation, and cooking. It begins again after cooking. The longer food sits on a receiving dock before being refrigerated or put in the freezer, the less time is left for preparation and cooking.

To manage both time and temperature, you need to monitor and control them. The thermometer may be the single most important tool you have to protect your food.

Choosing the Right Thermometer

There are many types of thermometers used in an establishment. Each is designed for a specific purpose. Some are used to measure the temperature of refrigerated or frozen storage areas. Others measure the temperature of equipment, such as ovens, hot-holding cabinets, and warewashing machines. Perhaps the most important types are thermometers that measure the temperature of food. The most common types used in establishments are the bimetallic stemmed thermometer, the thermocouple, and the thermistor. *(See Exhibit 5b.)* Infrared thermometers are also becoming increasingly popular.

Bimetallic Stemmed Thermometers

The most common and versatile type of thermometer used in the restaurant and foodservice industry is the **bimetallic stemmed thermometer.** *(See Exhibit 5c on the next page.)* This type of thermometer measures temperature through a metal probe with a sensor in the end. Bimetallic stemmed thermometers often have scales measuring temperatures from 0°F to 220°F (-18°C to 104°C). This makes them useful for measuring the temperatures of everything from incoming shipments of frozen food to hot-holding cabinets to sanitizing solution in warewashing equipment. When you select this type of thermometer, it should have:

○ An adjustable calibration nut to keep it accurate

○ Easy-to-read, numbered temperature markings

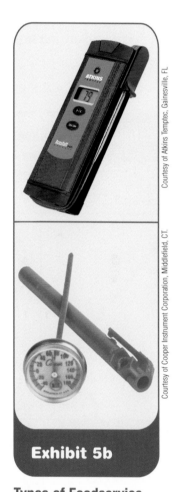

Exhibit 5b

Types of Foodservice Thermometers
Shown from top to bottom are a thermocouple and a bimetallic stemmed thermometer.

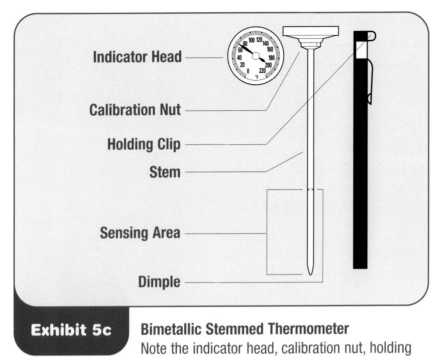

Exhibit 5c **Bimetallic Stemmed Thermometer**
Note the indicator head, calibration nut, holding clip, stem, sensing area, and dimple.

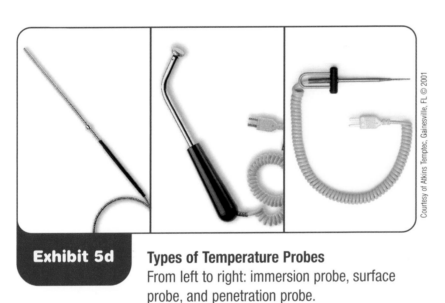

Courtesy of Atkins Temptec, Gainesville, FL © 2001

Exhibit 5d **Types of Temperature Probes**
From left to right: immersion probe, surface probe, and penetration probe.

○ A dimple to mark the end of the sensing area (which begins at the tip)

○ Accuracy to within ±2°F (±1°C)

Thermocouples and Thermistors

Thermocouples and thermistors measure temperatures through a metal probe or sensing area and display results on a digital readout. They come in a wide variety of styles and sizes, from small pocket models to panel-mounted displays. Many come with interchangeable temperature probes designed to measure the temperature of equipment and food.

Basic types of probes include immersion, surface, penetration, and air probes. *(See Exhibit 5d.)* Immersion probes are designed to measure temperatures of liquids, such as soups, sauces, or frying oil. Surface probes measure temperatures of flat cooking equipment like griddles. Penetration probes are used to measure the internal temperature of food. When measuring the internal temperature of thin food, such as meat and fish patties, small-diameter probes should be used. Air probes measure temperatures inside refrigerators or ovens.

Infrared (Laser) Thermometers

Infrared thermometers use infrared technology to produce accurate temperature readings of food and equipment surfaces. They are quick and easy to use. Infrared thermometers are noncontact thermometers that, when used properly, can reduce the risk of cross-contamination and damage to food products. *(See Exhibit 5e.)*

To gain the most accurate thermometer reading, remove any barriers between the thermometer and the product being measured and hold the thermometer as close as possible to the product without touching it.

When using infrared thermometers, remember the following:

○ Infrared thermometers should not be used to measure air temperature or the internal temperature of food. They are designed to measure surface temperature.

○ Do not take temperature measurements through glass or shiny or polished-metal surfaces, such as stainless steel or aluminum.

○ Always follow the manufacturer's guidelines for tips on obtaining the most accurate temperature reading with the infrared thermometer you are using.

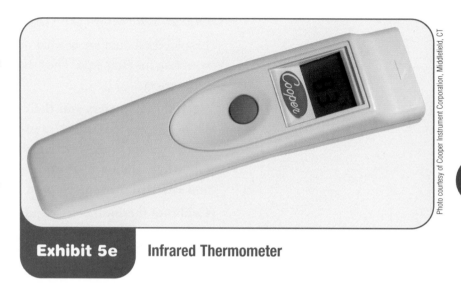

Exhibit 5e **Infrared Thermometer**

Time-Temperature Indicators (TTI) and Other Time-Temperature Recording Devices

Some instruments are designed to monitor both time and product temperature. The **time-temperature indicator (TTI)** is one example. Some suppliers attach these self-adhesive tags or sticks to a food shipment to determine if the product's temperature has exceeded safe limits during shipment or later storage. If the product's temperature has exceeded these limits, the TTI provides an irreversible record of the incident. A change in color inside the TTI indicators or windows notifies the receiver that the product has undergone time-temperature abuse. *(See Exhibit 5f.)*

More suppliers are using recording devices in their delivery trucks that continuously monitor temperatures. When delivered products appear to have suffered time-temperature abuse, the recording device can be checked to see if the temperature in the delivery truck changed at any time during transit.

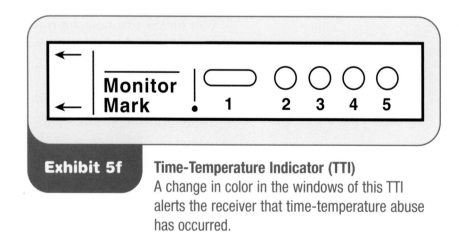

Exhibit 5f **Time-Temperature Indicator (TTI)**
A change in color in the windows of this TTI alerts the receiver that time-temperature abuse has occurred.

General Thermometer Guidelines

Employees should know what different thermometers are used for and how to care for each type. They should follow a few simple rules with all thermometers.

○ **Keep thermometers and their storage cases clean.** Thermometers should be washed, rinsed, sanitized, and air-dried before and after each use to prevent cross-contamination. Use an approved food-contact surface sanitizing solution to sanitize them. Have an adequate supply of clean and sanitized thermometers on hand.

○ **Calibrate thermometers regularly to ensure accuracy.** This should be done before each shift or before each day's deliveries. Thermometers should also be recalibrated any time they suffer a severe shock (for example, after being dropped or after an extreme change in temperature). Thermometers that hang or sit in refrigerators or freezers can be damaged easily. To make sure these thermometers are accurate, place a calibrated, stemmed thermometer in a cup of water, place the cup next to the hanging thermometer, and check the temperature. Hanging thermometers usually cannot be recalibrated and must be replaced if they are not accurate.

○ *Never* **use glass thermometers filled with mercury or spirits to monitor the temperature of food.** They can break and pose a serious danger to employees and customers.

○ **Measure internal temperatures of food by inserting the thermometer stem or probe into the thickest part of the product (usually the center).** It is good practice to take at least two readings, which should be in different locations because product temperatures may vary across the food portion. (Procedures for how to measure temperatures of packaged food delivered by suppliers will be discussed later in this chapter.)

○ **Wait for the thermometer reading to steady before recording the temperature of a food item.** Wait at least fifteen seconds from the time the thermometer stem or probe is inserted into the food.

How to Calibrate Thermometers

Calibration is the process of ensuring that a thermometer gives accurate recordings by adjusting it to a known standard. Most thermometers can be easily calibrated. Two accepted methods of calibration are the ice-point method *(see Exhibits 5g and 5h)* and the boiling-point method. *(See Exhibit 5i).* To calibrate your thermometers properly, follow one of these methods.

Cross-Contamination

Wash, rinse, sanitize, and air dry thermometers before and after each use to prevent cross-contamination. Use an approved food-contact surface sanitizing solution to sanitize all thermometers.

Exhibit 5g: Ice-Point Method for Calibrating a Thermometer

Step	Process	Notes
1	Fill a large container with crushed ice. Add clean tap water until the container is full.	Stir the mixture well.
2	Put the thermometer stem or probe into the ice water so the sensing area is completely submerged. Wait 30 seconds, or until the indicator stops moving.	Do not let the stem or probe touch the container's bottom or sides. The thermometer stem or probe must remain in the ice water.
3	Hold the calibration nut securely with a wrench or other tool and rotate the head of the thermometer until it reads 32°F (0°C).	On some thermocouples or thermistors, it may be possible to press a reset button to adjust the readout.

Exhibit 5g

Step 1

Step 2

Step 3

Exhibit 5h

Using the Ice-Point Method to Calibrate a Thermometer

Exhibit 5i: Boiling-Point Method for Calibrating a Thermometer

Step	Process	Notes
1	Bring clean tap water to a boil in a deep pan.	
2	Put the thermometer stem or probe into the boiling water so the sensing area is completely submerged. Wait 30 seconds, or until the indicator stops moving.	Do not let the stem or probe touch the pan's bottom or sides. The thermometer stem or probe must remain in the boiling water.
3	Hold the calibration nut securely with a wrench or other tool and rotate the head of the thermometer until it reads 212°F (100°C) or the appropriate boiling-point temperature for your elevation.	The boiling point of water is about 1°F (about 0.5°C) lower for every 550 feet (168 m) you are above sea level. On some thermocouples or thermistors, it may be possible to press a reset button to adjust the readout.

Exhibit 5i

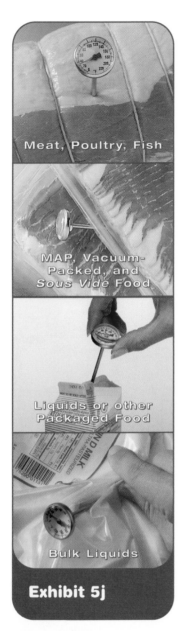

Meat, Poultry, Fish

MAP, Vacuum-Packed, and Sous Vide Food

Liquids or other Packaged Food

Bulk Liquids

Exhibit 5j

Checking the Temperatures of Various Food

How to Check Temperatures of Deliveries

Use the following guidelines when checking the temperature of deliveries. *(See Exhibit 5j.)*

○ **Check the temperature of meat, poultry, and fish by inserting the thermometer stem or probe into the thickest part of the product (usually the center).**

○ **Check the temperature of MAP, vacuum-packed, or *sous vide* food** *(see page 5-20)*, **by inserting the thermometer stem or probe between two packages.** Be careful not to puncture the product wrapping. Wait at least fifteen seconds for the reading to steady, and record the temperature.

○ **Check the temperature of liquids or other packaged food by opening the package and inserting the thermometer stem or probe into the food until the sensing area is immersed.** Do not let the thermometer stem or probe touch the container's sides. Wait until the indicator stops moving and record the temperature.

○ **Check the temperature of bulk liquids by folding the bag or pouch around the thermometer stem or probe.** Be careful not to puncture the bag or pouch.

○ **Check the temperature of live molluscan shellfish by inserting the stem or probe into the middle of the carton or case, between the shellfish, for an ambient reading.** Check the temperature of shucked shellfish by inserting the stem or probe into the container until the sensing area is immersed.

○ **When receiving eggs, check the ambient air temperature of the delivery truck, as well as the truck's temperature chart recorder for extreme temperature fluctuations during transport.** Temperature fluctuations, high humidity, and warm temperatures may result in the growth of harmful microorganisms.

○ **Always be sure to use a clean, sanitized thermometer each time you check a temperature.** If you don't have extra thermometers at the back door, clean and sanitize your thermometer after each use. Keep a bucket of sanitizing solution on the dock, or use approved sanitizing wipes.

RECEIVING AND INSPECTING FOOD

Every food product delivered to your establishment should be inspected carefully for damage or potential contamination. Internal temperatures of products should be checked and recorded. While receiving temperatures for fresh food are product-specific, frozen food should always be received frozen. In addition, note each food's appearance, texture, smell, and (in some cases) taste. *(See Exhibit 5k.)*

Exhibit 5k: Accept and Reject Criteria for Receiving Seafood, Meat, Poultry, and Eggs

Use this chart to determine whether to accept or reject deliveries.

Food	Accept	Reject
FRESH FISH Receive at 41°F (5°C) or lower	COLOR: bright red gills; bright, shiny skin ODOR: mild ocean or seaweed smell EYES: bright, clear, and full TEXTURE: firm flesh that springs back when touched	COLOR: dull, gray gills; dull, dry skin ODOR: strong fishy or ammonia smell EYES: cloudy, red-rimmed, sunken TEXTURE: soft, leaves an imprint when pressed
FRESH SHELLFISH (clams, mussels, oysters) Receive at 45°F (7°C) or lower for live shellfish	ODOR: mild ocean or seaweed smell SHELLS: closed and unbroken (indicates shellfish are alive) CONDITION: received alive; identified by shellstock identification tag. Retain tags for 90 days after product is used.	ODOR: strong fishy smell SHELLS: open shells that do not close when tapped (indicates shellfish are dead); broken shells CONDITION: dead on arrival TEXTURE: slimy, sticky, or dry
FRESH CRUSTACEANS (lobster, shrimp, and crabs) Receive at 45°F (7°C) or lower for live lobsters and crabs	ODOR: mild ocean or seaweed smell SHELLS: hard and heavy for lobsters and crabs CONDITION: shipped alive; packed with seaweed and kept moist	ODOR: strong fishy smell SHELLS: soft CONDITION: dead on arrival, tail fails to curl when lobster is picked up
FRESH MEAT Receive at 41°F (5°C) or lower	BEEF COLOR: bright, cherry red LAMB COLOR: light red PORK COLOR: pink lean meat, white fat TEXTURE: firm and springs back when touched	COLOR: brown or greenish; brown, green, or purple blotches; black, white, or green spots TEXTURE: slimy, sticky, or dry PACKAGING: broken cartons, dirty wrappers, or torn packaging ODOR: sour odor
FRESH POULTRY Receive at 41°F (5°C) or lower	COLOR: no discoloration TEXTURE: firm and springs back when touched PACKAGING: should be surrounded by crushed, self-draining ice	COLOR: purple or green discoloration around the neck; dark wing tips (red wing tips are acceptable) TEXTURE: stickiness under the wings or around joints ODOR: abnormal, unpleasant odor
FRESH EGGS (Shell) Receive at an air temperature of 45°F (7°C) or lower	ODOR: none SHELLS: clean and unbroken CONDITION: firm, high yolks that are not easy to break and whites that cling to yolks	ODOR: sulfur smell or off odor SHELLS: dirty or cracked

Exhibit 5k

Seafood

Fish and shellfish are very sensitive to rough treatment and time-temperature abuse. Both fresh and frozen seafood deteriorate quickly if improperly handled. Time-temperature abuse can result in the rapid growth of microorganisms, which can lead to foodborne illness.

Fish

Fresh fish should be packed in self-draining crushed or flaked ice. Upon delivery, it should be received at a temperature of 41°F (5°C) or lower. Fresh fish in good condition should meet the following standards:

○ Clear eyes

○ Firm flesh

○ Pleasant, mild scent of ocean or seaweed

○ Bright red and moist gills

○ Bright skin

Carelessly handled fish is not appetizing, let alone safe to eat. The following conditions are grounds for rejecting a shipment of fish:

○ Strong fishy or ammonia smell

○ Cloudy, red-rimmed, sunken eyes

○ Dark, dull red gills

○ Dry skin

○ Soft skin that leaves an imprint when pressure is applied with a finger

○ Tumors, abscesses, and cysts on the skin

See *Exhibit 5l* for examples of acceptable and unacceptable fish.

Acceptable   Unacceptable

Exhibit 5l   Acceptable vs. Unacceptable Fish

Frozen fish should be received frozen. If there is any indication it has been allowed to thaw, do not accept it. Fish that has thawed and then been refrozen before reaching your establishment may have a sour odor and be off-color. Fillets often turn brown at the edges when they have been refrozen. Other signs include large amounts of ice or liquid in the bottom of the shipping box and moist, discolored, or slimy wrapping paper.

Shellfish

Shellfish fall into two categories: *crustacea*, such as shrimp, crab, and lobster, and *molluscan bivalves* (mollusks), such as clams, oysters, and mussels. Shellfish can be shipped live, fresh, frozen, in the shell, or shucked.

Interstate shipping of shellfish is monitored by the FDA. Shellfish must be bought only from suppliers listed in the National Shellfish Sanitation Program Guide for the Control of Molluscan Shellfish, or from sources included in the Interstate Certified Shellfish Shippers List. *(Found on FDA Seafood Information and Resources Web site; see page 5-35.)* Shucked shellfish must be packaged in nonreturnable containers clearly labeled with the name, address, and certification number of the packer. Packages containing less than one-half gallon must have a sell-by date. Packages containing more than one-half gallon should list the date the shellfish was shucked.

Both live and shucked molluscan shellfish may be received at temperatures of 45°F (7°C) or lower. When shipped live, shellfish must be delivered alive in nonreturnable containers. The FDA requires that live molluscan shellfish carry **shellstock identification tags.** *(See Exhibit 5m.)* Restaurant and foodservice operators must write the date of delivery on the tags. The tag should remain attached to the container the

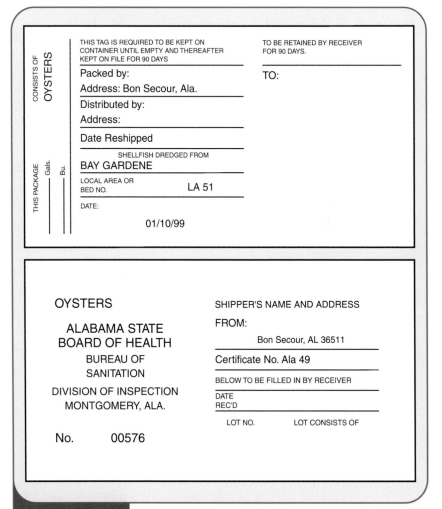

Exhibit 5m

Shellstock Identification Tags
Note that the tags must be dated when you receive the product, then kept on file for 90 days after the last shellfish has been used.

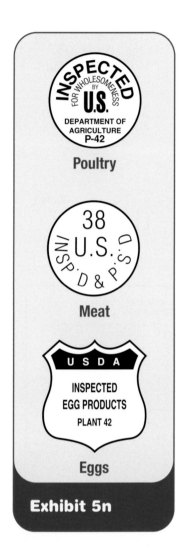

Poultry

Meat

Eggs

Exhibit 5n

USDA Inspection Stamps for Meat, Poultry, and Eggs Inspection is a mandatory process.

shellfish came in until the container is empty. Operators then must keep the tags on file for 90 days after the last shellfish has been used. Never mix shellfish from one shipment with another.

Shells of clams, mussels, and oysters will be closed if alive. Partly open shells may mean they are dead. To find out, tap on the shells. If they close, the mollusks are still alive. If the shells do not close or are badly cracked or broken, they should be discarded.

Fresh lobsters or crabs in good condition are easy to tell from those that have been poorly handled. A fresh lobster or crab will meet the following standards:

○ Show signs of movement

○ Have a hard and heavy shell

○ React when its eyes are pinched

○ Curl its tail under when turned on its back (lobsters)

Lobsters or crabs that show weak signs of life should be cooked right away. Dead ones must be discarded or returned to the vendor for credit.

Fresh Meat and Poultry

We often hear news stories of foodborne illnesses caused by *Salmonella* and *Campylobacter* from poultry or *E. coli* O157:H7 from beef or other meat. While fresh meat and poultry are carefully inspected for wholesomeness by government agencies, it is impossible to eliminate all microorganisms present during processing. Almost all cases of foodborne illness can be prevented, however, if food is properly handled and prepared.

Fresh meat and poultry should be delivered at 41°F (5°C) or lower. When it arrives, inspect it closely, checking temperature, color, odor, texture, and packaging. Meat and poultry must be purchased from plants inspected by the USDA or state department of agriculture. Meat products that have been inspected will be stamped with abbreviations for "inspected and passed" by the inspecting agency, along with the number identifying the processing plant. *(See Exhibit 5n.)* Stamps will not appear on every cut of meat, but one should be present on every inspected carcass and on packaging.

Meat and poultry inspection is mandatory. During the inspection process, products are checked for wholesomeness. USDA inspectors examine the carcass and viscera of each animal for possible signs of illness and check processing plants for sanitary conditions. The USDA inspection stamp means that both product and processing plant have met certain standards. It does not mean the product is free of microorganisms that can cause foodborne illness. Operators still are responsible for properly handling and preparing this food to be sure it is safe for consumers to eat.

Most meat and poultry also carry a stamp indicating its "grade" or palatability and level of quality. Grading is a voluntary service offered by the USDA and is paid for by processors and packers. USDA grades are printed inside a shield-shaped stamp. *(See Exhibit 5o.)*

When receiving meat and poultry, consider the following factors:

Meat

○ **Beef should be a bright, cherry red.** Aged beef may be darker in color, and the type of packaging used sometimes affects color. Any beef turning brown or green should be rejected. Beef usually spoils on or near the surface first. An off color, slimy texture, or sour odor are signs the meat has begun to deteriorate. Inspect ground beef very carefully. Ground beef spoils more easily than solid muscle cuts. Check for broken cartons, dirty meat wrappers, and torn or leaking bags. Do not accept any product that arrives in this condition. See *Exhibit 5p* for examples of acceptable and unacceptable beef.

○ **Lamb is light red when fresh and properly exposed to air.** Do not accept any fresh lamb that is brown or has a whitish surface covering the lean meat.

○ **Fresh pork is light pink in color with firm, white fat portions.** An excessively dark color, soft or rancid fat, and a sour odor all indicate the meat is spoiled and should be rejected.

Exhibit 5o

USDA Grading Stamps for Meat, Poultry, and Eggs Grading is a voluntary service that processors and packers pay for.

Exhibit 5p Acceptable vs. Unacceptable Beef

Poultry

Fresh poultry should be shipped in self-draining crushed ice and delivered at a temperature of 41°F (5°C) or lower, or chill-packed. Poultry shipped and stored at temperatures of 28°F (-2°C) will likely have a significantly longer shelf life.

Mishandled poultry is easy to spot by its appearance. *(See Exhibit 5q.)* Poultry that has started to spoil may have the following characteristics:

○ Purplish or greenish color
○ Abnormal odor
○ Stickiness under the wings and around joints
○ Dark wing tips (red tips are acceptable)

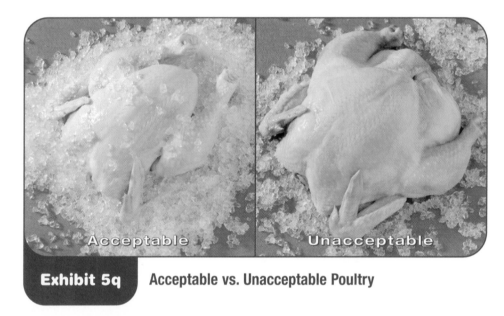

| Exhibit 5q | Acceptable vs. Unacceptable Poultry |

Poultry is inspected by federal or state agriculture agencies in much the same way as meat. As with meat, the grading system is voluntary and paid for by processors.

Eggs

Purchase fresh eggs from approved, government-inspected suppliers. The USDA inspection stamp on egg cartons indicates federal regulations are enforced to maintain quality and reduce contamination. As with meat and poultry, grading is voluntary and is provided by the USDA. The official grade stamp certifies the eggs have been graded for quality under federal and/or state supervision. Cases and cartons of shell eggs for direct sale to the consumer must display safe-handling instructions on them.

Choose suppliers who can deliver eggs within a few days of the packing date. Eggs must be delivered in refrigerated trucks. These trucks should be capable of documenting air temperature during transportation. When the eggs arrive, the truck's air temperature should be 45°F (7°C) or lower and the eggs must be stored immediately in refrigeration units that will hold them at an ambient (air) temperature of 45°F (7°C) or lower.

Exhibit 5r

Acceptable Eggs
Eggs should be clean, dry, and free of cracks.

Shells should be clean, dry, and free of cracks. *(See Exhibit 5r.)* To test freshness, break an egg into a flat dish. Acceptable eggs have firm, high yolks that are not easy to break, and the whites should cling to the yolks. In a very high-quality egg, the white will stand up on its own. Fresh eggs should have no odor. Refrigerate fresh eggs in their original containers. Cases or cartons of shell eggs for direct sale to the consumer must display safe-handling instructions on them.

Liquid, frozen, and dehydrated eggs must be pasteurized as required by law, and bear the USDA inspection mark. When delivered, they should be refrigerated or frozen at the proper temperature. Check packages for damage or indication of refreezing and note use-by dates to be sure the product is in good condition and still usable.

Dairy Products

Purchase only pasteurized dairy products. Unpasteurized milk and other dairy products are potential sources of microorganisms such as *Salmonella* spp., *Campylobacter jejuni,* and *Listeria monocytogenes,* all of which can cause serious illness. All milk and milk products should be labeled "Grade A." This means they meet standards for quality and sanitary processing methods set by the FDA and U.S. Public Health Service. Dairy products with the Grade A label—such as cream, dried milk, cottage cheese, cream cheese, butter, cheese—and frozen products like ice cream, are made with pasteurized milk.

Like other refrigerated products, milk and dairy products should be received at 41°F (5°C) or lower, unless the law governing their distribution specifies a different temperature. Check with the proper regulatory agency for temperature requirements.

Fresh milk has a sweetish taste. Any milk that tastes sour, bitter, or moldy should be rejected. Milk has a sell-by date stamped on the container. Milk delivered after that date, as well as milk with any off odor, should be rejected.

Key Point

Purchase pasteurized dairy products only.

Butter should have a sweet flavor, uniform color, and firm texture. Check for signs of mold, specks, or other foreign matter. Make sure containers are clean and undamaged. Don't accept any butter that is rancid or has absorbed odors.

In the United States, all cheeses must meet certain standards of identity. In order for the product to be called "cheddar" or "mozzarella," for example, the government specifies ingredients that must be used, maximum moisture content, minimum fat content, and general characteristics. When cheeses are delivered, check to see each type has its typical flavor, texture, and color. If the cheese has a rind, it should be clean and unbroken. Cheese should have no signs of mold or off odors. As always, check for proper temperature and clean, undamaged packaging.

Fresh Produce

Fresh fruit and vegetables have different temperature requirements for transportation and storage, so they may be held at different temperatures. No specific temperature is mandated by regulation for the transportation and storage of fresh produce, with a few exceptions. Cut melons, a potentially hazardous food, must be held at 41°F (5°C) or lower. Fresh-cut produce is best held at 33°F to 41°F (1°C to 5°C) to maintain quality.

Fresh fruit and vegetables are highly perishable and should be put into storage quickly. Most produce should not sit out at room temperature, let alone on a warm dock. Have a system in place to get fresh-cut and highly perishable items into a cooler.

In general, produce should not be washed before it is stored. While washing would not hurt leafy green items, many other products are likely to decay faster if washed before storage. This is especially true for mushrooms and berries. Wash produce just before preparing and serving it.

All fruit and vegetables should be handled with care. If they are pinched, squeezed, or roughly handled, they will bruise and spoil more quickly.

Check products being delivered for signs of mishandling and insect infestation, including insect eggs and egg cases. Visually inspect produce for quality. *(See Exhibit 5s.)* Spoilage will show up in a variety of ways, including mold, blemishes, cuts, mushiness, discoloration, wilting, or dull appearance. What applies to one fruit or vegetable may not apply to another. For example, peaches with cuts in them could be considered poor quality, but potatoes and carrots with cuts would be considered acceptable. Discoloration in produce may vary as well. For example, oranges may actually revert to a green color without affecting the quality of the orange or its juice.

Key Point

In general, produce should not be washed before it is stored.

Use smell and taste to help determine product quality. Unpleasant odors will tell you when a product is not acceptable. With fruit, sometimes taste is the best test. Outer peels or skins can be blemished without affecting flavor or quality. Be sure to wash or peel fruit and vegetables carefully before tasting them.

Since there are so many ways produce can show signs of spoilage, employees need to learn not only how to identify obviously unacceptable produce, but also produce that will spoil quickly in storage.

Refrigerated and Frozen Processed Food

More and more establishments are using prepared food that is either refrigerated or frozen. This includes precut meats, Individually Quick Frozen (IQF) poultry, frozen or refrigerated entrées that only require heating, and fresh-cut fruit and vegetables (including salads). Processed food can save time and money, but only if it is treated with the same care given to other food products. Mishandled products that end up causing a foodborne illness can put an establishment out of business.

The temperature of refrigerated processed food should be 41°F (5°C) or lower when delivered. While these products are usually fully cooked or ready to eat, they still require careful handling. Inspect packaging for tears or holes and check use-by dates.

All frozen food should be delivered frozen, with the exception of ice cream. Ice cream may be delivered and stored at temperatures of 6°F to 10°F (-14°C to -12°C) without affecting product safety or quality.

Frozen food should be checked for signs of thawing and refreezing. Simply because a product is frozen upon receipt does not mean it has not thawed and been refrozen during prior handling. Obvious signs are blocks of ice or liquid at the bottom of the case or large ice crystals on the product itself. Other signs include product discoloration or dryness and stains on the outer packaging. Food should be wrapped in airtight packaging, and boxes and outer cartons should be clean and undamaged.

Exhibit 5s

Inspecting the Quality of Fruit
Spoilage will show up in a variety of ways, including mold, blemishes, mushiness, discoloration, wilting, or dull appearance.

Key Point

Check frozen food for signs of thawing and refreezing, such as large ice crystals on the product.

MAP, Vacuum-Packed, and Sous Vide Food

MAP Food

MAP stands for **modified atmosphere packaging**. By this method, air is removed from a food package and replaced with gases, such as carbon dioxide and nitrogen. These gases help extend the shelf life of the product. Many fresh-cut produce items are packaged using MAP methods.

Vacuum-Packed Food

Vacuum-packed food is processed by removing the air around a food product sealed in a package. Bacon is one example.

Sous Vide Food

Sous vide (soo veed) is a French term meaning "under vacuum." Food processed by this method is vacuum-packed in individual pouches, partially or fully cooked, and then chilled. This food is then heated for service in the establishment.

Sous vide products may be received either refrigerated or frozen, depending on the manufacturer. Some frozen, reduced-calorie meals are packaged using this method.

Receiving MAP, Vacuum-Packed, and Sous Vide Food

Removing oxygen from packaged food can reduce or prevent the growth of some microorganisms that need oxygen to grow. However, the same conditions can promote the growth of anaerobic microorganisms, such as those producing the botulism toxin.

For this reason, the FDA doesn't allow establishments to package food on site using a MAP method except under certain conditions. To prepare *sous vide* or MAP food, operators must have a HACCP plan in place and limit the foods being packaged to those that can't support the growth of *Clostridium botulinum*. Refrigerated products must be delivered and stored at 41°F (5°C) or lower unless otherwise specified by the manufacturer. *(See Exhibit 5t.)*

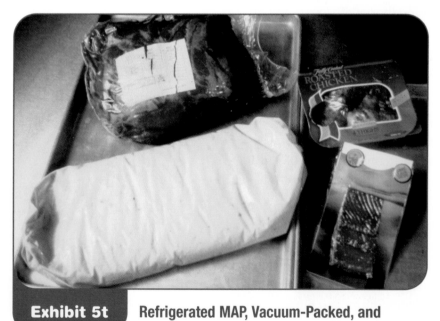

Exhibit 5t   Refrigerated MAP, Vacuum-Packed, and *Sous Vide* Products

When you receive MAP, vacuum-packed, and *sous vide* food, use the following guidelines:

○ Make sure the supplier has a HACCP plan in place.

○ Frozen products should be frozen when they arrive.

○ Reject packages with leaks.

○ Reject products that appear slimy or have bubbles.

○ Reject products that are an unacceptable color.

○ Reject packages that contain product with an expired code date.

Dry and Canned Products

Dry and canned products seem to pose little threat to consumers. Most have a fairly long shelf life and are usually used long before they have a chance to spoil. However, canned products provide a good environment for the microorganisms that cause botulism. Dry food can be contaminated by a variety of sources that can cause foodborne illnesses.

Dry food must be kept dry. Most microorganisms need moisture to grow and multiply, which is why dry food has a much longer shelf life than fresh food. Check both outer cases and inner packaging for dampness or moisture. Reject the shipment if it is damp or shows signs of prior wetness (moisture stains).

Dry food often attracts pests. Because it can be stored at room temperature, dry food might not be sealed and stored as securely as fresh, refrigerated, or frozen food. Insects and rodents have an easier time getting into dry-food packages. Carefully inspect packages for holes, tears, or punctures. Check products themselves for signs of infestation. You can spot insects or insect eggs in cereal or flour by sprinkling some of the product on brown paper. Rodents leave signs such as chewed packages and droppings in and around cartons.

Use sight and smell to inspect dry food, too. Off colors and odors, spots of mold, or a slimy appearance are signs of spoilage.

Canned food must also be checked carefully for damage. *(See Exhibit 5u on next page.)* Check can exteriors first, looking for the following signs of contamination:

○ **Swollen ends.** One or both ends of a can may bulge from gas produced by the presence of chemicals or the growth of foodborne bacteria inside. If one end bulges out when the other is pressed, discard it, because the can has not gone through the proper heat-treating process to eliminate foodborne microorganisms.

○ **Leaks and flawed seals.** If there are any leaks, flaws, or irregularities along the top or side seals, reject the can.

Key Point

Reject dry foods if there are moisture stains on the packaging.

Key Point

Reject packages with holes, tears, or punctures.

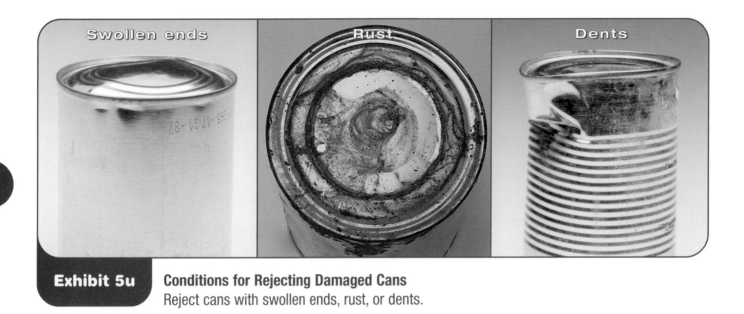

Exhibit 5u   **Conditions for Rejecting Damaged Cans**
Reject cans with swollen ends, rust, or dents.

Never taste
canned foods
you are unsure of.

○ **Rust.** If a can is rusted, the contents may be too old or the rust may have eaten holes in the can, which can lead to contamination.

○ **Dents.** Don't accept cans with dents along side or top seams or dents large enough to make it impossible to open the can with a can opener. The seams may be broken. (Check with your regulatory agency regarding dented cans; some do not allow any dents.)

Any cans received without labels should be rejected. Once the exteriors of the cans have been checked, spot-check the contents. Any food that does not have a normal color, texture, or odor, that are foamy, or that contain a milky-colored liquid should be thrown out immediately. *Never* taste canned food you are unsure of. Botulism, a foodborne illness associated with canned food, is so dangerous that people have died from just tasting and spitting out contaminated food.

Aseptically Packaged and Ultra-High Temperature (UHT) Pasteurized Food

Aseptically packaged and **ultra-high temperature (UHT)** pasteurized food such as milk, juice, and puddings can also be received and stored at room temperature. These products are pasteurized (heat-treated at very high temperatures for a short time) to kill microorganisms that can cause illness. They are then packaged under sterile conditions to keep them from being contaminated. Once opened, however, the products should be refrigerated at 41°F (5°C) or lower. Check packaging and seals to make sure they are intact. Reject these products if packaging is punctured or broken.

Potentially Hazardous Hot Food

Occasionally, operators may receive a shipment of hot food. Hot food must be properly cooked as required by local or federal codes. *(See Chapter 7 for guidelines.)* If you purchase hot food, make sure the supplier has a HACCP plan or other means of documenting proper cooking methods and temperatures. Potentially hazardous hot food must be delivered at a temperature of 140°F (60°C) or higher. This food should be delivered in appropriate containers that can maintain these temperatures.

SUMMARY

Even though federal and state agencies regulate and monitor the production and transportation of food such as meats, poultry, seafood, eggs, dairy products, and canned goods, it is your responsibility to check the quality and safety of food that comes into your establishment.

Receiving safe food starts with the careful selection of approved and reputable suppliers. By working with suppliers, operators can take steps to ensure the food they purchase is safe.

Operators must plan delivery schedules so products can be handled promptly and correctly. Employees assigned to receive deliveries should be trained to inspect food properly, as well as to distinguish between products that are acceptable and those that are not. They should also be authorized to reject products that don't meet company standards and to sign for products that do.

Thermometers are the most important tools operators have to prevent time-temperature abuse. Every facility should have an adequate supply of thermometers on hand. Thermometers must be cleaned and sanitized before and after each use. Managers should make sure employees know what different thermometers are used for and how to calibrate and use them properly.

All products arriving at the establishment should meet agreed-upon standards. Packaging should be clean and undamaged. Code dates should be current. Food should show no signs of mishandling.

Products must be delivered at the proper temperature. All products—especially meat, poultry, and fish—should be checked for proper color, texture, and odor. Live molluscan shellfish and crustacea must be delivered alive. Eggs should be inspected for freshness and for dirty and cracked shells. Dairy products must be checked for freshness. Produce should be fresh and wholesome. Frozen food should be inspected for signs of thawing and refreezing. MAP, vacuum-packed, and *sous vide* food should not bubble or appear slimy, its packaging should be intact, and code dates should not be expired. Canned food must be carefully examined for signs of damage. Dry food should be inspected for pest infestation and moisture. Potentially hazardous hot food must be delivered at 140°F (60°C) or higher.

A CASE IN POINT I

Case Study

On Monday, a large food delivery arrived at the Sunnydale Nursing Home during the busy lunch hour. All the food products were different: cases of frozen ground beef patties, canned vegetables, frozen shrimp, fresh tomatoes, a case of potatoes, and fresh chicken.

Betty, the new assistant manager, thought the best thing to do was to put everything away and check it later, since she was very busy. She told Ed, in charge of receiving, to sign for the delivery and put the food into storage. Ed asked her if it would be better to ask the delivery driver to come back later. Since she needed the chicken for that night's dinner, Betty asked Ed to accept the delivery now and went back to the front of the house.

Ed put the frozen shrimp and ground beef patties in the freezer and the fresh chicken in the refrigerator. Then he put the fresh tomatoes, potatoes, and canned vegetables in dry storage. When he was finished, he went back to work in the kitchen.

What was done incorrectly? What could be the possible result?

A CASE IN POINT II

Case Study

ABC Seafood makes its usual Thursday afternoon delivery to The Fish House. John, a prep cook, is the only person in the kitchen when the driver rings the bell at the back door. The kitchen manager, who is in charge of receiving, is in a managers' meeting. The chef is out on an errand, and the rest of the kitchen staff is on break, though some are still in the restaurant.

John follows the driver onto the dock, where the driver unloads two crates of ice-packed fresh fish, bags of live mussels, two buckets of live oysters, a case of shucked oysters in plastic containers, and a case each of frozen shrimp and frozen lobster tails. John goes back into the kitchen to get a bimetallic stemmed thermometer and remembers to look in the chef's office for a copy of the order form. He takes both out to the dock and begins to inspect the shipment.

John checks the products against both the order sheet and the invoice, then begins to check product temperatures. First he puts the thermometer stem between the packages in the cases of frozen shrimp and lobster. Then he checks the temperature of the shucked oysters by taking the cover off one container and inserting the thermometer stem into it. The thermometer reads 45°F (7°C), which worries him. He remembers being taught that refrigerated products should be 41°F (5°C) or lower. He decides not to say anything.

After wiping the thermometer stem on his apron, John checks the internal temperature of a whole fish packed in ice. Finally, he reaches into one of the buckets of live oysters with his hand to see if it feels cold. He notices a few mussels and oysters with open and broken shells. He removes those with broken shells, knowing the chef won't use them.

John records all his findings on the order sheet he took from the chef's office, signs for the delivery, and starts putting the products away. He first puts away the live shellfish, dumping the few still left in the refrigerator into the new containers so there is room. Then he puts the shucked oysters and fresh fish in the refrigerator and the shrimp and lobster into the freezer.

What was done wrong? What could be the result?

TRAINING TIPS

1.   Thermometer Calibration Demonstration

Purpose: *After completing this activity, trainees should be able to calibrate a bimetallic stemmed thermometer using the ice-point method.*

Time: 30 minutes

Directions: Calibrate a bimetallic stemmed thermometer using the ice-point method for your trainees. Allow time for trainees to practice the calibration process.

2.   Thermometer Group Discussion

Purpose: *After completing this activity, trainees should be able to explain how each thermometer functions and determine which ones would be best suited for the establishment.*

Time: 30 minutes

Directions: Discuss the primary uses, advantages, and disadvantages of the following thermometers:

- ○ Bimetallic stemmed thermometer
- ○ Thermistor
- ○ Thermocouple
- ○ Hanging thermometer

3.   Receiving Game

Purpose: *After completing this activity, trainees should be able to determine when food items should be accepted or rejected during receiving.*

Time: 30 minutes

Directions: Divide trainees up into four or five teams and give each team a bell.

Explain to each team that you will show a picture (or provide a verbal description) of a product with certain characteristics (for example, a fish with sunken eyes, a dented can, discolored poultry, a properly packaged dry product).

Ask each team to ring its bell when they have decided whether the product should be accepted or rejected. Then ask them to give their answer and tell the reasoning behind it. If a team provides the correct answer, it receives one point. It receives two points if it can identify conditions creating an *opposite* result. For example, if the team has decided a box of frozen steaks can be accepted, and it can identify the conditions resulting in the frozen steaks being rejected, it receives two points. Give bonus points to any team that can identify an item for which receiving is a critical control point.

4. Receiving Role-Play

Purpose: *After completing this activity, trainees should be able to determine when to accept or reject a food item.*

Time: 30 minutes

Directions: Ask two volunteers to role-play a delivery skit. One volunteer is assigned the role of a delivery driver bringing several different food items to an establishment, while the other plays the role of a receiving employee. Give the volunteers a list of specific food products to be delivered and allow them five or ten minutes to prepare a role play in which they act out both right and wrong practices during the receiving process.

Ask the other trainees to identify and write down each wrong practice they spot during the skit. Afterwards (after a round of applause!), ask the group to call out the "wrongs" they have spotted. Award prizes to both the role-play volunteers and to the person who spotted the correct number of wrongs. Discuss proper receiving practices with the trainees, including accept-and-reject criteria for specific food items.

5. Guest Lecturer: Thermometer Manufacturer Representative

Purpose: *After completing this activity, staff members will be able to properly calibrate, use, and sanitize thermometers found in the establishment.* *Note:* A sales representative from the manufacturer will add credibility to the presentation.

Time: 30 minutes

Directions: Bring in a knowledgeable and interesting sales representative from the company that manufactures the thermometers you use. Have the representative explain how the thermometers are made, how they are calibrated, how they are used, and how they should be cleaned and sanitized.

Make sure you bring in someone enthusiastic and knowledgeable. Provide the representative with information about your audience, how long the presentation should be, and what audio-visual equipment is available. Make it clear that the presentation should not be a sales pitch.

6. Field Trip to a Manufacturer or Supplier

Purpose: *After completing this activity, managers and supervisors will be able to identify the food safety practices used by their manufacturers or suppliers to keep products safe.*

Time: ½ day

Directions: Take your managers, supervisors, and any other interested employees on a field trip to a food manufacturer or distributor you do business with and tour its facility.

Consider these places:

- Dairy plant
- Poultry plant
- Meat packing house
- Cannery
- Seafood supplier
- Bakery
- Food distribution warehouse

Plan ahead. Group sizes may be limited, and access to different parts of these facilities is often restricted. With the cooperation of a plant manager, these tours can be remarkably informative.

7. Receiving Checklist

Purpose: *To have your receiving personnel create a checklist of accept-and-reject conditions for food items they receive at your establishment working directly with your suppliers. This process will help create an understanding between both parties and will create buy-in for your personnel through empowerment.*

Time: several weeks

Directions: Ask your staff to break down food items at your establishment by category, then by specific item. Certain types of products may be listed as a general group, such as canned or dry food. For each item, have them work with the supplier to create a list of accept-and-reject conditions.

For example:

Whole fresh chicken

Accept	**Reject**
Shipped in crushed ice	Shipped without ice
Temperature of 41°F (5°C) or lower	Temperature above 41°F (5°C)
Free of discoloration	Green or purple discoloration
Free of off-odors	Abnormal odor
Free of stickiness	Stickiness under wings
USDA inspection stamp	

DISCUSSION QUESTIONS

1. What are some general guidelines for receiving food safely?

2. How is a thermometer calibrated using the ice-point method?

3. What are proper methods for checking the temperature of frozen food, fresh poultry delivered on ice, and bulk milk? What should the temperature be for each?

4. What are three conditions that would result in rejecting a shipment of fresh poultry?

5. What four types of external damage to cans are cause for rejection?

MULTIPLE-CHOICE STUDY QUESTIONS

1. Which is most important in choosing a food supplier?
   A. It meets your food safety standards.
   B. Its prices are the lowest.
   C. Its warehouse is close to your establishment.
   D. It offers a convenient delivery schedule.

2. When you receive a delivery of food, it is important that you
   A. refrigerate it before inspecting it.
   B. inspect it before accepting it.
   C. put it with other recent deliveries.
   D. take it out of its original packaging before storing it.

3. Which of the following shipments should be rejected?
   A. Beef that is bright, cherry red in color
   B. Lamb with flesh that is firm and springs back
      when touched
   C. Fish that arrives with sunken eyes
   D. Chicken received at 41°F (5°C)

4. How should whole, fresh salmon be packaged for delivery and storage?

   A. Layered with salt
   B. Vacuum sealed
   C. Wrapped in dry, clean cloth
   D. Packed in self-draining crushed ice

5. Which condition would cause you to reject a shipment of live oysters?

   A. Most of the oysters have closed shells.
   B. The oysters have a mild seaweed scent.
   C. Most of the oysters shells are open and don't close when they're tapped.
   D. The oysters are delivered with shellstock identification tags.

6. A box of sirloin steaks carries both a USDA inspection stamp and a USDA choice grading stamp. What do these stamps tell you?

   A. The farm that supplied the beef uses only USDA-certified animal feed.
   B. The meat and processing plant have met USDA standards, and the meat quality is acceptable.
   C. The meat wholesaler meets USDA quality-grading standards.
   D. The steaks are free of disease-causing microorganisms.

7. Which condition would cause you to reject a shipment of fresh chicken?

   A. The flesh is firm and springs back when touched.
   B. The wing tips are brown.
   C. The grading stamp is missing.
   D. The chicken is odorless.

8. To check the freshness of a delivery of shell eggs, you crack one onto a plate. What about the egg would tell you that you can accept the delivery?

   A. It has a sulfurous smell.
   B. The egg white doesn't cling to the yolk.
   C. It comes out of its shell completely.
   D. The yolk is high and firm.

9. Statements from a dairy supplier's sales brochure are listed below. Which statement should tell you not to hire this supplier?

   A. We make our cheese with only the freshest unpasteurized milk.
   B. From farm to you, our milk is kept at temperatures below 41°F (5°C).
   C. Money-back guarantee—our prices are the lowest in the area.
   D. We deliver according to your schedule and needs.

10. Which of the following food *does not* have to be received at 41°F (5°C) or lower?

    A. Beef
    B. Live shellfish
    C. Pork
    D. Fish

11. All of the following would be grounds for rejecting a case of frozen food *except:*

    A. There are large ice crystals on the frozen food inside the case
    B. The outside of the case is water-stained
    C. The food in the box is frozen solid
    D. There is frozen liquid at the bottom of the case

12. Which delivery should be rejected?

    A. Several cans in a case of peaches have torn labels.
    B. Several cans in a case of tomato soup have swollen ends.
    C. A bag of oatmeal is delivered at 60°F (16°C).
    D. A case of rice is missing a USDA Inspection Stamp.

13. Which step for calibrating a thermometer using the ice-point method is *incorrect?*

    A. First, insert the thermometer stem or probe into a container of ice water.
    B. Second, wait thirty seconds or until the temperature indicator stops moving.
    C. Third, remove the thermometer stem or probe from the ice water.
    D. Finally, adjust the thermometer until it reads 32°F (0°C).

14. You have a thermocouple with several different types of probes. Which probe should you use to check the temperature of a large stock pot of soup?

A. Penetration probe
B. Surface probe
C. Air probe
D. Immersion probe

15. Your manager has asked you to purchase a new thermometer for the restaurant. Which type would *not* be a proper choice?

A. Bimetallic stemmed thermometer accurate to ±2°F (±1°C)
B. Thermistor
C. Thermocouple
D. Mercury-filled glass thermometer

For answers, please turn to Appendix A.

ADDITIONAL RESOURCES

Books and Periodicals

Cichy, R. F. 1993. *Sanitation management.* East Lansing, MI: Educational Institute of the American Hotel & Motel Association.

Cross out cross-contamination. 1998. *Best Practices.* 2 (1):6-9.

Equipped for food safety. 1998. *Best Practices.* 2 (4):6-10.

Kotschevar, L. H. 1994. *Quantity food purchasing,* 4th ed. New York: Macmillan.

Loken, J. K. 1995. *The HACCP food safety manual.* New York: Wiley.

Longree, K. 1996. *Quantity food sanitation,* 5th ed. New York: Wiley.

National Restaurant Association Educational Foundation. 1998.
A Practical Approach to HACCP: Coursebook. Chicago: National Restaurant Association Educational Foundation.

The right way to…accept deliveries. 1998. *Best Practices.* 2(1):12-13.

The right way to…calibrate equipment. 1997. *Best Practices.* 1(3):12.

The right way to…prepare beef. 1998. *Best Practices.* 2(3):12-13.

The right way to…prepare chicken. 1998. *Best Practices.* 2(1):10.

The right way to…prepare eggs. 1998. *Best Practices.* 2(2):10-11.

The right way to…prepare pork. 1997. *Best Practices.* 1(3):10-11.

The right way to…preparing fruits and vegetables. 1998.
Best Practices. 2(4), 12-13.

The right way to…seafood safety. 1999. *Best Practices.* 3(1):16.

Secrets of self-inspection. 1998. *Best Practices.* 2(2):6-9.

Warfel, M. C. (1996). *Purchasing for food service managers,* 3rd ed. Berkeley, CA: McCutchan.

Web Sites

Alaska Seafood Marketing Institute

http://www.alaskaseafood.org

The Alaska Seafood Marketing Institute enhances the appeal and popularity of Alaska seafood to the restaurant and foodservice industry and to the general public. This useful Web site contains information on how to market seafood to your customers, how to purchase Alaskan seafood, detailed information on types of Alaskan seafood, and how to handle seafood to maintain safety and quality.

American Egg Board (AEB)

http://www.aeb.org

As the egg industry's promotional arm, the AEB is the U.S. egg producer's link to the consumer for communicating the value of the egg. This Web site offers a wealth of information about eggs and egg safety.

American Meat Institute

http://www.meatami.com

The American Meat Institute is a membership trade association representing the interests of the U.S. meat and poultry industry to the federal government, Congress, the media, and the customer. This Web site provides up-to-date information about what is happening in the meat and poultry processing industry.

Atkins Temptec

http://www.atkinstemptec.com

Atkins is a manufacturer of digital thermometers and probes for the restaurant and foodservice industry. This Web site provides information on how thermometers work, how to get the most out of your thermometer, and how to hold food safely during service. The site also provides an interesting demonstration on how food color is a poor indicator of internal temperature. Consult this site for useful tips related to thermometers.

Bunzl Distribution, Inc.

http://www.bunzldistribution.com

Bunzl Distribution is a supplier of disposable paper and plastic packaging supplies, as well as specialty items for service and packaging for the restaurant and foodservice industry.

Cooper Instrument Corporation

http://www.cooperinstrument.com

Cooper Instrument Corporation is a supplier of temperature, time, and humidity instruments for a variety of global markets. This Web site contains information on how to calibrate thermometers properly and maintain food temperatures during cooking, holding, cooling, etc.

Daydots International

http://www.daydots.com

Makers and distributors of many sanitation and food safety tools for the restaurant, most of which can be ordered online. The site also contains food safety-related articles and information.

FDA–Center for Food Safety and Applied Nutrition (CFSAN)

http://www.cfsan.fda.gov

As the center within FDA responsible for food safety and nutrition, CFSAN promotes and protects public health by researching and implementing guidelines, policies, and standards to ensure food is safe, nutritious, wholesome, and properly labeled. This Web site provides a wealth of information on food safety and sanitation, including corresponding guidelines, policies, and standards.

FDA Food Code

http://vm.cfsan.fda.gov/~dms/foodcode.html

As the basis for many local sanitation codes, as well as the basis for information in this coursebook, the FDA Food Code, available at this Web address, is a useful resource for information relating to food safety for the restaurant and foodservice industry.

FDA Seafood Information and Resources

http://vm.cfsan.fda.gov/seafood1.html

The FDA operates an oversight compliance program for fishery products under which responsibility for product safety, wholesomeness, identity, and economic integrity rests with the processor or importer, who must comply with regulations under the Federal Food, Drug and Cosmetic (FD&C) Act. This Web site houses information on the seafood program, foodborne pathogens and contaminants associated with seafood, and HACCP compliance.

Food Marketing Institute (FMI)

http://www.fmi.org

The FMI is a nonprofit association conducting programs in research, education, industry relations, and public affairs on behalf of its members, which include large multistore chains, small regional firms, and independent supermarkets.

International Dairy-Deli-Bakery Association (IDDBA)

http://www.iddba.org

IDDBA is a resource for information and services across all food channels for the dairy, deli, and bakery industry. This Web site provides information relevant to all people who work in the dairy, deli, and bakery industry.

International Fresh-cut Produce Association (IFPA)

http://www.fresh-cuts.org

IFPA serves commercial producers, suppliers, and distributors of fresh-cut produce. It provides members with a forum for representation, education, technical support, networking, information, and resources. This site contains information about recent industry happenings and regulations, and provides guidance documents for purchasing, receiving, and storing fresh-cut produce.

The Mushroom Council

http://www.mushroomcouncil.com

This interesting and useful Web site from the Mushroom Council contains recipes and information about how to handle and store mushrooms to ensure safety and quality.

National Cattlemen's Beef Association (NCBA)

http://www.beef.org

NCBA is a consumer-focused, producer-directed organization representing the largest segment of the nation's food and fiber industry. This Web site, as well as a companion site for foodservice (www.beeffoodservice.com), provides excellent information on purchasing, cooking, and serving beef. Both sites also provide important food safety information.

The National Chicken Council

http://www.eatchicken.com

This Web site is brought to you by the National Chicken Council and the U.S. Poultry & Egg Association, which represent a wide spectrum of companies and individuals in the chicken business. This fun and interesting Web site contains chicken safety tips and nutritional and statistical information.

National Food Processors Association (NFPA)

http://www.nfpa-food.org

The NFPA is the voice of the food processing industry on scientific and public policy issues involving food safety, nutrition, technical and regulatory matters, and consumer affairs. Its Web site offers many resources on food safety and food processing for the entire food industry.

National Frozen Food Association (NFFA)

http://www.nffa.org

NFFA's mission is to promote the sales and consumption of frozen food through education training, research, sales planning, and menu development, and to provide a forum for industry dialogue. Publications and information on frozen foods and how to market them to consumers are available at this Web site.

National Pork Producers Council

http://www.nppc.org

Billing itself as the place for pork lovers, pork producers, and anyone who wants to learn more about pigs and pork, the National Pork Producers Council's Web site provides information on pork safety and quality, nutrition and industry information, and publications.

National Restaurant Association

http://www.restaurant.org

The National Restaurant Association is the leading business association for the restaurant industry. Together with the National Restaurant Association Educational Foundation, the Association's mission is to represent, educate, and promote the rapidly growing restaurant and foodservice industry. This Web site should be your starting place for all issues and concerns related to your establishment. This Web site has it all, from tips for running your establishment to vital data on your customers' spending habits.

National Turkey Federation

http://www.turkeyfed.org

The National Turkey Federation advocates for all segments of the turkey industry, providing services and conducting activities which increase demand for its members' products by protecting and enhancing their ability to profitably provide wholesome, high-quality, nutritious products. Its Web site contains a special area specifically for the restaurant and foodservice industry, including recipe and menu ideas, cooking demonstrations, and a reference on purchasing, preparing, and promoting turkey.

Produce Marketing Association

http://www.pma.com

The Produce Marketing Association is a not-for-profit trade association serving members who market fresh fruit, vegetables, and floral products worldwide. Its members are involved in the production, distribution, retail, and foodservice sectors of the industry. This Web site provides promotional ideas and information about hot topics in the produce industry.

SYSCO Corporation

http://www.sysco.com

Sysco is North America's leading marketer and distributor of food and foodservice products. In addition to providing information about Sysco brands and products, this Web site contains a recipe database, food safety information, and information about building your business.

Tyson Foods, Inc.

http://www.tyson.com

Tyson Foods is the world's largest poultry producer. This Web site is a great resource about chicken products for the restaurant and foodservice industry. It includes recipes and menu ideas, information on new products, and information about Tyson Foods' efficient distribution and transportation system.

USDA-FSIS

http://www.fsis.usda.gov

The Food Safety and Inspection Service (FSIS) is the public health agency of the U.S. Department of Agriculture responsible for ensuring that the nation's commercial supply of meat, poultry, and egg products is safe, wholesome, and correctly labeled and packaged. This Web site contains a wealth of information on food safety relating to meat, poultry, and eggs.

U.S. Foodservice™

http://www.usfoodservice.com

U.S. Foodservice is one of the largest broadline foodservice distributors in the United States, distributing food and related products to restaurants and institutional foodservice establishments across the country.

Vitsab TTI

http://www.vitsab.com

Vitsab time temperature indicator (TTI) labels are inexpensive monitoring devices that can be placed on temperature-sensitive containers to track their possible exposure to out-of-range, high-temperature conditions. The indicator tracks time, as well as temperature, so the result is a true indication of temperature exposure. This Web site provides technical information on TTIs and Vitsab.

Chapter 6
Keeping Food Safe in Storage

Knowledge

TEST YOUR FOOD SAFETY KNOWLEDGE

1. **True or False:** Potentially hazardous, ready-to-eat food stored in refrigeration at 41°F (5°C) must be discarded if not used within three days of preparation. *(See page 6-12.)*

2. **True or False:** Foodservice chemicals may be transferred to sturdy, clearly labeled containers. *(See 6-12.)*

3. **True or False:** Freezing destroys all harmful microorganisms in food. *(See page 6-4.)*

4. **True or False:** Raw chicken should be stored below ready-to-eat food, such as pumpkin pie, if it is stored in the same walk-in refrigerator. *(See page 6-5.)*

5. **True or False:** If stored food has passed its expiration date, you should cook and serve it at once. *(See page 6-2.)*

For answers, please turn to Appendix A.

Key Terms

First in, first out (FIFO)
Refrigerated storage
Frozen storage
Dry storage

Chemical storage
Shelf life
Hygrometer

Table of Contents

Learning Objectives

After completing this chapter, you should be able to

○ Label and store specific types of refrigerated food.

○ Label and store frozen food.

○ Properly store dry and canned food.

○ Apply first in, first out (FIFO) practices.

○ Properly store raw food to prevent cross-contamination.

○ Properly transfer food from original containers to storage containers.

○ Follow the proper procedures for using refrigerators, freezers, and dry-storage areas.

Key Point

Use the first in, first out (FIFO) stock rotation method. Products closest to their expiration date should be used first.

When food is stored improperly and not used in a timely manner, quality and safety suffer. Poor storage practices can cause food to spoil quickly with potentially serious results.

GENERAL STORAGE GUIDELINES

Every facility has a wide variety of products that need to be stored. A few general rules can be applied to most storage situations.

○ **Follow the first in, first out (FIFO) method.** Write the date on each product, indicating when it was received or prepared. *(See Exhibit 6a.)* Train employees to store products with the earliest use-by or expiration dates in front of products with later dates. *(See Exhibit 6b.)* Once items have been properly shelved, use those stored in front first. Make sure employees check the use-by or expiration date and discard products that have passed this date.

○ **Keep potentially hazardous food out of the temperature danger zone.** Store deliveries as soon as they have been inspected. Take out only as much food as you can prepare at one time. Put prepared food away until needed. Properly cool and store cooked food as soon as it is no longer needed. *(See Chapter 7 for more information on cooling cooked food.)*

○ **Check temperatures of stored food and storage areas.** Temperatures should be checked at the beginning of the shift. Many establishments use a preshift checklist to guide employees through this process.

○ **Label food.** Always label potentially hazardous, ready-to-eat food with the date it should be sold, consumed, or discarded. All food should clearly note the contents of the package. If you take food out of its original package, put it in a clean, sanitized food container with a tight-fitting lid. (Eggs and produce should be stored in their original containers.)

Courtesy of Daydots International, Fort Worth, TX, 800-321-3687

Exhibit 6a   **Labeling System**
There are several products on the market that help simplify the labeling process for storing food.

- **Keep all storage areas clean and dry.** Floors, walls, and shelving in refrigerators, freezers, dry storerooms, and heated holding cabinets should be properly cleaned on a regular basis. Clean up spills and leaks right away to keep them from contaminating other food.
- **Clean dollies, carts, transporters, and trays often.**
- **Store food only in designated storage areas.** Do not store food products near chemicals or cleaning supplies *(see Exhibit 6c)*; in restrooms, locker rooms, janitor closets, furnace rooms or vestibules; or under stairways or pipes of any kind. Food can easily be contaminated in any of these areas.

TYPES OF STORAGE

An ideal facility has two types of storage areas: food storage and chemical storage. Food-storage areas include refrigerators, freezers, and dry storerooms. Chemical-storage areas are used to store cleaning supplies and other chemicals.

Each storage area has a specific purpose. Operators should consider the design and management of storage space a priority since improperly used storage areas can cause problems. For example, an overstocked refrigerator may not be able to hold temperatures properly. Stock may not get rotated correctly, and there is a greater chance of cross-contamination.

Courtesy of Daydots International, Fort Worth, TX. 800-321-3687.

Exhibit 6b

Follow FIFO
Train employees to store products with the earliest use-by or expiration dates in front of products with later dates.

Exhibit 6c

Improper Storage
Never store food near chemicals or cleaning supplies.

Key Point

Potentially hazardous food should be stored in refrigerators at an internal temperature of 41°F (5°C) or lower.

Storage spaces should also be located to prevent food contamination. Food should be stored away from warewashing areas and garbage rooms. To ease the flow of food through the operation, storage areas must be accessible to receiving, food-preparation, and cooking areas.

Most restaurants and foodservice establishments have several types of storage areas in their facilities. The most common include:

○ **Refrigerated storage.** These areas are typically used to hold potentially hazardous food at 41°F (5°C) or lower. Refrigeration slows the growth of microorganisms and helps keep them from multiplying to levels high enough to cause illness. (Some jurisdictions allow food to be held at an internal temperature of 45°F [7°C] or lower. Check with the local regulatory agency for specific regulations.)

○ **Frozen storage.** These areas are typically used to hold frozen food at 0°F (-18°C) or lower. Freezing does not kill all microorganisms, but it does slow their growth substantially. Frozen food should be used as quickly as possible.

○ **Dry storage.** These areas are used to hold dry and canned food. To maintain the quality of this food, dry-storage areas should be kept at the appropriate temperature and humidity levels. Storerooms should be clean, well-ventilated, and well-lighted.

○ **Chemical storage.** Cleaning tools and chemicals should be kept in their own storage area. Store these supplies away from food in clean, dry rooms or in cabinets that can be locked.

STORAGE TECHNIQUES

A few commonsense rules apply to each of these storage areas. Make sure employees follow these rules to keep food safe.

Refrigerated Storage

In general, the colder food is, the safer it is. Keeping food as cold as possible without freezing it also extends its shelf life. Ideal storage temperatures will vary depending on the food. Fruit and vegetables will freeze if stored at temperatures ideal for fish. Meat and poultry will have a shorter **shelf life** if stored at temperatures better suited for produce. If possible, store food such as meat and poultry in separate refrigerators to hold them at optimal temperatures. If this is impractical, store meat, poultry, fish, and dairy products in the coldest part of the unit, away from the door.

Key Point

Generally, the colder food is, the safer it is.

While there are many types of refrigeration equipment available to operators, from walk-in refrigerators to refrigerated drawers, some general guidelines apply when using all of them.

○ **To hold food at a specific internal temperature, refrigerator air temperature usually must be approximately 2°F (1°C) lower.** For example, to hold poultry at an internal temperature of 41°F (5°C), the air temperature in the refrigerator should be 39°F (4°C). Use hanging thermometers in the warmest part of the refrigerator. Some units have a readout panel outside to check the temperature without opening the door. These should also be checked for accuracy. At least once during each shift, check the temperature of the unit.

○ **Monitor food temperature regularly.** Randomly sample the temperature of food stored inside using a calibrated thermometer. *(See Exhibit 6d.)* You may also want to monitor food temperatures by using a product-mimicking device.

○ **Never place hot food in the refrigerator.** This can warm the interior enough to put other food in the temperature danger zone.

○ **Do not overload the unit.** Storing too many products prevents good airflow and makes the unit work harder to stay cold.

○ **Use open shelving.** Lining shelves with aluminum foil or paper restricts circulation of cold air in the unit.

○ **Keep the refrigerator door closed as much as possible.** Frequent opening lets warm air inside, which can affect food safety and make the refrigerator work harder. Consider using cold curtains to help maintain walk-in refrigerator temperatures.

○ **Wrap all food properly.** Leaving food uncovered can lead to cross-contamination.

○ **Store raw meat, poultry, and fish separately from cooked and ready-to-eat food to prevent cross-contamination.**

○ **Store cooked or ready-to-eat food above raw meat, poultry, and fish if these items are stored in the same unit.** This will prevent raw-product juices from dripping onto the prepared food and causing a foodborne illness. Store raw meat, poultry, and fish in the following order from top to bottom: whole fish, whole cuts of beef and pork, ground meat and fish, whole and ground

Cross-Contamination

Store raw meat, poultry, and fish separately from cooked and ready-to-eat food whenever possible.

Exhibit 6d **Monitoring Food Temperature** Randomly sample the internal temperature of refrigerated food using a calibrated thermometer.

poultry. This order is based on the required minimum internal cooking temperature of each food. *(See Exhibits 6e and 6f.)*

Frozen Storage

Some general guidelines for using freezers include:

○ **Check unit and food temperatures regularly.** Use a calibrated thermometer to check the accuracy of hanging thermometers or unit readout.

○ **Keep freezer temperatures at 0°F (-18°C) or lower unless the food you are storing requires a different temperature.**

○ **Place frozen food deliveries in the freezer as soon as they have been inspected.** Clearly label the food, identifying the package's contents, date of delivery, and use-by date, if there is one. If products are going to be used immediately, they can be stored in a refrigerator. Never hold frozen food at room temperature.

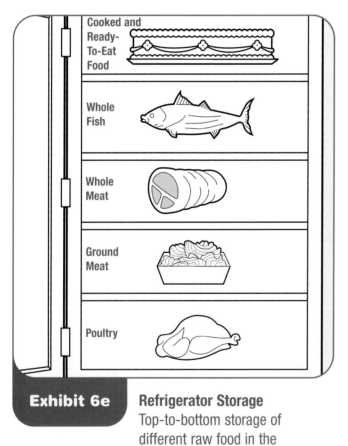

Exhibit 6e **Refrigerator Storage**
Top-to-bottom storage of different raw food in the same refrigerator.

Exhibit 6f **Improper Storage of Different Raw Products in the Same Refrigerator**
Never store raw products this way.

○ **Use caution when placing food into a freezer.** Warm food can raise the temperature in the unit and partially thaw the food inside. Preparing several small batches of food is a better idea than freezing leftovers. Store food to allow good air circulation. Overloading a freezer makes it work harder and makes it more difficult to find and rotate food properly.

○ **Defrost freezer units on a regular basis.** They will operate more efficiently when free of frost. Move contents to another freezer when defrosting, or use them immediately.

○ **Never refreeze thawed food unless it has been thoroughly cooked.** Refreezing damages food quality. Thawed food is also more likely to support the growth of microorganisms, which refreezing will not kill.

○ **Rotate frozen food using the FIFO method.** Check use-by dates to be sure they haven't expired.

○ **Store food in its original package.** If food is removed from its packaging, wrap it in moisture-proof material or place it in a cleaned and sanitized container. Clearly label all packages and containers identifying the contents. Always label potentially hazardous, ready-to-eat food with the date it should be sold, consumed, or discarded.

○ **Open the unit as infrequently as possible.** Use cold curtains to help maintain temperatures.

Dry Storage

Dry food remains safe and retains quality if held in the right conditions. Moisture and heat are the biggest problems. The storeroom temperature should be between 50°F and 70°F (10°C to 21°C). Keep relative humidity around 50 to 60 percent. If high humidity is a problem, consider using a dehumidifier. Some general guidelines for dry storage include:

○ **Monitor temperature and humidity regularly.** Use a hanging thermometer to record temperatures and a **hygrometer** to measure relative humidity. Units are available that combine both instruments.

○ **Store dry food at least six inches off the floor and away from walls.** Keep it out of direct sunlight.

○ **Store food in its original packaging whenever possible.** Once packages are opened, store product in tightly covered containers. This will prevent insects, rodents, and microorganisms from contaminating the food.

○ **Keep the area clean and dry.** Clean up any spills immediately.

○ **Make sure storerooms are well-ventilated.** This will help keep temperature and humidity more even and constant throughout the storage area.

Key Point

Fresh raw meat, poultry, and fish must be stored at an internal temperature of 41°F (5°C) or lower.

Time & Temperature

Fish to be served raw or partially cooked must be frozen by the processor to specific temperatures prior to shipment.

STORING SPECIFIC FOOD

The general storage guidelines discussed apply to most food. However, certain foods have special requirements.

Meat

○ Store meat immediately after delivery in its own storage unit or in the coldest part of the refrigerator. Fresh meat must be held at an internal temperature of 41°F (5°C) or lower. Frozen meat should be stored at a temperature that will keep it frozen.

○ Wrap raw cuts of meat, especially ground beef, so they are airtight. Meat will turn brown when exposed to air. Frozen meat should be wrapped in airtight, moisture-proof material or placed in containers to prevent freezer burn.

○ Primal cuts, quarters or sides of raw meat, and slab bacon can be hung on clean, sanitized hooks or can be placed on sanitized racks. To prevent cross-contamination, do not store meat above any other food.

○ Throw out any meat showing signs of spoilage.

Poultry

○ Store raw, fresh poultry at an internal temperature of 41°F (5°C) or lower. Frozen poultry should be stored at temperatures that will keep it frozen. If it has been removed from its original packaging, place it in airtight containers or wrap it in airtight material.

○ Ice-packed poultry can be stored in a refrigerator as is. Containers must be self-draining. Change the ice and sanitize the container often.

Fish

○ Store fresh fish at an internal temperature of 41°F (5°C) or lower. Keep fillets and steaks in original packaging or wrap them in moisture-proof materials. Fresh, whole fish can be packed in flaked or crushed ice, and ice beds must be self-draining. Change the ice and sanitize the container regularly.

○ Store frozen fish at temperatures that will keep it frozen. Wrap fish in moisture-proof packaging.

○ Fish that will be served raw or partially cooked should be frozen by the processor, as follows, prior to shipment to kill any parasites that may be present.

Fish should be frozen by the processor at:

- -4°F (-20°C) or lower for seven days (168 hours) in a storage freezer; or
- -31°F (-35°C) or lower for fifteen hours in a blast freezer

The only exceptions are shellfish and certain species of tuna not susceptible to parasites. Ask your suppliers whether the fish you are purchasing meet these conditions.

Shellfish

- Store live shellfish in their containers at an internal temperature of 45°F (7°C) or as low as 35°F (2°C). You can store molluscan shellfish (clams, oysters, mussels, scallops) in a display tank under one of two conditions:
  - The tanks carry a sign stating that the shellfish are for display only
  - You obtain a variance from the health department to serve the shellfish on display
- To obtain a variance, you must submit a HACCP-based plan showing the following:
  - The water in the tank will not come into contact with any other fish
  - Using the tank won't affect product quality or safety
  - You retain shellstock identification tags as required

Eggs

- Eggs received at an air temperature of 45°F (7°C), in compliance with laws governing their shipment from suppliers, must be placed immediately upon their receipt in refrigeration equipment capable of maintaining an ambient (air) temperature of 45°F (7°C) or lower. Maintain constant temperature and humidity levels in refrigerators used to store eggs. Do not wash eggs before storing them since they are washed and sanitized at the packing facility.
- Use the FIFO method of storage. Plan to use all eggs within four to five weeks of the packing date.
- Keep shell eggs in cold storage right up until the time they are used. Take out only as many eggs as are needed for immediate use.
- Do not combine cracked eggs in a bowl (pool them) unless you intend to use them right away.
- Store frozen egg products at temperatures that will keep them frozen.
- Liquid egg products should be refrigerated in their original containers at 41°F (5°C) or lower. Do not freeze.

Store dairy products at 41°F (5°C) or lower.

○ Dried egg products can be stored in a cool, dry storeroom away from light. Storing them in a refrigerator is best. Once they are reconstituted (mixed with water), store them in the refrigerator at 41°F (5°C) or lower.

Dairy Products

○ Store dairy products at 41°F (5°C) or lower.

○ Frozen dairy products such as ice cream and frozen yogurt can be stored at 6°F to 10°F (-14°C to -12°C).

○ Dairy products used for cooking should not be held at room temperature longer than two hours. They can sour quickly. Like all potentially hazardous food, dairy products must be thrown away if they have been in the temperature danger zone for more than four hours. Do not put products that have been held at room temperature back into refrigeration.

○ Always use the FIFO method of storage. Discard products if they have passed their use-by or expiration dates.

Fresh Produce

○ Fruit and vegetables have various temperature requirements for storage. While many whole, raw fruit and vegetables can be stored at 41°F (5°C) or lower, not all will be stored at these temperatures. Whole, raw produce and raw, cut vegetables—such as celery, carrots, and radishes—delivered packed in ice can be stored that way. The containers must be self-draining, and ice should be changed regularly.

○ Fruit and vegetables kept in the refrigerator can dry out quickly. Keep the relative humidity at 85 to 95 percent.

○ Though most produce can be stored in the refrigerator, avocados, bananas, pears, and tomatoes ripen best at room temperature.

○ Produce should not be washed before storage. Moisture promotes the growth of mold in many instances. Instead, wash produce before preparing or serving it.

○ Store whole citrus fruit, hard-rind squash, eggplant, and root vegetables—such as potatoes, sweet potatoes, rutabagas, and onions—in a cool, dry storeroom. Temperatures of 60°F to 70°F (16°C to 21°C) are best. Make sure containers are well-ventilated. Store onions away from other vegetables that might absorb odor.

While many whole, raw fruit and vegetables can be stored at 41°F (5°C) or lower, not all will be stored at these temperatures.

MAP, Vacuum-Packed, and Sous Vide Food

○ Always store MAP, vacuum-packed, and *sous vide* food at temperatures recommended by the manufacturer. Most should be stored at 41°F (5°C) or lower. Frozen MAP, vacuum-packed, and *sous vide* food should be stored at temperatures that will keep it frozen. Store and handle these products carefully.

○ Vacuum packaging will not stop the growth of anaerobic microorganisms. MAP fish, for example, is especially susceptible to the growth of *Clostridium botulinum*. Discard product if the package is torn or slimy, if it contains excessive liquid, or if the product bubbles, indicating the possible growth of *Clostridium botulinum*.

○ Always check the expiration date before using MAP, vacuum-packed, and *sous vide* products. Labels should clearly list contents, storage temperature, preparation instructions, and a use-by date.

○ Operators who produce MAP food on site must follow specific labeling rules. The FDA rules for processing MAP food on site are very strict.

UHT and Aseptically Packaged Food

○ Food that has been pasteurized at ultra-high temperatures (UHT) and aseptically packaged (the packaging is free of microorganisms) can be stored at room temperature. Since much of this food, such as milk and pudding, are served cold, you might want to store them in the refrigerator.

○ Once it is opened, store UHT and aseptically packaged food in the refrigerator at 41°F (5°C) or lower.

○ UHT products not aseptically packaged must be stored at an internal temperature of 41°F (5°C) or lower.

Canned Goods

○ Store canned goods and other dry food at a temperature between 50°F and 70°F (10°C to 21°C). Even canned food spoils over time. Higher storage temperatures may shorten shelf life. Acidic food, such as canned tomatoes, does not last as long as food low in acid. The acid can also form pinholes in the metal over time.

○ Keep storerooms dry. Too much moisture will cause cans to rust.

○ Wipe cans clean with a sanitized cloth before opening them to help prevent dirt from falling into the contents of the can.

Key Point

Storage temperatures higher than 70°F (21°C) may shorten the shelf life of canned goods.

Dry Food

○ Keep flour, cereal, and grain products such as pasta or crackers in airtight containers. They can quickly become stale in a humid room and can become moldy if there is too much moisture.

○ If dry food is removed from its original packaging and stored in containers, these containers should be clearly labeled. Many foods, such as salt and sugar, or flour and baking mixes, often look alike.

○ Before using dry food, check containers or packages for damage from insects or rodents. Cereal and grain products are favorite targets for these pests.

○ Salt and sugar, if stored in the right conditions, can be held almost indefinitely.

Chemicals

○ Never use empty food containers to store chemicals, and never store food in empty chemical containers. Keep chemicals in their original containers. If it is necessary to transfer chemicals, store them in sturdy containers clearly labeled with the contents and their hazards.

Shelf Life of Food

Pay close attention to each food's use-by date. Discard any food that has exceeded this time period.

All potentially hazardous, ready-to-eat food stored in refrigeration at 41°F (5°C) should be discarded if not consumed within seven days of preparation. Potentially hazardous, ready-to-eat food that has been frozen should be discarded if not consumed within twenty-four hours of being thawed.

SUMMARY

When food is stored improperly, quality and safety will suffer. Though different food has different storage needs, some commonsense rules apply. Food should be stored in designated areas using the FIFO method. Store it in its original packaging, or in cleaned and sanitized containers with tight-fitting lids. Wrap food in clean, moisture-proof materials. Label all stored food with the contents of the package. Always label potentially hazardous, ready-to-eat food with the date it should be sold, consumed, or discarded. Check the temperatures of stored food and the storage area regularly, and keep these areas clean and dry to prevent contamination.

Hold potentially hazardous food in storage areas refrigerated at 41°F (5°C) or lower, temperatures which slow the growth of microorganisms. To

Health Alert

Never use empty food containers to store chemicals, and never store food in empty chemical containers.

Key Point

Remember: when in doubt, throw it out.

hold food at the proper temperature, keep the air temperature of the unit approximately 2°F (1°C) lower than the desired food temperature. Never place hot food in the refrigerator, which could raise the temperature inside. Don't line refrigerator shelves, overload the unit, or open the door too often. These practices make the unit work harder to keep food cold. If possible, raw meat, poultry, and fish should be stored *separately* from cooked and ready-to-eat food to prevent cross-contamination. If not, store these items below cooked or ready-to-eat food in the appropriate order.

While freezing does not kill microorganisms, it does slow their growth substantially. Freezers should be kept at 0°F (-18°C) or lower unless the food being stored requires a different temperature. Unit and food temperatures should be checked often. Dry storage areas should be kept at the appropriate temperature and humidity levels and should be clean and well-ventilated to maintain food quality. Food in dry storage should be stored at least six inches off the floor and away from walls. Foodservice chemicals should be kept in their own storage area, away from food and food-preparation areas. Empty food containers should never be used to store chemicals. Chemicals should be kept in their original containers or in sturdy containers that have been clearly marked with the contents and their hazards.

Fresh meat, poultry, fish, and dairy products should be stored at 41°F (5°C) or lower. Fish and poultry can be stored under refrigeration in crushed ice as long as the containers are self-draining, the ice is changed, and the container is sanitized regularly. Eggs should be refrigerated at an ambient (air) temperature of 45°F (7°C) or lower right up until they are used, and should not be pooled unless they will be used immediately. Shellfish should be stored in their original containers at 45°F (7°C). Fresh produce has various temperature requirements for storage. Produce should not be washed before storage, because it can promote mold growth. MAP, vacuum-packed, and *sous vide* food should be stored at temperatures recommended by the manufacturer. Packages should be checked for signs of contamination, including bubbling, excessive liquid, and torn or slimy packaging. Once opened, UHT and aseptically packaged foods should be stored in the refrigerator at 41°F (5°C) or lower. Dry and canned food should be stored at temperatures between 50°F and 70°F (10°C to 21°C).

TRAINING TIPS

1. "What's This?"

Purpose: *To illustrate the importance of keeping products in their original containers, labeling products in secondary containers, and storing cleaning supplies and chemicals away from food.*

Time: 10 minutes

Directions: Place a number of food and nonfood products in unlabeled clear containers, or otherwise conceal their identity. Products should include items that are difficult to identify or distinguish from other products, such as:

○ Sugar   ○ Baking soda   ○ Caustic solvent

○ Salt   ○ Bleach   ○ Window cleaner

○ Flour   ○ Vinegar   ○ Pan spray

○ Cleanser   ○ Oil   ○ Stainless-steel cleaner

Have kitchen employees try to identify each product without smelling, touching, and certainly without tasting. Offer a reward to the person who can correctly identify the most products. Discuss the importance of proper labeling.

2. "Where Do I Go?" Team Contest

Purpose: *After completing this activity, trainees will be able to identify proper storage areas and handling guidelines for specific types of food.*

Time: 20 minutes

Directions: Create a worksheet for the activity by dividing a piece of paper into three columns with the following headings:

Item   **Location**   **Storage Practices**

In the **Item** column, list twenty-five to thirty specific food products that might be stored in a restaurant or foodservice establishment. Be sure to include a wide variety of food: fresh and frozen meat, poultry, fish, seafood, shellfish, dairy products, eggs, MAP and sous vide food, fresh produce, UHT products, dry food, and canned food.

Divide trainees into teams of two to four people and provide each team with a copy of the worksheet. Allow ten minutes for the teams to identify the proper storage area for each item, as well as a proper storage practice.

Then, permitting each team to score its own worksheet, go down the list of food, coming to agreement among the group about where each product should be stored and storage practices that should be followed. The team identifying the most correct storage locations for each item and the best storage practices wins.

3. "Rooms for Rent" Group Activity

Purpose: *After completing this activity, trainees will be able to identify the most effective storage procedures for freezers, refrigerators, dry-storage rooms, and chemical-storage rooms.*

Time: 30 minutes

Directions: Divide trainees into four groups and assign one of the following storage areas to each group:

○ Freezer   ○ Dry-storage room

○ Refrigerator   ○ Chemical-storage room

Give each group five to ten minutes to come up with as many storage procedures as possible for its assigned storage area. Discuss the storage procedures each group identifies and record the most important ones.

After discussing procedures for each storage area, create a list of general storage procedures from the procedures common to all four areas.

4. Developing SOPs for Storing Food

Purpose: *After completing this activity, trainees will be able to develop standard operating procedures (SOPs) for storage of various types of food.*

Time: 30 minutes

Directions: Divide trainees into groups and provide them with the following list of food:

○ Live oysters   ○ Fresh hamburger patties

○ MAP chicken breasts   ○ Fresh shell eggs

○ Prepared shrimp salad   ○ *Sous vide* chicken cordon bleu

○ Whole snapper   ○ Canned meats

Ask each group to develop SOPs for storing these products while considering the following questions:

1. What is the ideal storage temperature for each product?

2. Where and how should it be stored?

3. Under what conditions should the product be discarded?

4. Are there other storage requirements for the product?

Ask each group to present the storage procedures it has identified and then discuss them with the other groups.

5. "What's Wrong with This Picture?" Contest

Purpose: *To involve your kitchen staff in the assessment of storage practices in your establishment and to implement corrective actions based on their assessment.*

Time: several days

Directions: Equip your employees with clipboards, pencils, blank paper, and thermometers. Have them inspect the establishment's food-storage areas and note any storage errors or practices that violate safe-storage principles. The inspection should include your freezers, refrigerators, dry storage areas, and any other areas used to store food. When your staff has finished its inspection, collect its reports and make copies. The "inspector" who has found the most storage mistakes wins a prize.

Hold a follow-up meeting to discuss the inspection findings. Let your staff know the inspection was just the first step. Ask them to help create an action plan for correcting problems they found in the storage areas. Tell them you will need their help in implementing the plan. As incentive, let your staff know you will be performing the next inspection. If you give storage an A+, you'll treat them all to a party!

6. Developing Food Storage Job Aids

Purpose: *To involve your kitchen management team in the development of food storage job aids that can be used by employees and management as easy reference tools.*

Time: several days

Directions: Ask your kitchen management team to develop usable charts, checklists, and diagrams documenting the storage guidelines and procedures in your establishment.

These reference tools might include:

○ A storage temperature list for food stored in freezers, refrigerators, and dry-storage areas.

○ Storage guidelines for each storage area.

○ Time-temperature logs for checking food temperatures during each shift (with corrective actions listed).

○ Use-by or discard dates for food products.

When developing these job aids, ask your management team to:

○ Keep them simple. Less is more.

○ Make them colorful. You may want to color code them by storage area.

○ Print them in several languages, if necessary.

○ Laminate all materials.

○ Post them in the appropriate areas.

○ Use pictures or graphics, and fewer words, whenever possible.

Important: Ask your managers to make sure the standards set in the job aids meet or exceed codes enforced by your local regulatory agency. The job aids should also include any recommendations provided by the supplier or manufacturer.

A CASE IN POINT I

Case Study

On Monday afternoon, the kitchen staff at the Sunnydale Nursing Home was busy cleaning up after lunch and preparing for dinner. Pete, a kitchen assistant, put a large stockpot of hot, leftover vegetable soup in the refrigerator to cool. Angie, a cook, began deboning the chicken breasts that were stored earlier. When she finished, she put the chicken on an uncovered sheet pan and stored the pan in the refrigerator. She carefully placed the raw chicken on the top shelf, away from the hot soup. Next, Angie iced a carrot cake she had baked that morning. She put the carrot cake in the refrigerator on the shelf directly below the chicken breasts.

What storage errors were made? What food items are at risk?

A CASE IN POINT II

Case Study

Anticipating a slow dinner service, Ed, the owner of Big City Diner, gave his kitchen manager the afternoon off. Some time later, Acme Distributors delivered an order of canned and dry food and supplies, including canned tomato sauce, canned soup, crackers, pasta, paper napkins, and cleaning supplies, such as warewashing detergent and sanitizer.

After checking the order, Ed stacked the canned and dry food onto a dolly and wheeled it into the storeroom. There was an open case of tomato sauce on the shelf with one can left in it and a full case behind it. Ed removed the open case and replaced it with the case on the dolly. He put the single can in an open spot on another shelf. By shifting some boxes around, Ed was able to put the cases of soup, napkins, and pasta on shelves in front of cases already stored there. While moving cases, he accidentally knocked over a container of flour, spilling some on the floor. There was no room on the shelves for the crackers, so he left the case on the floor. He was sweating from exertion and wondered if the storeroom was too hot. He checked the hanging thermometer, which read 70°F (21°C).

Ed went back to the dock to get the cleaning supplies. He wheeled them into the storeroom and stacked them against the wall next to several cases of pasta. When he returned the dolly to the back door, a truck from Mike's Produce pulled up to deliver cases of potatoes and several bags of onions. Ed knew the produce room was overstocked, so he stored the produce in a space sometimes used for storage underneath the dining room stairs.

What did Ed do wrong? What are the possible consequences?

DISCUSSION QUESTIONS

1. In top-to-bottom order, how would you store the following food in a refrigerator: raw trout, an uncooked beef roast, raw chicken, and raw ground beef?

2. How should food that has been taken out of its original package be labeled?

3. What is the proper storage temperature for frozen dairy products?

4. What temperature should the air inside a walk-in refrigerator be to hold products at an internal temperature of 41°F (5°C)?

5. How should products be stored when following FIFO?

MULTIPLE-CHOICE STUDY QUESTIONS

1. At what internal temperature would stored ground beef most likely become unsafe to use?
   A. 0°F (-17°C)   C. 41°F (5°C)
   B. 30°F (-1°C)   D. 60°F (16°C)

2. Dry-storage rooms should be kept at
   A. 35°F to 41°F (2°C to 5°C).   C. 50°F to 70°F (10°C to 21°C).
   B. 45°F to 50°F (7°C to 10°C).   D. 70°F to 80°F (21°C to 27°C).

3. Under which condition could you use a tank to display live mussels you will be serving to customers?

   A. You have obtained a variance from the health department.
   B. You clean the display tank at least once a month.
   C. You will mix them with other mussels that have not been on display.
   D. You have removed the shellstock identification tags as required by law.

4. Which of the following is *not* a good storage practice?

   A. Shelving food based on its expiration date

   B. Storing raw poultry at temperatures between 41°F and 140°F (5°C and 60°C)

   C. Storing live shellfish at 45°F (7°C) or lower

   D. Storing raw meat below ready-to-eat food

5. All of the following statements about the FIFO method of stock rotation are correct *except:*

   A. Food items should be used by their expiration date.

   B. Products with the earliest use-by date should be stored in front of products with later dates.

   C. The oldest products should be used first.

   D. Products that have passed their expiration dates should be used first.

6. Improperly stored vacuum-packed food can be a problem because

   A. loss of moisture can lessen food quality.

   B. heat can cause vacuum-packaged cans to explode.

   C. air leakage can contaminate neighboring food.

   D. bacteria that cause botulism can grow.

7. If you see that a package of *sous vide* food is torn, you should

   A. cook and serve the product.

   B. discard the product.

   C. mend the tear and freeze the package.

   D. measure the internal temperature of the food.

8. To keep refrigerated food at an internal temperature of 41°F (5°C), the air temperature in the refrigerator should be approximately

   A. 39°F (4°C).

   B. 32°F (0°C).

   C. 26°F (-3°C).

   D. 0°F (-18°C).

9. All of the following are incorrect storage practices *except*

A. lining refrigerator shelving with aluminum foil.

B. cooling hot food in a refrigerator.

C. storing products with the earliest expiration dates in front of products with older dates.

D. storing fresh poultry over ready-to-eat food.

10. Which storage practice is incorrect?

A. Storing fresh poultry at 41°F (5°C)

B. Storing eggs at room temperature

C. Storing raw clams in their shipping crate at 45°F (7°C)

D. Storing raw ground beef in its original packaging

For answers, please turn to Appendix A.

ADDITIONAL RESOURCES

Books and Periodicals

Cichy, R. F. 1993. *Sanitation management.* East Lansing, MI: Educational Institute of the American Hotel & Motel Association.

Cross out cross-contamination. 1998. *Best Practices.* 2 (1):6-9.

Equipped for food safety. 1998. *Best Practices.* 2 (4):6-10.

Kotschevar, L. H. 1994. *Quantity food purchasing,* 4th ed. New York: Macmillan.

Loken, J. K. 1995. *The HACCP food-safety manual.* New York: Wiley.

Longree, K. 1996. *Quantity food sanitation,* 5th ed. New York: Wiley.

National Restaurant Association Educational Foundation. 1998. *A Practical Approach to HACCP: Coursebook.* Chicago: National Restaurant Association Educational Foundation.

The right way to…prepare beef. 1998. *Best Practices.* 2 (3):12-13.

The right way to…prepare chicken. 1998. *Best Practices.* 2 (1):10.

The right way to…prepare eggs. 1998. *Best Practices.* 2 (2):10-11.

The right way to…prepare fruits and vegetables. 1998. *Best Practices.* 2 (4):12-13.

The right way to…prepare pork. 1997. *Best Practices.* 1 (3):10-11.

The right way to…seafood safety. 1999. *Best Practices.* 3 (1):16.

Secrets of self-inspection. 1998. *Best Practices.* 2 (2):6-9.

Stefanelli, J. M. 1997. *Purchasing: Selection and procurement for the hospitality industry.* New York: Wiley.

Websites

Alaska Seafood Marketing Institute

http://www.alaskaseafood.org

The Alaska Seafood Marketing Institute enhances the appeal and popularity of Alaska seafood to the restaurant and foodservice industry and to the general public. This useful Web site contains information on how to market seafood to your customers, how to purchase Alaskan seafood, detailed information on types of Alaskan seafood, and how to handle seafood to maintain safety and quality.

American Egg Board (AEB)

http://www.aeb.org

As the egg industry's promotional arm, the AEB is the U.S. egg producer's link to the consumer for communicating the value of the egg. This Web site offers a wealth of information about eggs and egg safety.

American Meat Institute

http://www.meatami.com

The American Meat Institute is a membership trade association representing the interests of the U.S. meat and poultry industry to the federal government, Congress, the media, and the customer. This Web site provides up-to-date information about what is happening in the meat and poultry processing industry.

Atkins Temptec

http://www.atkinstemptec.com

Atkins is a manufacturer of digital thermometers and probes for the restaurant and foodservice industry. This Web site provides information on how thermometers work, how to get the most out of your thermometer, and how to hold food safely during service. The site also provides an interesting demonstration on how food color is a poor indicator of internal temperature. Consult this site for useful tips related to thermometers.

Cooper Instrument Corporation

http://www.cooperinstrument.com

Cooper Instrument Corporation is a supplier of temperature, time, and humidity instruments for a variety of global markets. This Web site contains information on how to calibrate thermometers properly and maintain food temperature during cooking, holding, cooling, etc.

FDA–Center for Food Safety and Applied Nutrition (CFSAN)

http://www.cfsan.fda.gov

As the center within the FDA responsible for food safety and nutrition, CFSAN promotes and protects public health by researching and implementing guidelines, policies, and standards to ensure that food is safe, nutritious, wholesome, and properly labeled. This Web site provides a wealth of information on food safety and sanitation, including corresponding guidelines, policies, and standards.

FDA Food Code

http://vm.cfsan.fda.gov/~dms/foodcode.html

As the basis for many local sanitation codes, as well as the basis for information in this textbook, the FDA Food Code, available at this Web address, is a useful resource for information relating to food for the restaurant and foodservice industry.

FDA Seafood Information and Resources

http://vm.cfsan.fda.gov/seafood1.html

The FDA operates an oversight compliance program for fishery products, under which responsibility for product safety, wholesomeness, identity, and economic integrity rests with the processor or importer, who must comply with regulations under the Federal Food, Drug and Cosmetic (FD&C) Act. This Web site houses information on the seafood program, foodborne pathogens and contaminants associated with seafood, and HACCP compliance.

International Dairy-Deli-Bakery Association (IDDBA)

http://www.iddba.org

IDDBA is a resource for information and services across all food channels for the dairy, deli, and bakery industry. This Web site provides information relevant to all people who work in the dairy, deli, and bakery industry.

International Fresh-cut Produce Association (IFPA)

http://www.fresh-cuts.org

IFPA serves commercial producers, suppliers, and distributors of fresh-cut produce. It provides members with a forum for representation, education, technical support, networking, information, and resources. This site contains information about recent industry happenings and regulations, and provides guidance documents for purchasing, receiving, and storing fresh-cut produce.

The Mushroom Council

http://www.mushroomcouncil.com

This interesting and useful Web site from the Mushroom Council contains recipes and information about how to handle and store mushrooms to ensure safety and quality.

National Cattlemen's Beef Association (NCBA)

http://www.beef.org

NCBA is a consumer-focused, producer-directed organization representing the largest segment of the nation's food and fiber industry. This Web site, as well as its companion site for foodservice (www.beeffoodservice.com), provides excellent information on purchasing, cooking, and serving beef. Both sites also provide important food safety information.

The National Chicken Council

http://www.eatchicken.com

This Web site is brought to you by the National Chicken Council and the U.S. Poultry & Egg Association, representing a wide spectrum of companies and individuals in the chicken business. This fun and interesting Web site contains chicken safety tips and nutritional and statistical information.

National Frozen Food Association (NFFA)

http://www.nffa.org

NFFA's mission is to promote the sales and consumption of frozen food through education, training, research, sales planning, and menu development, as well as to provide a forum for industry dialogue. Publications and information on frozen food and how to market it to consumers are available at this Web site.

National Pork Producers Council

http://www.nppc.org

Billing itself as the place for pork lovers, pork producers, and anyone who wants to learn more about pigs and pork, the National Pork Producers Council Web site provides information on pork safety and quality, nutrition and industry information, and publications.

National Restaurant Association

http://www.restaurant.org

The National Restaurant Association is the leading business association for the restaurant industry. Together with the National Restaurant Association Educational Foundation, the Association's mission is to represent, educate, and promote the rapidly growing restaurant and foodservice industry. This Web site should be your starting place for all issues and concerns related to your restaurant. This Web site has it all, from tips for running your establishment to vital data on customers' spending habits.

National Turkey Federation

http://www.turkeyfed.org

The National Turkey Federation advocates for all segments of the turkey industry, providing services and conducting activities that increase demand for its members' products by protecting and enhancing their ability to profitably provide wholesome, high-quality, nutritious products. Its Web site contains a special area specifically for the restaurant and foodservice industry, including recipe and menu ideas, cooking demonstrations, and a reference on purchasing, preparing, and promoting turkey.

Produce Marketing Association

http://www.pma.com

The Produce Marketing Association is a not-for-profit trade association serving members who market fresh fruit, vegetables, and floral products worldwide. Its members are involved in the production, distribution, retail, and foodservice sectors of the industry. This Web site provides promotional ideas and information about hot topics in the produce industry.

Tyson Foods, Inc.

http://www.tyson.com

Tyson Foods is the world's largest poultry producer. This Web site is a great resource about chicken products for the restaurant and foodservice industry. It includes recipes and menu ideas, information on new products, and information about Tyson Foods' efficient distribution and transportation system.

USDA-FSIS

http://www.fsis.usda.gov

The Food Safety and Inspection Service (FSIS) is the public health agency of the U.S. Department of Agriculture responsible for ensuring that the nation's commercial supply of meat, poultry, and egg products is safe, wholesome, and correctly labeled and packaged. This Web site contains a wealth of information on food safety relating to meat, poultry, and eggs.

Chapter 7
Protecting Food During Preparation

Knowledge

TEST YOUR FOOD SAFETY KNOWLEDGE

1. **True or False:** After the loose dirt has been brushed off the outside of a melon, it can be safely cut open. *(See page 7-11.)*

2. **True or False:** Cooked vegetables must never be left out or held at room temperature. *(See page 7-16.)*

3. **True or False:** Cooked, potentially hazardous food must be cooled from 140°F to 70°F (60°C to 21°C) within two hours and from 70°F to 41°F (21°C to 5°C) or lower within an additional four hours. *(See page 7-17.)*

4. **True or False:** When potentially hazardous food is reheated for hot-holding, it must be heated to 145°F (63°C) for fifteen seconds within two hours. *(See page 7-19.)*

5. **True or False:** Poultry must be cooked to a minimum internal temperature of 155°F (68°C) or higher for fifteen seconds. *(See page 7-13.)*

For answers, please turn to Appendix A.

Key Terms

Four-hour rule
Minimum internal
     temperature
Slacking

Two-stage cooling
Ice-water bath
Cold paddle

Table of Contents

Learning Objectives

After completing this chapter, you should be able to

○ Use time and temperature controls at each step in the flow of food.

○ Thaw frozen food properly.

○ Handle and prepare raw food safely.

○ Prevent contamination and microbial growth in the flow of food.

○ Identify the minimum internal cooking temperatures of food.

○ Cool, store, and reheat cooked food properly.

○ Make corrections if time and temperature standards are not met.

Serving safe food requires constant attention to the factors that can cause foodborne illness. Even food that is safe when delivered and properly stored can quickly become unsafe if handled improperly during preparation and cooking.

SAFE FOODHANDLING

The key to serving safe food is to handle it safely. Teach employees how important it is to control time and temperature and to prevent cross-contamination. Then put commonsense guidelines in place for safe foodhandling.

Time and Temperature Control

Probably the biggest factor in foodborne-illness outbreaks is time-temperature abuse. Disease-causing microorganisms grow and multiply at temperatures between 41°F and 140°F (5°C and 60°C), which is why this range is known as the temperature danger zone. Microorganisms grow much faster in the middle of the zone, at temperatures between 70°F and 125°F (21°C and 52°C). *(See Exhibit 7a.)* Whenever food is held in the temperature danger zone, it is being abused.

As you learned in Chapter 2, time also plays a critical role in food safety. Microorganisms need both time and temperature to grow. The longer food stays in the temperature danger zone, the more time microorganisms have to multiply and make food unsafe. When heating or cooling food, it is important to pass it through the danger zone as quickly and as infrequently as possible.

To keep food safe during preparation and cooking, you must follow the **four-hour rule**. Never let food remain in the temperature danger zone for more than four hours.

Exposure time adds up during each stage of handling, from the time food arrives at the receiving dock to the time it is cooked. Once food is properly cooked to the required **minimum internal temperature**, the number of microorganisms in it will have been reduced to a safe level. However, you must continue to handle the food safely. There are further opportunities for time-temperature abuse, including:

○ Not cooling food properly

○ Failing to reheat food to 165°F (74°C) for fifteen seconds within two hours

○ Failing to hold food at a minimum internal temperature of 140°F (60°C) or higher

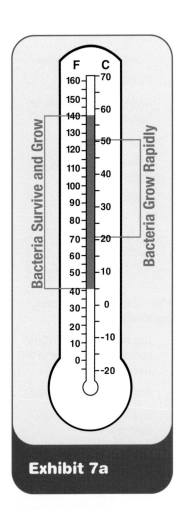

Exhibit 7a

Temperature and Bacterial Growth

Foodborne microorganisms grow most rapidly at temperatures between 70°F and 125°F (21°C and 52°C).

You can see how four hours in the danger zone easily add up if you don't handle food carefully. The best way to avoid a problem is to establish procedures employees must follow and then monitor them. Build time and temperature controls into your establishment's HACCP plan and standard operating procedures. Make time and temperature control part of every employee's job. Some suggestions include:

○ **Decide the best way to monitor time and temperature in your establishment.** Determine which foods should be monitored, how often, and who should check them. Then assign responsibilities to employees in each area. Make sure employees understand exactly what you want them to do, how to do it, and why it is important.

○ **Make sure the establishment has the right kind of thermometers available in the right places.** Give employees their own calibrated thermometers. Have them use timers in prep areas to monitor how long food is being kept in the temperature danger zone.

○ **Regularly record temperatures and the times they are taken.** Print simple forms employees can use to record temperatures and times throughout the shift. Post these forms on clipboards outside of refrigerators and freezers, near prep tables, and next to cooking and holding equipment. *(See Exhibit 7b.)*

○ **Incorporate time and temperature controls into standard operating procedures for employees.** These might include:

- Not removing from the refrigerator more food than can be prepared in a short period of time

- Refrigerating ingredients and utensils before preparing certain recipes, such as tuna or chicken salad

- Cooking potentially hazardous food to minimum internal cooking temperatures

○ **Develop a set of corrective actions.** Decide what action should be taken if time and temperature standards are not met. For example, an establishment that holds egg rolls

Time & Temperature

The time that food remains in the temperature danger zone accumulates at each step in the flow of food. This accumulated time cannot exceed four hours.

Exhibit 7b **Time and Temperature Control**
Print simple forms employees can use to record temperatures.

Reprinted with permission from Roger Bonifield and Dingbats.

on a steam table might throw them out if their internal temperature falls below 140°F (60°C) for more than four hours, or they might reheat the egg rolls to 165°F (74°C) for at least fifteen seconds within two hours.

Preventing Cross-Contamination

The other major hazard for food in your operation is cross-contamination, which is the transfer of microorganisms from one food or surface to another. Microorganisms move around easily in a kitchen. They can be transferred from food or unwashed hands to prep tables, equipment, utensils, cutting boards, dish towels, sponges, or other food.

Cross-contamination can occur at almost any point in an operation. When you know where and how microorganisms can be transferred, cross-contamination is fairly simple to prevent. Prevention starts with the creation of barriers between food products. Barriers can be physical or procedural. For example, in Chapter 6 you learned to store food in covered containers and to make sure that raw meat or poultry is not stored above cooked and ready-to-eat food in a refrigerator.

Prevention is a matter of common sense. Here are some simple barriers you should consider using in your establishment.

○ **Prepare raw meat, fish, and poultry in separate areas from produce or cooked and ready-to-eat food.** This can be as easy as using separate prep tables in the same area. If you don't have sufficient space for separate tables, prepare these items at different times so they don't cross paths.

○ **Assign specific equipment to each type of food product.** For example, use one set of cutting boards, utensils, and containers just for poultry, another set for meat, and a third set for produce. Some manufacturers make colored cutting boards and utensils with colored handles. Color-coding can tell employees which equipment to use with what products, such as green for produce, yellow for chicken, and red for meat.

○ **Use specific containers for each type of food product.** Clearly label containers with their contents—such as "Raw Chicken" or "Tuna Salad"—so the containers are less likely to be used for something else. Make sure you have an adequate supply of containers on hand and keep them clean and sanitized between uses.

○ **Clean and sanitize all work surfaces, equipment, and utensils after each task.** After cutting up raw chicken, for example, it is not enough to simply rinse the cutting board. Wash, rinse, and sanitize cutting boards and utensils in a three-compartment sink, or run them through a warewashing machine. Make sure employees know which cleaners and sanitizers to use for each job. Sanitizers used on food-contact surfaces

Designating different colored cutting boards for different food reduces the risk of cross-contamination.

Manufactured by KatchAll/San Jamar

must meet local or state department codes conforming to the Code of Federal Regulations (21CFR178.1010). *(See Chapter 11 for more information on cleaning and sanitizing.)*

○ **Cloths or towels used for wiping food spills must not be used for any other purpose.** There are a few ways to prevent the spread of microorganisms from cleaning cloths to equipment and utensils. One way is to use disposable towels. Another is to use color-coded cleaning cloths that match a specific food-preparation area or task. After each use, cloths or towels must be rinsed and stored in a clean sanitizing solution. *(See Chapter 11 for more information on how to make sanitizing solutions.)*

○ **Consider requiring employees to wear single-use gloves while preparing or serving ready-to-eat food in your establishment.** If you adopt this policy, teach employees to use gloves properly. Employees must wash their hands before putting on gloves. Gloves should be used only for the job at hand and changed each time a new task is started. Hands must be washed before putting on a new pair of gloves. If punctured or ripped, gloves must be changed.

○ **Employees should watch what they touch after handling raw food and should practice good personal hygiene.** Hands are the most common cause of cross-contamination in the kitchen. Employees should be trained to wash their hands properly and at the appropriate times. *(See Chapter 4.)*

PREPARING FOOD

Controlling time and temperature and preventing cross-contamination will help keep food from becoming unsafe during preparation. However, some food items and methods of preparation require a bit more care.

Thawing Food Properly

Often, the first step in food preparation is thawing what you intend to cook. If frozen food is thawed improperly, foodborne microorganisms can rapidly grow to unsafe levels.

Suppose a cook wants to prepare a twenty-pound frozen turkey. The cook is in a hurry, so instead of putting the turkey in the refrigerator he puts it in a pan on a worktable to thaw overnight. The air temperature in the kitchen is about 72°F (22°C).

There are two problems with this method of thawing. First, when the bird begins to thaw, the skin and outer layers are exposed to the temperature danger zone even though the bird's core is still frozen. Microorganisms may multiply to

Cross-Contamination

If gloves are used, they should be changed before starting a new task, and hands must be washed before putting on a new pair.

a point where cooking the turkey may not be sufficient to kill them. The cook may end up serving a potentially dangerous dish to customers.

Second, microorganisms on the turkey may contaminate everything in the area: the cook's hands, the worktable, the pan, the cutting board, knives, and other utensils. If the area and equipment aren't completely cleaned and sanitized before the cook starts another task, microorganisms could spread throughout the entire kitchen.

There are only four acceptable ways to thaw frozen food safely. *(See Exhibit 7c.)*

○ **Thaw food in the refrigerator at 41°F (5°C) or lower.** This method requires advance planning. A twenty-pound turkey can take from two to four days to thaw completely in the refrigerator. Employees need to take food out of frozen storage far enough in advance to make sure there is adequate thawed product on hand when it's needed.

○ **Submerge the frozen product under running potable water at a temperature of 70°F (21°C) or lower.** The water has to flow fast enough to wash loose food particles into the overflow drain. Make sure the thawed product does not drip water onto other products or food-contact surfaces. Clean and sanitize the sink and work area before and after thawing food this way. Remember, the product cannot remain in the temperature danger zone for more than four hours, which includes both the time the food is thawed under running water and the time needed for preparation prior to cooking.

○ **Food may be thawed in a microwave oven only if it will be cooked immediately afterward.** Microwave thawing can actually start cooking the product, so don't use this method unless you intend to continue cooking the food immediately. Large items such as roasts or turkeys don't thaw well in the microwave.

○ **Food may be thawed as part of the cooking process as long as the product reaches the required minimum internal cooking temperature.** Frozen hamburger patties, for example, can go straight from the freezer onto a grill without being thawed first. Frozen chicken can go straight into a deep fryer.

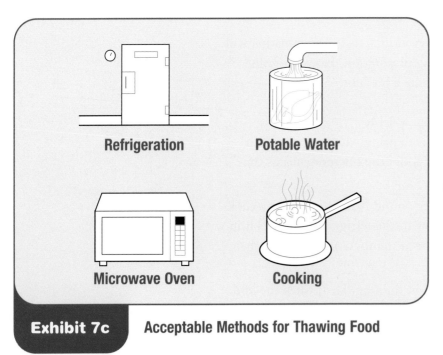

Refrigeration   **Potable Water**

Microwave Oven   **Cooking**

Exhibit 7c   **Acceptable Methods for Thawing Food**

These products cook quickly enough from the frozen state to pass through the temperature danger zone without harm. However, always make sure you verify the final internal cooking temperature with a calibrated thermometer.

Some frozen food may be *slacked* before cooking. **Slacking** is the process of gradually thawing frozen food in preparation for deep frying, allowing even heating during cooking. For example, you might allow a large block of frozen spinach to warm from -10°F (-23°C) to 25°F (-4°C). Since food should never be refrozen once it begins to thaw, slack food just before you cook it. Don't let food get any warmer than 41°F (5°C). If slacking at room temperature is permitted in your jurisdiction, a system must be in place to ensure the product does not exceed 41°F (5°C).

PREPARING SPECIFIC FOOD

Meat, Fish, and Poultry

The sources of most cross-contamination in an operation are raw meat, poultry, and seafood. Your staff should follow safe procedures when handling these products.

○ **Use clean and sanitized work areas, cutting boards, knives, and utensils.** Prepare raw meat, poultry, and seafood separately or at a different time from fresh produce.

○ **Be sure hands are washed properly.** If gloves are used, change them before starting each new task. Wash hands again before putting on a new pair of gloves.

○ **Remove from the refrigerator only as much product as can be prepared at one time.** When cubing beef for stew, for example, take out and cube one roast and refrigerate it before taking out another roast.

○ **Put raw, prepared meat away or cook it as quickly as possible.** If put back into storage, wrap raw, prepared meat first or put it in covered containers that are clearly labeled. Store it on bottom shelves in the refrigerator. Never store it above cooked or ready-to-eat food, such as fresh produce, because raw meat juices may drip and cross-contaminate other food.

Salads Containing Potentially Hazardous Food

Chicken, tuna, egg, pasta, and potato salads all have been known to cause foodborne-illness outbreaks. These salads are typically made from food that can easily support the rapid growth of microorganisms. Since they usually will not be cooked after preparation, there is no chance to kill microorganisms that could be introduced during preparation. Therefore,

Cross-Contamination

Raw meat, poultry, and seafood are the sources of most cross-contamination in an establishment.

care must be taken to both control time and temperature and prevent cross-contamination when preparing these salads. Handle all ingredients with care and follow these precautions:

○ Make sure leftover meat and poultry have been properly cooked, held, cooled, and stored. Leftover meat and poultry should be discarded after seven days if stored in the refrigerator at 41°F (5°C) or lower. (Four days if stored at 45°F [7°C].) Salads made with these items should be discarded when this storage period expires.

○ Make sure foodhandlers wash their hands before preparing food and, if required by local health codes, have them wear clean, single-use gloves. Tongs, deli tissue, or other barriers can also be used to protect food.

○ Prepare other ingredients away from raw meat, seafood, and poultry. Always prepare raw vegetables using clean, sanitary equipment and utensils.

○ Leave food in the refrigerator until all ingredients are ready to be mixed. For example, don't take out chicken and leave it on the prep table while you chop the vegetables.

○ Consider chilling all ingredients before making salads. Refrigerate unopened cans of tuna and jars of mayonnaise the day before you make tuna salad, for example. Consider chilling mixing bowls and utensils, too. Taking these steps will provide additional barriers that might prevent the growth of harmful microorganisms that could be present.

○ Prepare food in small batches. This will prevent large amounts of food from sitting out at room temperature for long periods of time. Set a working time limit of twenty minutes and make sure you take out only enough ingredients to use within that time period. When time is up, refrigerate what you've made, then get ingredients to make more. *(See Exhibit 7d.)*

Eggs and Egg Mixtures

Historically, the contents of whole, clean, uncracked shell eggs were considered free of bacteria. Now it is known that, in rare cases, a certain bacteria, *Salmonella* enteritidis, can be found inside eggs. *Salmonella* enteritidis can live inside a laying hen and can be deposited in the white of an egg before the shell is formed. Although only a small number of eggs produced in the United States are likely to carry this bacteria, all untreated shell eggs are considered to be a potentially hazardous food because they are able to support the rapid growth of microorganisms. As a consequence, eggs should be purchased fresh and should be stored and handled properly. Eggs should be kept at an ambient

Exhibit 7d

Preparing Salads Containing Potentially Hazardous Food
Prepare these salads in small batches so large amounts of food don't sit out at room temperature for too long.

(air) temperature of 45°F (7°C) or lower until immediately before use. Special handling precautions for eggs are described below.

○ Use clean bowls, whisks, blenders, and other equipment when working with eggs and other potentially hazardous food items. To prevent cross-contamination, promptly clean and sanitize all equipment and utensils used with eggs. Place freshly made batters or sauces in clean containers. Never reuse containers that held raw eggs, even to replenish the same recipe.

○ Food containing eggs that receive little or no cooking—such as Caesar salad dressing, meringue, mayonnaise, hollandaise sauce, and eggnog—should be prepared with great care. Follow the directions precisely and monitor temperatures using a thermometer. Many jurisdictions now require operators to post a notice to consumers when raw eggs are used as an ingredient in these types of food. The use of pasteurized shell eggs or pasteurized egg products may be a safer alternative when preparing these types of dishes.

○ Eggs that are cracked open and combined in a container (pooled eggs) must be handled with special care because bacteria present in one egg can be spread to the rest. Cross-contamination from one pooled egg mixture to another may also be a hazard if the food is not prepared or handled properly. Shell eggs must be kept at an ambient (air) temperature of 45°F (7°C) or lower until immediately before use. Pooled eggs, if used, must be cooked promptly after mixing or stored at 41°F (5°C) or lower. Containers that held pooled eggs must be washed and sanitized before being used again. There should be no carryover of egg residue from one food batch to another, even if it is the same recipe.

○ Federal public health officials require that operations serving highly susceptible populations—such as those in nursing homes and hospitals—use only pasteurized shell eggs or pasteurized egg products. This is recommended for all types of eggs and egg dishes served to these people and is required when liquid egg mixtures are prepared, when eggs are cooked and held for service, and when egg-containing food will receive little or no cooking.

Batters and Breading

If prepared with eggs or milk, batters and breading are at risk from time-temperature abuse and cross-contamination, so they should be handled with care. In some cases, it may be better to buy frozen breaded items that can be taken directly from the freezer and then cooked thoroughly in the oven or fryer. If you prefer to make breaded or battered food from scratch, follow some simple rules:

Key Point

Batters and breading can be potentially hazardous and face the same risks of time-temperature abuse and cross-contamination as other food.

○ Consider making batters with pasteurized egg products instead of raw shell eggs whenever possible.

○ Prepare batters in small batches and keep what you don't need in a covered container in the refrigerator. Using small amounts prevents time-temperature abuse of both the batter and the food being coated.

○ When making a batch of batter or breading for later use, store it in the refrigerator at 41°F (5°C) or lower. Don't combine it with other batches already stored in the refrigerator.

○ If you are breading food to cook later, put the breaded food in the refrigerator as quickly as possible.

○ Cook battered and breaded food thoroughly. The coating acts as an insulator, which can prevent food from being thoroughly cooked. When deep-frying food, make sure the temperature of the oil recovers before loading each batch. Overloading the basket also slows cooking time, which means product could be removed from the fryer before it is thoroughly cooked. Be sure to monitor oil and food temperatures using calibrated thermometers, and watch cooking time.

○ Throw out any used batter or breading left over after each shift. Whenever a potentially hazardous food is dipped in batter or dredged in breading, consider the batter or breading contaminated. Never use a batter or breading for more than one product. For example, dipping raw chicken into batter then using the same batter for onion rings may contaminate the onion rings and possibly make someone sick.

Fruit and Vegetables

For the most part, fresh fruit and vegetables are not as likely to carry pathogens as are some other food. But viruses such as hepatitis A, bacteria such as *Listeria* spp., and parasites such as *Cryptosporidium parvum* can survive on produce, especially cut produce. The risk from such microorganisms can be minimized or eliminated by simple preparation safeguards.

○ Start with a clean, sanitized work space. Prepare vegetables away from raw meat, poultry, and eggs, as well as from cooked and ready-to-eat food, to prevent cross-contamination.

○ Surfaces that come in contact with raw meat—such as hands, gloves, knives, prep tables, or cutting boards—should never make contact with food that will be eaten raw, such as salad or fruit, unless they have been cleaned and sanitized.

Cross-Contamination

When preparing raw vegetables, start with a clean, sanitized work space. Prepare vegetables away from raw meat, poultry, and eggs, and from cooked and ready-to-eat food.

Reprinted with permission from Tony Soluri and Charlie Trotter

○ Wash fruit and vegetables thoroughly under running water to remove dirt or other contaminants just before cutting, combining with other ingredients, or cooking. Failure to wash fruit before cutting it may result in contamination of the fruit's interior.

○ Pay particular attention to leafy items such as lettuce and spinach, because dirt and microorganisms can get into inner leaves. Remove the outer leaves, pull lettuce and spinach completely apart, and rinse thoroughly.

○ Cut away bruised or damaged areas when preparing fruit and vegetables.

○ Hold cut melons at 41°F (5°C) or lower. Discard after four hours if not held at this temperature.

○ Do not add sulfites to food products.

○ Juice prepared on-site for service or sale to high-risk populations must be prepared following a HACCP plan.

Ice

People often forget that ice is a food, too. It is used to chill beverages and to chill or dilute food. In Chapter 6, you learned how to store products on ice, such as poultry, fish, fruit, and vegetables. Many establishments also use ice to chill food on display, such as canned beverages or fruit. Any time ice has been used to cool food in this way, you may not reuse it as a food.

Ice used as a food, or used to chill other food, must always be made with potable water (water that is safe to drink). Remember, ice can become contaminated just as easily as other food. When transferring ice from an ice machine to an ice bin or to a display, use a clean, sanitized scoop and container. Store ice scoops outside of the ice machine in a sanitary, protected location. Never hold ice in containers that have been used to store raw meat, poultry, fish, or chemicals.

Cross-Contamination

When transferring ice from an ice machine, use a clean, sanitized scoop and container.

COOKING FOOD

Whatever cooking method you choose, temperature is key to killing microorganisms that can cause foodborne illness. The food's internal temperature must reach a certain level for a specific amount of time before you can be sure all foodborne microorganisms have been sufficiently reduced in number. The only way to be certain a food has reached the required minimum internal cooking temperature is to check it using a calibrated thermometer.

The minimum internal temperature at which foodborne microorganisms are destroyed varies from product to product. Minimum standards have been

While cooking can kill microorganisms, it cannot destroy the spores or toxins they might create. Handling food safely before and after it is cooked will prevent microorganisms from growing or producing spores or toxins.

developed for most cooked food and are included in local and state health codes. *(See Exhibit 7e.)*

It is important to remember that potentially hazardous food—such as meat, eggs, seafood, and poultry—should be cooked to the minimum internal temperatures specified in this chapter unless otherwise ordered by the customer. Potentially hazardous food not cooked to these temperatures—including over-easy eggs, raw oysters, sashimi, or rare hamburgers—generally do not pose an unacceptable risk of foodborne illness to the healthy customer. However, if a customer is in a group at risk for foodborne illness *(see Chapter 1)*, consuming raw or undercooked, potentially hazardous food could possibly increase the risk of illness, sometimes seriously.

In all cases, high-risk customers should be advised, if they ask about or specifically request undercooked food or any potentially hazardous food (or ingredient) that is raw or not fully cooked, of this potential risk. They may want to consult with a physician before regularly consuming these types of food.

Here are general guidelines to follow when cooking:

○ **Specify cooking time and required minimum internal cooking temperature in all recipes.**

Exhibit 7e: Minimum Internal Cooking Temperatures

Adapted from the FDA Food Code.

Product	Temperature
POULTRY **STUFFING, STUFFED MEAT, AND DISHES COMBINING RAW AND COOKED FOOD** (including soups and casseroles)	165°F (74°C) for 15 seconds
GROUND MEATS (beef, pork, or other meat or fish)	155°F (68°C) for 15 seconds
INJECTED MEATS (including brined ham and flavor-injected roasts)	155°F (68°C) for 15 seconds
PORK, BEEF, VEAL, LAMB	Steaks/Chops: 145°F (63°C) for 15 seconds Roasts: 145°F (63°C) for 4 minutes
FISH	145°F (63°C) for 15 seconds
FRESH SHELL EGGS FOR IMMEDIATE SERVICE	145°F (63°C) for 15 seconds
ANY POTENTIALLY HAZARDOUS FOOD COOKED IN A MICROWAVE OVEN	165°F (74°C); let food stand for 2 minutes after cooking

Exhibit 7e

○ **Use properly calibrated thermometers with a suitable size probe, accurate to within 2°F (1°C), to measure food temperatures.** Check internal temperatures in several places, in the thickest part of the food. Clean and sanitize the thermometer after each use.

○ **Avoid overloading ovens, fryers, and other cooking equipment.** The equipment or oil temperature might drop, and the food might not cook properly. Crowding food also might lead to cross-contamination.

○ **Let the cooking equipment's temperature recover between batches.**

○ **Use utensils or gloves to handle food after cooking and taste food correctly to avoid cross-contamination.** The safest and most sanitary way to taste food is to ladle a small amount into a dish. Taste the food in the dish with a clean utensil. When finished, remove the dish and utensil from the area and have them cleaned and sanitized.

COOKING REQUIREMENTS FOR SPECIFIC FOOD

Poultry, Stuffing, Stuffed Meats, and Casseroles

Poultry, stuffing, and stuffed meats should be cooked to an internal temperature of 165°F (74°C) or higher for fifteen seconds. *(See Exhibit 7f.)* Poultry tends to have more types and higher counts of microorganisms than other meat and, therefore, should be cooked more thoroughly.

Stuffing poses a hazard for several reasons. It can be made with potentially hazardous food, such as eggs or oysters, which need to be fully cooked. It also acts as insulation, preventing heat from reaching the center of meat or poultry. For example, when poultry is stuffed, the thighs, wings, and even the breast can reach temperatures of 165°F (74°C), while the stuffing may still be at an unsafe temperature. If the bird is taken out of the oven too soon, microorganisms in the stuffing will not be killed, creating a potential hazard.

Exhibit 7f

Checking the Temperature of Poultry
Poultry should be cooked to an internal temperature of 165°F (74°C) or higher for fifteen seconds.

For this reason, stuffing should be cooked separately, particularly when cooking whole, large birds or large cuts of meat. Smaller cuts of meat, such as pork tenderloins or veal chops, may be stuffed before cooking, but verify that

Exhibit 7g	**Checking the Temperature of Pork**

Pork, such as chops or medallions of tenderloin, should be cooked to an internal temperature of 145°F (63°C) or higher for fifteen seconds.

Temperature	Hold for (in minutes)
130°F (54°C)	112
133°F (56°C)	56
135°F (57°C)	36
136°F (58°C)	28
138°F (59°C)	18
140°F (60°C)	12
142°F (61°C)	8
144°F (62°C)	5
145°F (63°C)	4

Chart: Adapted from the FDA Food Code.

Exhibit 7h	**Alternative Minimum Internal Temperatures for Cooking Beef and Pork Roasts**

the internal temperature of both the meat and the stuffing has reached 165°F (74°C) for at least fifteen seconds.

Casseroles and dishes combining raw and cooked food, such as soups, should also be cooked to 165°F (74°C) for at least fifteen seconds.

Pork

Cook pork, such as chops or medallions of tenderloin, to an internal temperature of 145°F (63°C) or higher for fifteen seconds. *(See Exhibit 7g.)* This temperature is high enough to destroy any *Trichinella* spp. larvae that might have contaminated the pork. Pork roasts, on the other hand, must hold the same internal temperature for at least four minutes. Depending on the type of roast and the oven used, however, pork can be cooked at different internal temperatures. *(See Exhibit 7h.)*

Beef

Steaks must reach and hold an internal temperature of at least 145°F (63°C) for fifteen seconds. Roasts, on the other hand, must hold the same internal temperature for at least four minutes. Depending on the type of roast and the oven used, however, beef can be cooked at different internal temperatures. *(See Exhibit 7h.)*

Ground Meats

Ground beef, pork, and other meat or fish must be cooked to an internal temperature of 155°F (68°C) for at least fifteen seconds. Most whole-muscle cuts of meat are likely to have microorganisms only on their surface. When meat or seafood is ground, microorganisms on the surface are mixed throughout the product. To make sure microorganisms are destroyed, thorough cooking is a must. Check with your local or state regulatory agency for additional requirements. *(See Exhibit 7i.)*

Temperature	Hold for
145°F (63°C)	3 minutes
150°F (66°C)	1 minute
155°F (68°C)	15 seconds
158°F (70°C)	< 1 second

Chart: Adapted from the FDA Food Code

Injected Meats

Injected and pinned meats, such as brined ham or flavor-injected beef roasts, must be cooked to an internal temperature of 155°F (68°C) for at least fifteen seconds. When meats are injected or pinned, foodborne microorganisms on the surface can be carried to the interior.

Game and Ratites

Commercially raised and inspected game animals—such as elk, deer, bison, and rabbit—can be cooked in much the same way as beef. Small cuts, such as steaks, can be cooked to a minimum internal temperature of 145°F (63°C) or higher, for fifteen seconds. Ground meat must be cooked to an internal temperature of 155°F (68°C) or higher for fifteen seconds. Stuffed meats should be cooked to 165°F (74°C) or higher for fifteen seconds. Roasts can be cooked in the same way and to the same temperatures as beef and pork roasts.

Though they are birds, ratites (ostrich, emu, and rhea) are not cooked the same way as poultry. They will have a metallic taste when cooked to internal temperatures higher than 165°F (74°C). Ratites are fully cooked when their internal temperature reaches 155°F (68°C) for fifteen seconds. If portions or cutlets of these birds are stuffed, however, cook them to an internal temperature of 165°F (74°C) for fifteen seconds.

Exhibit 7i

Photo: Courtesy of Atkins Temptec, Gainesville, FL

Alternative Minimum Internal Temperatures for Cooking Ground Meats

Fish

Cook fish to an internal temperature of 145°F (63°C) or higher for fifteen seconds. *(See Exhibit 7j.)* Stuffed fish should be cooked to 165°F (74°C) or higher, for fifteen seconds. This is also true for stuffing containing fish. If fish has been ground, chopped, or minced, it should be cooked to an internal temperature of 155°F (68°C) or higher for fifteen seconds.

Exhibit 7j **Checking the Temperature of Fish**
Cook fish to an internal temperature of 145°F (63°C) or higher for fifteen seconds.

Key Point

Shell eggs cooked to order should be cooked to 145°F (63°C) or higher for fifteen seconds.

Key Point

Never hold brewed tea at room temperature for more than one day.

Eggs and Egg Mixtures

In general, shell eggs cooked for immediate service should be cooked to 145°F (63°C) or higher for fifteen seconds. When eggs are cooked this way, the white is set and the yolk begins to thicken. Properly cooked scrambled eggs are firm, with no visible liquid. To hold eggs for later service, cook them to 155°F (68°C) or higher for fifteen seconds, then hold them at 140°F (60°C).

When cooking egg dishes, cook them precisely as directed and make sure cooking is complete before serving the item. Egg dishes, especially those prepared in deep dishes, such as casseroles or stuffing, or previously frozen dishes must be cooked to an internal temperature of 165°F (74°C) for fifteen seconds.

When cooking eggs, remove from storage only as many eggs as you need for immediate use. Never stack egg trays (flats) near the grill or stove.

Vegetables

Although most vegetables can be eaten raw, many are cooked before they are served. When cooking fruit or vegetables for hot-holding, cook them to 140°F (60°C) or higher. Cooked vegetables must never be left out or held at room temperature. Hold cooked vegetable dishes at 140°F (60°C) or higher, otherwise discard them after four hours.

Tea

Dry tea leaves contain low levels of bacteria, yeast, and mold (like most plant-derived food). Using improper brewing temperatures to prepare tea and storing it at room temperature for long periods of time can cause these microorganisms to grow to high levels. Improperly cleaned and sanitized equipment can also promote growth of microorganisms. When handling tea, follow these recommendations:

○ Brew only as much tea as you reasonably expect to sell within a few hours.

○ Never hold brewed tea at room temperature for more than one day. Discard any unused tea at the end of the day.

○ To protect tea flavor and avoid microbial contamination and growth, clean and sanitize tea brewing, storage, and dispensing equipment at least once a day. Equipment should be disassembled, washed, rinsed, and sanitized. Urn spigots should be replaced at the end of each day with freshly sanitized ones.

○ For any brewing method, use a thermometer to make sure brewing water in your equipment meets the specified temperature.

   • 195°F (91°C) for automatic iced tea and automatic coffee machine equipment. Tea leaves should remain in contact with the water for a minimum of one minute.

- 175°F (80°C) minimum when using the traditional steeping method. Tea leaves must be exposed to the water for approximately five minutes by this method.

Microwave Cooking

Microwave ovens tend to cook food more unevenly than other methods of cooking. For this reason, there are special rules for using microwave ovens to cook meat, poultry, and fish.

○ Rotate or stir food halfway through the cooking process to distribute heat more evenly.

○ Cover food to prevent the surface from drying out.

○ Let food stand for at least two minutes after cooking to let product temperature equalize.

Meat, poultry, or fish cooked in a microwave oven must be heated to 165°F (74°C) or higher. Check the internal temperature of the food in several places to make sure it has cooked through.

COOLING FOOD

You have already seen how important it is to keep food out of the temperature danger zone. When cooked food will not be served immediately, it is essential to hold it properly or to cool it as quickly as possible.

The FDA Food Code recommends **two-stage cooling**. Cooked food must be cooled from 140°F (60°C) to 70°F (21°C) within two hours and from 70°F (21°C) to 41°F (5°C) or lower in an additional four hours, for a total cooling time of six hours. (Some jurisdictions use one-stage cooling, by which food must be cooled to 41°F [5°C] or lower in less than four hours.)

Keep in mind this is a two-stage process (two hours plus four hours). While you have learned that microorganisms grow best in the temperature danger zone, the temperature range between 125°F (52°C) and 70°F (21°C) is ideal for the growth of pathogenic microorganisms. Food must pass through this temperature range quickly during cooling in order to minimize the growth of these microorganisms. Because only two hours are allowed to cool food from 140°F (60°C) to 70°F (21°C), the two-stage cooling process passes potentially hazardous food through this temperature range quickly—and safely.

When using two-stage cooling, if the food has not reached 70°F (21°C) within two hours, the food must be discarded or properly reheated. Reheat the food to 165°F (74°C) for fifteen seconds within two hours and then cool it properly.

Key Point

Meat, poultry, or fish cooked in a microwave oven must be heated to 165°F (74°C) or higher.

Time & Temperature

Using two-stage cooling, food must be cooled from 140°F (60°C) to 70°F (21°C) within two hours and from 70°F (21°C) to 41°F (5°C) or lower in an additional four hours, for a total cooling time of six hours.

Reduce size of food

Ice-water bath

Blast chiller

Exhibit 7k

Safe Methods for Cooling Food

Methods for Cooling Food

While common sense may suggest that the quickest way to cool food is to put it in the refrigerator, it isn't. Refrigerators are designed to keep cold food cold. They usually don't have the capacity to cool hot food quickly.

In general, the thickness or density of food is the biggest factor in how quickly it cools. The denser the food product, the more slowly it cools. For example, refried beans take longer to cool than vegetable broth since the beans are more dense.

The container in which food is stored also affects how fast it will cool. Stainless steel transfers heat from food faster than plastic. Shallow pans allow the heat from food to disperse faster than deep ones.

There are a number of methods you can use for cooling food quickly and safely. *(See Exhibit 7k.)*

○ **Reduce the quantity or size of the food you are cooling.** Cut large food items into smaller pieces or divide large containers of food into smaller containers or shallow pans.

○ **Use ice-water baths to bring food temperature down quickly.** After dividing food into smaller quantities, put the pots or pans into a sink or large pot filled with ice water. *(See Exhibit 7l.)*

○ **Use blast chillers to cool food before placing it into refrigeration.** Since blast chillers are now more affordable, you might consider one for your establishment.

○ **If properly equipped, steam-jacketed kettles can be used to cool food.** Simply run cold water through the jacket.

○ **Stir food as it cools.** This allows the product to cool faster and more evenly. Some manufacturers make plastic paddles you can fill with water and freeze. Stirring food products with these **cold paddles** cools food quickly. *(See Exhibit 7m.)* Make sure all utensils used to stir food are cleaned and sanitized between each use.

○ **Add ice or cool water as an ingredient.** This method works for recipes that require water as an ingredient, such as soup or stew. The recipe can initially be prepared with less water than is required. Cold water or ice can then be added after cooking to cool the product and to provide the remaining water required by the recipe.

Once food has been cooled to at least 70°F (21°C), the refrigerator can handle it more easily. Food will continue to chill faster if you do the following:

○ **Keep food in shallow stainless-steel pans.** You may want to put dense foods, like chili or stew, into pans that are two inches deep. Thinner liquids can be stored in three-inch pans.

- **Always place pans on top shelves in the cooler.** Leave them uncovered while they cool if they are protected from overhead contaminants. Otherwise, cover them loosely with foil or plastic wrap. Stir the contents of the pans occasionally, using caution while stirring.

- **Position pans so air can circulate around them.** This will help transfer heat from the pan.

All cooked food should be stored in containers labeled with the date the food was prepared and stored.

REHEATING POTENTIALLY HAZARDOUS FOOD

When previously cooked, potentially hazardous food is reheated for hot-holding, take it through the temperature danger zone again as quickly as possible. Food must be reheated to an internal temperature of 165°F (74°C) for fifteen seconds, within two hours. When adding previously cooked food as an ingredient to another food, such as adding ground beef to spaghetti sauce, the whole mixture must be reheated to 165°F (74°C) for fifteen seconds.

When using the microwave oven to reheat previously cooked food, follow the same rules used for microwave cooking. Cover the product. Rotate or stir it midway through cooking. Allow it to stand for two minutes. Check its internal temperature in several places to see if it has reached at least 165°F (74°C).

Food that is reheated for immediate service to a customer, such as a roast beef sandwich, may be served at any temperature as long as the food was cooked properly.

Exhibit 7l | **Cooling Food with an Ice-Water Bath**
Ice-water baths can be used to cool food quickly.

Exhibit 7m | **Using a Cold Paddle To Cool Food**
Stirring products with cold paddles chills food quickly.

If food has not reached 165°F (74°C) for fifteen seconds *within two hours,* it must be discarded.

SUMMARY

To protect food during preparation, you must handle it safely. The keys are time and temperature control and the prevention of cross-contamination.

All potentially hazardous food should be kept out of the temperature danger zone as much as possible. The longer food is exposed to these temperatures, the greater the risk that microorganisms will grow.

Thaw frozen food in the refrigerator or under cool running water, never at room temperature. Have employees prepare food in small batches, use chilled utensils and bowls, and record product temperatures and preparation times.

To prevent contamination, prepare raw food separately from cooked or ready-to-eat food. Assign specific equipment to each type of food. Always clean and sanitize equipment, utensils, and containers before and after each use. Wash your hands and consider using gloves when handling food products, especially ready-to-eat food that will not be cooked.

Cooking can reduce the number of microorganisms in food to safe levels. To ensure that microorganisms are destroyed, food must be cooked to required minimum internal temperatures for a specific amount of time. These temperatures vary from product to product. Cooking does not kill the spores or toxins some microorganisms produce. That is why it is so important to inspect product once it arrives and handle it safely during preparation.

Once food is cooked, it should be served as quickly as possible. If it is going to be stored and served later, it must be cooled rapidly. Cooked food must be cooled from 140°F (60°C) to 70°F (21°C) within two hours and from 70°F (21°C) to 41°F (5°C) or lower in an additional four hours (for a total cooling time of six hours) unless otherwise required by your local health code. Placing large containers of hot food into the refrigerator can put all other stored food in danger. Ways to cool large quantities of cooked food quickly include dividing it into smaller portions, putting it in shallow stainless steel pans, using an ice-water bath or blast chiller, and stirring it often. When the food is cold enough, store it properly in the refrigerator.

Previously cooked, potentially hazardous food that is to be held hot, must be reheated to an internal temperature of 165°F (74°C) for fifteen seconds within two hours before it can be served.

A CASE IN POINT I

Case Study

On Friday, John went to work at The Fish House knowing he had a lot to do. After changing clothes and punching in, he took a case of frozen raw shrimp out of the freezer. To thaw it quickly, he put the boxes of frozen shrimp into the prep sink and turned on the hot water. While waiting for the shrimp to thaw, John took several fresh whole fish out of the walk-in refrigerator. He brought them back to the prep area and began to clean and fillet them. When he finished, he put the fillets in a pan and returned them to the walk-in refrigerator. He rinsed off the boning knife and cutting board in the sink, and wiped off the worktable with a dish towel.

Next, John transferred the shrimp from the sink to the worktable using a large collander. On the cutting board he peeled, deveined, and butterflied the shrimp using the boning knife. He put the prepared shrimp in a covered container in the refrigerator, then started preparing fresh produce.

What did John do wrong?

A CASE IN POINT II

Case Study

By 7:30 P.M., all the residents at Sunnydale Nursing Home had eaten dinner. As she began cleaning up, Angie realized she had a lot of chicken breasts left over. Betty, the new assistant manager, had forgotten to inform Angie that several residents were going to a local festival and would miss dinner.

"No problem," Angie thought. "We can use the leftover chicken to make chicken salad."

Angie left the chicken breasts in a pan on the prep table while she started putting other food away and cleaning up the kitchen. At 9:45 P.M., when everything else was clean, she put her hand over the pan of chicken breasts and decided they were cool enough to handle. She covered the pan with plastic wrap, and put it in the refrigerator.

Three days later, she came in to work on the early shift. Angie decided to make chicken salad from the leftover chicken breasts. After she hung up her coat and put on her apron, Angie took all the ingredients she needed for chicken salad out of the refrigerator and put them on a worktable. Then she started breakfast.

First, she cracked three dozen eggs into a large bowl, added some milk, and set the bowl near the stove. Then she took bacon out of the refrigerator and put it on the worktable next to the chicken salad ingredients. She peeled off strips of bacon onto a sheet pan and put the pan into the oven. After wiping her hands on her apron, she went back to the stove to whisk the eggs and pour them onto the griddle. When they were almost done, Angie scooped the scrambled eggs into a hotel pan and put it in the steam table.

As soon as breakfast was cooked, Angie went back to the prep table to wash and cut up celery and cut up the chicken for chicken salad.

What did Angie do wrong?

TRAINING TIPS

1. "Cook-Kill" Temperature Quiz

Purpose: *After completing this activity, trainees will be able to identify safe minimum internal cooking temperatures for different food.*

Time: 20 minutes

Directions: Give trainees a printed list of twelve to fifteen varied menu items, with a blank space next to each item. Ask them to fill in the required minimum internal cooking temperature for each item. Put in some trick menu items, such as rare roast beef, stuffed ostrich cutlets, and buffalo burgers.

Limit this activity to five minutes. Review the different temperatures with the group and see if anyone got a perfect score.

Use this activity as a springboard to a group discussion on safe cooking temperatures for different food.

2. SOPs for Menu Items

Purpose: *After completing this activity, trainees will be able to develop standard operating procedures (SOPs) for preparing food.*

Time: 30 minutes

Directions: Make a list of six to eight different menu items, including at least one item from each major food category discussed in Chapter 7 under "Preparing Specific Food." A representative list might include:

○ Fruit and Vegetables: chef's salad
○ Meat, Fish, and Poultry: fresh whitefish, roast turkey, loin of pork
○ Protein Salads and Sandwiches: chicken salad
○ Eggs: Denver omelette
○ Batters and Breading: hand-breaded fried chicken

Divide trainees into groups of about three people each. Give each group a sheet of paper with one menu item printed at the top. Each group will work on a different menu item.

Ask each group to create a recipe for its assigned menu item. The recipe should include: ingredients, preparation steps, and, especially, food safety instructions. Note: *Accurate weights and measures are not necessary for these recipes, since this activity should focus on food safety guidelines rather than culinary accuracy.*

Assign a time limit to develop the recipe. Offer a reminder that the recipe should focus on SOPs for safe foodhandling.

When time is up, have each group present its recipe. Allow time for discussion. Make necessary changes and additions. Distribute copies of the recipes to the trainees.

3. "Chill Out!" Team Contest

Purpose: *After completing this activity, trainees will be able to identify the various methods of cooling food, discuss their advantages and disadvantages, and identify the food best suited for each cooling method. (Note: Use this activity after reviewing the critical limits for cooling food, but before a discussion of cooling methods.)*

Time: 30 minutes

Directions: Print a form that asks trainees to list as many cooling methods as they can think of, food types best suited for each method, and advantages and disadvantages of using each method.

Divide trainees into teams of two and give each team one form. Allow about seven minutes for the teams to list cooling methods, food types, and advantages and disadvantages.

After time is up, let the teams grade their sheets as follows:

○ For each cooling method recognized as safe, award   5 points

○ For each type of food to be cooled by this method, award   2 points

○ For each cooling method advantage, award   3 points

○ For each cooling method disadvantage, award   3 points

The team with the most points wins.

Allow time for discussion, and let trainees make additions and correct errors on their papers. Create a master list of cooling methods from this activity as a reference handout.

4. "Cool It" Cooling Activity

Purpose: *To examine the methods used to cool food in your operation and to identify any problems with your current practices.*

Time: one day

Directions: Invite the kitchen staff to participate in a competition. Select a menu item your establishment prepares in advance, then cools and stores for future use. Chili is a good example. The next time you're scheduled to make chili, have your cooks take the temperature of the chili just before taking it off the stove. Have them record this temperature, then tell them to cool the chili as they normally do.

Ask each staff member to estimate what the temperature of the chili will be two hours after removing it from the stove, and again four hours later. Record their estimates on index cards.

Have someone check and record the temperature of the chili every hour. After the first two hours, and again after six hours, verify the final temperature yourself, then check the index cards. List the different temperatures the staff estimated and announce a winner.

Discuss with your staff the importance of cooling food properly. Solicit their input to develop a system for cooling chili and other food products from 140°F (60°C) to 70°F (21°C) within two hours and from 70°F (21°C) to 41°F (5°C) or lower in an additional four hours, for a total cooling time of six hours. Brainstorm and discuss different methods, then implement the best method as your standard cooling process.

5. Work Station Safety Checklists

Purpose: *To involve kitchen staff members in the creation of food safety checklists for specific work areas in your establishment.*

Time: several weeks

Directions: Create a food safety team from a group of volunteers from the kitchen staff. Ideally, all different shifts and work areas should be represented.

Assign team members to study different work areas in your operation (cold prep, grill, stove, oven, bakery, cleanup, storage, and so on) and ask them to develop food safety checklists for those specific work areas. They should keep in mind the major food safety concerns: monitoring time and temperature, preventing cross-contamination, and proper personal hygiene.

After receiving the checklists, edit them to make them consistent and easy to read. Suggest ways to make them colorful or color-coded, and translate them into several languages if necessary. Review your work with the team. Have the checklists professionally printed and laminated and post them in the appropriate work areas for easy reference.

6. "Cooking Pays" Temperature Quiz

Purpose: *To reinforce the importance of monitoring temperature when handling food in your operation.*

Time: ongoing activity

Directions: Every day, carry a number of tokens in your pocket, along with your calibrated thermometer or thermocouple. Whenever you walk through the kitchen and see an employee using a thermometer to check the temperature of food, give him or her a token. If employees can tell you the temperature of the food product and are accurate within 2°F (1°C), give them two more tokens. During your regular staff meetings, give out prizes based on the number of tokens collected. This activity verifies that employees are using thermometers.

As you conduct this activity, remind the cooks of the importance of monitoring both time and temperature when handling potentially hazardous food, and solicit their cooperation in minimizing the time food spends in the temperature danger zone.

DISCUSSION QUESTIONS

1. What are the required minimum internal cooking temperatures for poultry, fish, pork, and ground beef?

2. What are four proper methods for thawing food?

3. What are various methods for cooling cooked food?

4. What are the steps for properly cooking food in a microwave oven?

MULTIPLE-CHOICE STUDY QUESTIONS

1. An employee has just trimmed raw chicken on a cutting board and must now use the board to prepare vegetables. What should the employee do with the board prior to preparing the vegetables?

   A. Wash, rinse, and sanitize the cutting board.
   B. Dry it with a paper towel.
   C. Rinse it under very hot water.
   D. Turn it over and use the reverse side.

2. Which of the following is *not* a safe method for thawing frozen food?

   A. Thawing it by submerging it under running water at 70°F (21°C) or lower.
   B. Thawing it in the microwave and cooking it immediately afterwards.
   C. Thawing it at room temperature.
   D. Thawing it in the refrigerator overnight.

3. Casseroles and other dishes combining raw and cooked food should be cooked to

   A. 135°F (57°C) for 15 seconds.
   B. 145°F (63°C) for 15 seconds.
   C. 155°F (68°C) for 15 seconds.
   D. 165°F (74°C) for 15 seconds.

4. You are making omelettes to order for Sunday brunch at the hotel where you work. How should you handle the eggs to ensure you are serving safe omelettes?

   A. Crack all of the eggs you think you will use into a large container stored near the grill.
   B. Stack flats of eggs as close to the grill as possible so they can be cooked quickly.
   C. Cook omelette orders to 125°F (52°C) or higher.
   D. Keep all of the eggs you need in their shells and refrigerated right up until the time they are used.

5. Which of the following is a bad practice when handling vegetables?
   A. Washing vegetables under running water before cutting them
   B. Pulling leafy vegetables completely apart before rinsing them
   C. Holding cooked vegetables at 125°F (52°C) or higher
   D. Refrigerating cut vegetables at 41°F (5°C) or lower

6. Potentially hazardous cooked food cannot remain in the temperature danger zone for more than _____ before it must be reheated or discarded.
   A. one hour
   B. two hours
   C. three hours
   D. four hours

7. What is the proper way to cool a large stockpot of clam chowder?
   A. Allow the stockpot to cool at room temperature.
   B. Put the hot stockpot into the walk-in refrigerator to cool.
   C. Divide the clam chowder into smaller containers and place them in an ice-water bath.
   D. Put the hot stockpot into the walk-in freezer to cool.

8. When reheating potentially hazardous food for hot-holding, reheat the food to
   A. 135°F (57°C) for 15 seconds within two hours.
   B. 145°F (63°C) for 15 seconds within two hours.
   C. 155°F (68°C) for 15 seconds within two hours.
   D. 165°F (74°C) for 15 seconds within two hours.

9. Which of the following food has been safely cooked?
   A. A hamburger cooked to an internal temperature of 135°F (57°C) for fifteen seconds
   B. A pork chop cooked to an internal temperature of 145°F (63°C) for fifteen seconds
   C. A whole turkey cooked to an internal temperature of 155°F (68°C) for fifteen seconds
   D. A fish fillet cooked to an internal temperature of 135°F (57°C) for fifteen seconds

10. All of the following practices can help prevent cross-contamination during food preparation *except*
    A. preparing meat separately from ready-to-eat food.
    B. assigning specific equipment for preparing specific food.
    C. rinsing cutting boards between preparing raw food and ready-to-eat food.
    D. using specific storage containers for specific food.

For answers, please turn to Appendix A.

ADDITIONAL RESOURCES

Books and Periodicals

Bax, B. 1997. *Handbook for safe food service management.* Upper Saddle River, NJ: Prentice-Hall.

Cross out cross-contamination. 1998. *Best Practices.* 2(1):6-9.

Equipped for food safety. 1998. *Best Practices.* 2(4):6-10.

Laconi, D. 1995. *Fundamentals of professional food preparation: A laboratory text workbook.* New York: Wiley.

Loken, J. 1995. *The HACCP food safety manual.* New York: Wiley.

Longree, K. and G. Armbruster. 1996. *Quantity food sanitation.* New York: Wiley.

The right way to...prepare beef. 1998. *Best Practices.* 2(3):12-13.

The right way to...prepare chicken. 1998. *Best Practices.* 2(1):10.

The right way to...prepare eggs. 1998. *Best Practices.* 2(2):10-11.

The right way to...prepare pork. 1997. *Best Practices.* 1(3):10-11.

The right way to...prepare fruits and vegetables. 1998. *Best Practices.* 2(4):12-13.

The right way to…seafood safety. 1999. *Best Practices.* 3(1):16.

Secrets of self-inspection. 1998. *Best Practices.* 2(2):6-9.

Web Sites

Alaska Seafood Marketing Institute

http://www.alaskaseafood.org

The Alaska Seafood Marketing Institute enhances the appeal and popularity of Alaska seafood to the restaurant and foodservice industry, and to the general public. This useful Web site contains information on how to market seafood to your customers, how to purchase Alaskan seafood, detailed information on types of Alaskan seafood, and how to handle seafood to maintain safety and quality.

American Egg Board

http://www.aeb.org

As the egg industry's promotional arm, the American Egg Board is the U.S. egg producer's link to the consumer for communicating the value of the egg. Its Web site offers a wealth of information about eggs and egg safety.

American Meat Institute

http://www.meatami.com

The American Meat Institute is a membership trade association representing the interests of the U.S. meat and poultry industry to the federal government, Congress, the media, and the customer. This Web site keeps you up-to-date on what is happening in the meat and poultry processing industry.

Atkins Temptec

http://www.atkinstemptec.com

Atkins is a manufacturer of digital thermometers and probes for the restaurant and foodservice industry. This Web site provides information on how thermometers work, how to get the most out of your thermometer, and how to hold food safely during service. The site also provides an interesting demonstration on how food color is a poor indicator of internal temperature. Consult this site for useful tips related to thermometers.

Chlorine Chemistry Council®

http://www.c3.org

The Chlorine Chemistry Council, comprised of chlorine and chlorinated product manufacturers, is a business council of the American Chemistry Council. At this Web site, you will find the history of chlorine, its many uses at home, in the restaurant, and in the foodservice industry, and chlorine in the news.

Cooper Instrument Corporation

http://www.cooperinstrument.com

Cooper Instrument Corporation is a supplier of temperature, time, and humidity instruments for a variety of global markets. This Web site contains information on how to calibrate thermometers properly and maintain food temperatures during cooking, holding, cooling, etc.

ECOLAB, INC

http://www.ecolab.com

An excellent source of information on housekeeping and sanitation supplies for the restaurant and foodservice industry from the world's leading sanitation product supplier.

FDA Center for Food Safety and Applied Nutrition (CFSAN)

http://vm.cfsan.fda.gov

As the center within the FDA responsible for food safety and nutrition, CFSAN promotes and protects public health by researching and implementing guidelines, policies, and standards to ensure that food is safe, nutritious, wholesome, and properly labeled. This Web site provides a wealth of information on food safety and sanitation, including corresponding guidelines, policies, and standards.

FDA Food Code

http://vm.cfsan.fda.gov/~dms/foodcode.html

As the basis for many local sanitation codes, as well as the basis for information in this textbook, the FDA Food Code, available at this Web address, is a useful resource for information related to food safety for the restaurant and foodservice industry.

FDA Seafood Information and Resources

http://vm.cfsan.fda.gov/seafood1.html

The FDA operates an oversight compliance program for fishery products under which responsibility for product safety, wholesomeness, identity, and economic integrity rests with the processor or importer, who must comply with regulations under the Federal Food, Drug and Cosmetic (FD&C) Act. This Web site houses information on the seafood program, foodborne pathogens and contaminants associated with seafood, and HACCP compliance.

International Dairy-Deli-Bakery Association (IDDBA)

http://www.iddba.org

IDDBA is a resource for information and services across all food channels for the dairy, deli, and bakery industry. This Web site provides information relevant to people who work in the dairy, deli, and bakery industry.

International Food Information Council (IFIC) Foundation

http://ific.org

IFIC collects and disseminates scientific information on food safety, nutrition, and health. They work with an extensive roster of scientific experts to help translate research into understandable and useful information for opinion leaders and, ultimately, consumers. This Web site provides easy-to-understand information on foodborne illness and health-related stories circulating in the news.

International Fresh-cut Produce Association (IFPA)

http://www.fresh-cuts.org

IFPA serves commercial producers, suppliers, and distributors of fresh-cut produce. It provides members with a forum for representation, education, technical support, networking, information, and resources. This site contains information on recent industry happenings and regulations, and provides guidance documents for purchasing, receiving, and storing fresh-cut produce.

KatchAll/San Jamar

http://www.sanjamar.com

KatchAll is an industry leader in innovative food safety products for the restaurant and foodservice industry. The Web site contains information on all KatchAll products that control time and temperature and help avoid cross-contamination.

The Mushroom Council

http://www.mushroomcouncil.com

This interesting and useful Web site from the Mushroom Council contains recipes and information about how to handle and store mushrooms to ensure safety and quality.

National Cattlemen's Beef Association (NCBA)

http://www.beef.org

NCBA is a consumer-focused, producer-directed organization representing the largest segment of the nation's food and fiber industry. This Web site, as well as its companion site for foodservice (www.beeffoodservice.com), provides excellent information on purchasing, cooking, and serving beef. Both sites also provide important food safety information.

The National Chicken Council

http://www.eatchicken.com

This Web site is brought to you by the National Chicken Council and the U.S. Poultry & Egg Association, representing a wide spectrum of companies and individuals in the chicken business. This fun and interesting site contains chicken safety tips and nutritional and statistical information.

National Frozen Food Association (NFFA)

http://www.nffa.org

NFFA's mission is to promote the sales and consumption of frozen food through education, training, research, sales planning, and menu development, as well as provide a forum for industry dialogue. Publications and information on frozen food and how to market it to consumers are available at this Web site.

National Pork Producers Council

http://www.nppc.org

Billing itself as the place for pork lovers, pork producers, and anyone who wants to learn more about pigs and pork, the National Pork Producers Council Web site provides information on pork safety and quality, nutrition and industry information, and publications.

National Restaurant Association

http://www.restaurant.org

The National Restaurant Association is the leading business association for the restaurant industry. Together with the National Restaurant Association Educational Foundation, the Association's mission is to represent, educate, and promote the rapidly growing restaurant and foodservice industry. This Web site should be your starting place for all issues and concerns related to your restaurant. This Web site has it all, from tips for running your establishment to vital data on your customers' spending habits.

National Turkey Federation

http://www.turkeyfed.org

The National Turkey Federation advocates for all segments of the turkey industry, providing services and conducting activities that increase demand for its members' products by protecting and enhancing their ability to profitably provide wholesome, high-quality, nutritious products. Its Web site contains a special area specifically for the restaurant and foodservice industry, including recipe and menu ideas, cooking demonstrations, and a reference on purchasing, preparing, and promoting turkey.

Pasteurized Eggs, L.P.

http://www.davidsonseggs.com

This company is the developer of an all-natural process that eliminates the risk of foodborne illness by destroying *Salmonella,* which may be present in raw or soft-cooked eggs. This Web site explains the process of in-shell pasteurization, why in-shell pasteurization is an additional step restaurants and foodservice establishments can take to ensure the safety of food they serve, and where pasteurized eggs are available for purchase in the U.S.

Procter & Gamble

http://www.pg.com

Procter & Gamble produces cleaning and food and beverage products for the restaurant and foodservice industry. This Web site provides information on its products.

Produce Marketing Association

http://www.pma.com

The Produce Marketing Association is a not-for-profit trade association serving members who market fresh fruit, vegetables, and floral products worldwide. Its members are involved in the production, distribution, retail, and foodservice sectors of the industry. This Web site provides information on hot topics in the produce industry and promotional ideas.

Tyson Foods, Inc.

http://www.tyson.com

Tyson Foods is the world's largest poultry producer. This Web site is a great resource about chicken products for the restaurant and foodservice industry. It includes recipes and menu ideas, information on new products, and information about Tyson Foods' efficient distribution and transportation system.

USDA-FSIS

http://www.fsis.usda.gov

The Food Safety and Inspection Service (FSIS) is the public health agency of the U.S. Department of Agriculture responsible for ensuring that the nation's commercial supply of meat, poultry, and egg products is safe, wholesome, and correctly labeled and packaged. This Web site contains a wealth of information on food safety relating to meat, poultry, and eggs.

Chapter 8
Protecting Food During Service

Knowledge

TEST YOUR FOOD SAFETY KNOWLEDGE

1. **True or False:** Most hot-holding equipment can be used to safely reheat potentially hazardous food. *(See page 8-2.)*

2. **True or False:** Hot, potentially hazardous food items should be discarded after four hours if not held at 140°F (60°C) or higher. *(See page 8-3)*

3. **True or False:** Freshly prepared food should not be mixed with food being held for service. *(See page 8-3)*

4. **True or False:** Servers can contaminate food simply by handling the food-contact surface of a plate. *(See page 8-4)*

5. **True or False:** Unused plate garnishes that have been served to one customer, such as fruit or pickles, can be re-served to another customer. *(See page 8-7)*

For answers, please turn to Appendix A.

Key Terms

Hot-holding equipment
Cold-holding equipment
Food bar
Sneeze guard
Off-site service
Single-use items
Mobile unit
Temporary unit
Vending machine

Table of Contents

Learning Objectives

After completing this chapter, you should be able to

○ Identify procedures for holding hot and cold food properly.

○ Identify the proper procedures for setting up and replenishing food bars and buffets.

○ Identify food that can be re-served.

○ Identify the proper procedures for serving food off site.

○ List and apply the Eight Rules of Safe Foodhandling.

○ Identify proper and improper food-serving practices.

○ Identify measures to prevent contamination in self-service areas.

The job of protecting food continues even after it has been prepared and cooked properly, since microorganisms can still contaminate food before it is eaten. The key to serving safe food is to prevent time-temperature abuse and cross-contamination. Hold, display, and serve food at the right temperature and handle it safely. People do many things without knowing their actions can lead to contamination. Train employees to serve food properly, and make sure food safety rules are followed.

HOLDING FOOD FOR SERVICE

In many establishments, food is cooked to order. Food that is stored, prepared, and cooked properly, and then served immediately, is less likely to cause illness. Even in facilities that cook food to order, many menu items are cooked and held for service. A coffee shop might hold soup in a warming kettle. A steak house might keep prime rib warm on a steam table. Many establishments, such as cafeterias and buffets, hold almost all food they serve.

Kitchen staff might be tempted to hold hot food at a lower temperature that keeps it warm but does not affect quality. However, all employees must remember that microorganisms can grow at temperatures between 41°F (5°C) and 140°F (60°C). To ensure the safety of food that is held hot or cold, specific procedures must be followed.

Time & Temperature

Never use hot-holding equipment to reheat food if it is not designed to do so.

Hot Food

○ **Never use hot-holding equipment to reheat food if it is not designed to do so.** Reheat food to 165°F (74°C) for fifteen seconds within two hours, then transfer it to holding equipment. Most hot-holding equipment is incapable of passing food through the temperature danger zone quickly enough during the reheating process to prevent the growth of microorganisms.

○ **Only use hot-holding equipment that can keep food at an internal temperature of 140°F (60°C) or higher.** Hot-holding equipment includes steam tables, chafing dishes, double boilers, and heated cabinets. *(See Exhibit 8a.)*

Courtesy of Cooper Instrument Corporation, Middlefield, CT

Exhibit 8a Hot-Holding Equipment
Only use hot-holding equipment that can keep food at 140°F (60°C) or higher.

○ **Stir food at regular intervals to distribute heat evenly.**

○ **Keep food covered.** Covers retain heat and keep out contaminants.

○ **Check internal food temperature at least every two hours.** Use a thermometer. The holding equipment's thermostat measures the temperature of the equipment—not the food. Record temperatures in a log.

Exhibit 8b **Refilling Food in Holding Equipment**
Never mix freshly prepared food with food being held for service.

○ **Discard potentially hazardous food after four hours if it has not been held at or above 140°F (60°C).**

○ **Never mix freshly prepared food with food being held for service.** Mixing food may lead to cross-contamination. *(See Exhibit 8b.)*

○ **Prepare food in small batches so it will be used faster.** Do not prepare hot food any further in advance than necessary.

Cold Food

○ **Only use cold-holding equipment** that can keep food at 41°F (5°C) **or lower.** *(See Exhibit 8c.)*

○ **Do not store food directly on ice.** Whole fruit and vegetables and raw, cut vegetables are the only exceptions. Place food in pans or on plates first. Ice used on a display should be self-draining. Clean and sanitize drip pans after each use.

○ **Check internal food temperatures at least every two hours.** If above 41°F (5°C), take corrective action. Record temperatures (and corrective actions) in a log.

○ **Protect food from contaminants with covers or sneeze guards.**

Time & Temperature

Discard potentially hazardous food after four hours if it has not been held at or above 140°C (60°C).

Exhibit 8c **Cold-Holding Equipment**
Cold-holding equipment must keep food at 41°F (5°C) or lower.

Exhibit 8d **Properly Stored Utensils**
If stored in food, utensils should be stored with the handle extended above the rim of the container.

SERVING FOOD SAFELY

After handling food safely and cooking it properly, you don't want to risk contamination when serving it.

Kitchen Staff

Train your kitchen staff to follow these procedures to serve food safely.

○ **Store serving utensils properly.** Serving utensils can be stored in the food, with the handle extended above the rim. *(See Exhibit 8d.)* They can also be placed on a clean, sanitized food-contact surface. Spoons or scoops used to serve food such as ice cream or mashed potatoes can be stored under running water.

○ **Use serving utensils with long handles.** Long-handled utensils keep the server's hands away from food.

○ **Use clean and sanitized utensils for serving.** Use separate utensils for each food item, and properly clean and sanitize them after each serving task. Utensils should be cleaned and sanitized at least once every four hours during continuous use.

○ **Minimize bare-hand contact with food that is cooked or ready-to-eat.** Use gloves or utensils to handle this type of food. *(See Exhibit 8e.)*

○ **Practice good personal hygiene.** Wash hands after using the restroom, after hands have come into contact with food, and after handling dirty equipment or utensils.

Servers

Food servers need to be just as careful as kitchen staff. If they aren't careful, they can contaminate food simply by handling the food-contact surface of a plate. Servers should use the following guidelines to serve food safely. *(See Exhibit 8f on page 8-6.)*

○ **Glassware and dishes should be handled properly.** The food-contact area of plates, bowls, glasses, or cups should not be touched. Dishes should be held by the bottom or the edge. Cups should be held by their handles, and glassware should be held by the middle, bottom, or stem.

○ **Glassware and dishes should not be stacked when serving.** The rim or surface of one can be contaminated by the one above it. Stacking china and glassware also can cause it to chip or break.

○ **Flatware and utensils should be held by their handles.** Store flatware so servers grasp handles not food-contact surfaces.

○ **Minimize bare-hand contact with cooked and ready-to-eat food.** Serve this food with tongs or gloves, for example.

○ **Serve milk from refrigerated bulk dispensers or in single-serve cartons.** Serve cream and half-and-half in single-serve containers or covered pitchers.

○ **Use ice scoops or tongs to get ice.** The ice scoop should be stored in a sanitary location—not in the ice bin. A glass should never be used to scoop ice, because it might chip or break in the ice. Servers should not scoop ice with their bare hands.

○ **Practice good personal hygiene.** Servers should be neat and clean, and their hair should be pulled back and covered. They should avoid touching their hair or face when serving food and should refrain from habits such as chewing fingernails or licking their fingers.

○ **Never use cloths meant for cleaning food spills for any other purpose.** When tables are cleaned between guest seatings, spills should be wiped up with a disposable, dry cloth. Then clean the table with a moist cloth that has been stored in a fresh detergent solution.

Cross-Contamination

Use gloves or utensils to handle cooked or ready-to-eat food.

Exhibit 8e **Handling Cooked and Ready-to-Eat Food** Minimize bare-hand contact with cooked and ready-to-eat food by using gloves or tongs to serve them.

Exhibit 8f   **Serving Food Safely**
The right and wrong way to handle food, glassware, dishes, and utensils.

Division of Labor

To prevent cross-contamination, it's a good idea to schedule staff so they are not assigned to do more than one job during a shift. Serving food, setting tables, and busing dirty dishes are separate tasks, with different responsibilities. Since this division of labor is difficult to manage in most establishments, it's important for servers and busers who do double duty to wash their hands often and handle food safely. After wiping tables or busing dirty dishes, servers must wash their hands before handling food or place settings.

Re-serving Food Safely

Servers and kitchen staff should also know the rules about re-serving food that previously has been served to a customer. In general, only unopened, prepackaged food can be re-served. Condiment packets, wrapped crackers or wrapped breadsticks, and other sealed food generally offer adequate protection from contamination.

Never re-serve plate garnishes, such as fruit or pickles, to another customer. Served but unused garnishes must be discarded. Never re-serve uncovered condiments. Don't combine leftovers with fresh food. Opened portions of salsa, mayonnaise, mustard, or butter, for example, should be thrown away.

Uneaten bread or rolls may not be re-served to other customers. Linens used to line bread baskets must be changed after each customer.

Cross-Contamination

Never re-serve plate garnishes, uncovered condiments, or uneaten bread.

Self-Service Areas

Customers at **food bars** often unknowingly serve themselves in ways that can put themselves and other customers in danger. A customer may eat from her plate or nibble from the food bar while moving through the line. Another might pick up carrot sticks, pickles, and olives with her fingers, or dip a finger into salad dressing to taste it. Another might return unwanted food items, use a soiled plate for a second helping, or put his head under the sneeze guard to reach items in the back of the display.

Buffets and food bars should be monitored closely by employees trained in food safety procedures. Assign a staff member to replenish food-bar items and to hand out fresh plates for return visits. Post signs with polite tips about food-bar etiquette. These practices will go a long way toward keeping self-service areas more sanitary.

Cross-Contamination

Assign a trained staff member to monitor food bars.

Here are more rules for food bars:

○ **Protect food on display with sneeze guards or food shields.** *(See Exhibit 8g.)* These must be between fourteen and forty-eight inches above the food, in a direct line between the food and the mouth or nose of a customer of average height.

○ **Identify all food items in the foodbar by labelling containers** (but don't put probe-type tags into the food itself). Write names of salad dressings on ladle handles. *(See Exhibit 8h.)*

○ **Maintain proper food temperatures.** Keep hot food hot—140°F (60°C) or higher, and cold food cold—41°F (5°C) or lower. Check food temperatures at least every two hours and record them in a log.

○ **Replenish food on a timely basis.** Prepare and replenish small amounts at a time so food is fresher and has less chance of being exposed to contamination. Never mix fresh food with food being replaced to prevent cross-contamination. Practice the FIFO method of product rotation.

○ **Keep raw meat, fish, and poultry separate from cooked and ready-to-eat food.** Customers can easily spill food when serving themselves. Use separate displays or food bars for raw and cooked food (for example, at a Mongolian barbecue) to reduce the chance of cross-contamination.

○ **Do not let customers use soiled plates or silverware for refills.** Encourage all customers to take a clean plate for return trips to the food bar. Customers can use glassware for refills as long as beverage-dispensing equipment doesn't come in contact with the rim or interior of the glass.

Cross-Contamination

Never mix fresh food with food being replaced.

Exhibit 8g   Protecting Food on Display
Sneeze guards must be between fourteen and forty-eight inches above the food, in a direct line between the food and the mouth or nose of a customer of average height.

OFF-SITE SERVICE

Establishments use many different ways to deliver food to people. No matter how food is prepared and delivered, operators must follow the same food safety rules as permanent establishments. Food must be protected from contamination and time-temperature abuse, and facilities and equipment used to prepare food must be clean and sanitary. Menu items must lend themselves to safe service. Food must be handled safely. **Off-site services**—including delivery, mobile/temporary kitchens, and vending machines—all present special challenges.

Exhibit 8h **Identifying Food Items**
Label all items on a food bar.

Delivery

Many establishments—such as schools, hospitals, caterers, and even restaurants—use a central commissary to prepare food. Food from the central kitchen is then delivered to remote locations for service. The greater the time and distance from the point of preparation to the point of consumption, the greater the risk food will be exposed to contamination or time-temperature abuse. Equipment used to transport food—both containers and vehicles—must be designed to maintain safe food temperatures and be easy to keep clean.

When transporting food from a central kitchen, the following safety procedures should be followed:

○ **Use rigid, insulated food containers capable of maintaining food temperatures above 140°F (60°C) or below 41°F (5°C).** Containers should be sectioned so food doesn't mix, leak, or spill. *(See Exhibit 8i.)* Containers must

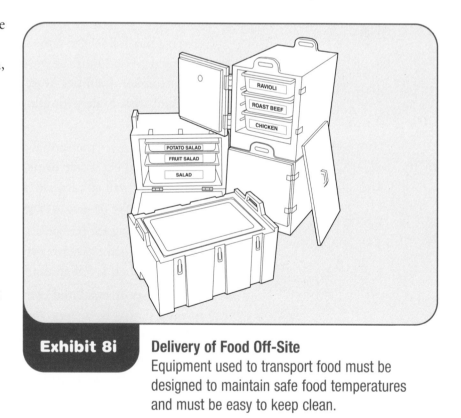

Exhibit 8i **Delivery of Food Off-Site**
Equipment used to transport food must be designed to maintain safe food temperatures and must be easy to keep clean.

When delivering food off-site, check temperatures regularly, and take corrective action if the food isn't at the proper temperature.

also allow air circulation to keep temperatures even. Containers should be kept clean and sanitized.

○ **Clean and sanitize the inside of delivery vehicles regularly.**

○ **Make sure employees practice good personal hygiene when distributing food.**

○ **Check internal food temperatures regularly.** Take corrective action if food isn't at the proper temperature. If containers or delivery vehicles aren't maintaining proper food temperatures at the end of each route, reevaluate the length of delivery routes or the efficiency of the equipment being used.

○ **Label food with storage, shelf life, and reheating instructions for employees at off-site locations.**

○ **Provide food safety guidelines for consumers.** If you or your employees won't be serving the food you deliver, provide customers with information on which items should be eaten immediately, which items may be saved for later, and how to serve all items.

Catering

Caterers provide food for private parties and events, as well as public and corporate functions. They might bring prepared food or cook food on-site in a mobile unit, a temporary unit, or in the customer's own kitchen.

Caterers must follow the same food safety rules as permanent establishments. Food must be protected from contamination and time-temperature abuse. Facilities must be clean and sanitary. Food must be prepared and served safely. Employees must follow good personal hygiene practices.

Caterers must meet the special challenges of off-site foodhandling. They must make sure there is safe drinking water for cooking, warewashing, and handwashing, as well as adequate power for holding and cooking equipment. Caterers must make proper arrangements for garbage disposal.

Outdoor catering for barbecues and cookouts may require special arrangements. When power or running water is not available, caterers may have to change their foodhandling procedures.

○ **Use ice chests or insulated containers for all potentially hazardous food.** Raw meat should be wrapped and stored on ice. Deliver milk and dairy products in a refrigerated vehicle or on ice.

○ **Serve cold food in containers on ice or in chilled, gel-filled containers.** If that is not possible, record the time when the potentially hazardous food was first taken out of cold storage and discard it after four hours. Note: *This method is not permitted in some jurisdictions.*

○ **Keep raw and ready-to-eat products separate during delivery and storage.** For example, store raw chicken separately from ready-to-eat salads.

○ **Use single-use items.** Make sure customers get a new set of disposable tableware for refills. Arrange for proper garbage disposal away from food-prep and serving areas.

○ **If leftovers are given to customers, provide instructions on how they should be handled.** This may include reheating and storage instructions for the products and shelf-life dates.

Mobile Units

Mobile units are portable facilities ranging from concession vans to elaborate field kitchens. Those serving only frozen novelties, candy, packaged snacks, and soft drinks have to meet basic sanitation requirements. Mobile kitchens preparing and serving potentially hazardous food, however, must follow the same rules required of permanent foodservice kitchens. Both might be required to apply for a special permit or license from the local regulatory agency.

Like permanent establishments, mobile kitchens have to be outfitted with adequate cooking equipment, mechanical cooling and hot-holding units, and warewashing and handwashing sinks with hot and cold potable water under pressure. Mobile kitchens must provide adequate ventilation, garbage storage and disposal facilities, and pest control. Operators should clean and maintain mobile units.

Key Point

Mobile units serving frozen novelties, candy, packaged snacks, and soft drinks have to meet basic sanitation requirements.

Temporary Units

Temporary units typically operate in one location for less than fourteen days. Foodservice tents or kiosks set up for food fairs, special celebrations, or sporting events may be temporary units. *(See Exhibit 8j.)* In some areas, the definition also extends to units set up for longer periods of time. Temporary units usually serve prepackaged food or food requiring limited preparation, such as hot dogs. It is best to keep the menu simple to limit the amount of on-site food preparation. Check with regulatory agencies for operating requirements.

Courtesy of the Illinois Restaurant Association

Exhibit 8j

Temporary Unit
It is best to keep the menu simple to limit the amount of on-site food preparation.

Temporary units should be constructed so dirt and pests are kept out. If floors are made of dirt or gravel, cover them with mats or platforms to control dust and mud. Construct walls and the ceiling with materials that will protect food from weather and windblown dust.

In addition, the same safe-handling rules previously discussed apply to food preparation in temporary units. If food is prepared on site, the unit must have adequate cooking, cold storage, and hot-holding equipment. If food is prepared off site, it must be transported and held at proper temperatures below 41°F (5°C) or above 140°F (60°C).

Safe drinking water must be available for cleaning, sanitizing, and handwashing. Since warewashing facilities will most likely be limited, it is best to use disposable, single-use items.

Vending Machines

Food prepared and packaged for **vending machines** has to be handled with the same care as any other food served to a customer. Vending operators also have to protect food from contamination and time-temperature abuse during transport, delivery, and service.

Keep food at the right temperature. Machines that contain potentially hazardous food must be able to maintain proper food temperatures of 41°F (5°C) or lower and 140°F (60°C) or higher. *(See Exhibit 8k.)* They must also have automatic shutoff controls that prevent food from being dispensed if the temperature stays in the danger zone for a specified amount of time.

Check product shelf life daily. Replace food with expired code dates. If refrigerated food isn't used within seven days after preparation, it must be discarded. Dispense potentially hazardous food, such as milk, in its original container. Fresh fruit with an edible peel should be washed and wrapped before being put into a machine.

Place machines in appropriate locations, away from garbage containers, sewage drains, and overhead pipes. Make sure the vending area is clean and well-lighted. Supply safe drinking water for beverage machines.

Clean and service vending machines regularly. Sanitize food-contact surfaces in machines each time food is replenished. Train employees to use good personal hygiene. Employees should wash their hands before and after servicing or refilling machines.

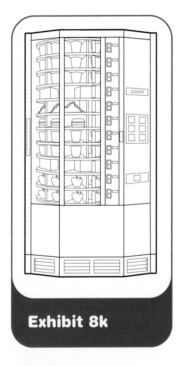

Exhibit 8k

Vending Machines
Replace food with expired code dates and discard food if it isn't used within seven days of preparation.

EIGHT RULES OF SAFE FOODHANDLING

All employees should know the Eight Rules of Safe Foodhandling and should be responsible for safe food practices in their assigned areas.

1. Practice strict personal hygiene.

2. Monitor time and temperature.

3. Prevent cross-contamination.

4. Clean and sanitize food-contact surfaces, equipment, and utensils before and after every use, and at least once every four hours during continuous use.

5. Cook food to its required minimum internal cooking temperature or higher.

6. Hold potentially hazardous, hot food at 140°F (60°C) or higher and cold food at 41°F (5°C) or lower.

7. Cool cooked food from 140°F (60°C) to 70°F (21°C) within two hours, and then from 70°F (21°C) to 41°F (5°C) or lower in an additional four hours, for a total cooling time of six hours.

8. Reheat potentially hazardous food for hot-holding to an internal temperature of 165°F (74°C) for fifteen seconds within two hours.

SUMMARY

Safe foodhandling doesn't stop once food is properly prepared and cooked. To make sure the food you serve is safe, you must continue to protect it from time-temperature abuse and contamination until it is eaten. When holding potentially hazardous food for service, keep hot food hot, above 140°F (60°C), and cold food cold, below 41°F (5°C). Use clean, sanitized utensils to serve food. Minimize bare-hand contact with cooked and ready-to-eat food.

Make sure all employees practice good personal hygiene. Train them to avoid cross-contamination when handling service items and tableware. Teach them about the potential hazards posed by re-serving plate garnishes, breads, or open dishes of condiments.

Customers can unknowingly contaminate food in self-service areas, such as food bars and buffets. Post signs to communicate self-service rules. Station employees in these areas to make sure customers follow those rules. Protect food in food bars and buffets with sneeze guards and make sure equipment can hold food at the proper temperature.

Take special precautions when preparing, delivering, or serving food off site. Catering, mobile kitchens, temporary units, and vending machines all pose unique challenges to food safety. Learn and follow all regulations in your jurisdiction. Be prepared with backup plans if off-site facilities aren't adequate or equipment breaks down. Finally, make sure all employees know and understand the Eight Rules of Safe Foodhandling.

A CASE IN POINT I

Case Study

Jill, a line cook on the morning shift at Memorial Hospital, was busy helping the kitchen staff put food on display for lunch in the hospital cafeteria. Ann, the kitchen manager who usually supervised lunch in the cafeteria, was at an all-day seminar on food safety. Jill was responsible for making sure meals were trayed and put into food carts for transport to the patients' rooms. The staff also packed two dozen meals each day for a neighborhood group that delivered them to homebound elderly people.

First, Jill looked for insulated food containers for the delivery meals. When she couldn't find them, she loaded the meals into cardboard boxes she found near the back door, knowing the driver would be there soon to pick them up. To help the cafeteria staff, Jill filled a soup baine with soup by dipping a two-quart measuring cup into the stockpot and pouring it into the baine. She carried the baine out to the cafeteria, put it into the steam table, and turned it on low.

The lunch hour was hectic. The cafeteria was busy, and the staff had many patient meals to tray and deliver. Halfway through lunch, a cashier came back to the kitchen to tell Jill the salad bar needed replenishing. Since she was busy, Jill asked a kitchen employee to take pans of prepared ingredients out of the refrigerator and put them on the salad bar. When she looked up a few moments later, she saw the kitchen employee send away two children who were eating carrot sticks from the salad bar. The employee then dumped the pans of lettuce and vegetables into the salad bar.

With lunch almost over, Jill breathed a sigh of relief. She moved down the cafeteria serving line, checking food temperatures. One of the casseroles was about 130°F (54°C). Jill checked the water level in the steam table and turned up the thermostat, then went to clean up the kitchen and finish her shift.

What did Jill do wrong? What should have been done?

A CASE IN POINT II

Case Study

Megan, a new server at The Fish House, reported for work ten minutes early on Thursday. Excited about her new job, she made sure to shower and wash her hair before going to work. When she arrived, she changed into a clean uniform, pulled her hair back tightly into a ponytail, and checked her appearance. She had used makeup sparingly and wore only small hoop earrings and a short chain necklace.

Her shift started off well. One of her customers ordered a menu item that Megan hadn't tried yet. When the order came up, Megan dipped her finger into the sauce at the edge of the plate for a taste. As the shift progressed, Megan's station got busier. When the hostess came to ask how soon one of Megan's tables could be cleared, Megan decided to do it herself. She took the dirty dishes to a bus station, then wiped down the table with a serving cloth she kept in her apron.

The buser, John, finished another task and came to help her set the table. While he put out silverware and linens, Megan filled water glasses with ice by scooping the glasses into the ice bin in the bus station. Only one slice of bread and one pat of butter were missing from the basket that had been on a previous customer's table, so she put that on a tray with the water glasses and took it to the table.

The table was reset in record time, and Megan soon had more guests. While taking their orders, Megan reached up to scratch a mosquito bite on her neck. Then she went to the kitchen to turn in the order and pick up a dessert order for another table.

What errors did Megan make? What should she have done?

TRAINING TIPS

1.   Group Activity: Food Contamination During Service

Purpose: *After completing this activity, trainees will be able to identify ways that food can be contaminated during service, as well as standard operating procedures (SOPs) that can minimize contamination. They will also determine corrective actions that should be taken for various modes of food service.*

Time: 40 minutes

Directions: Divide trainees into teams, and assign one team to each of the following topics about safe food handling as discussed in Chapter 8.

- Hot-holding
- Cold-holding
- Food bars, buffets
- Table service and re-service of food
- Catering
- Mobile units
- Temporary units
- Vending machines

Have the teams address the following questions about the assigned topics.

- How can food become contaminated at this stage or under these conditions?
- What SOPs must be in place to minimize food contamination?
- How do you verify that food safety procedures have been followed and standards have been met?
- What corrective actions and documentation must be performed, and who will perform them?

Allow about ten minutes for the teams to answer the questions, and allow time for each team to present its answers to the group. Then discuss the following questions with the entire group.

- Which team's assigned topics apply to your establishment?
- Does your establishment have these systems and procedures in place?
- If not, how might they be put in place?

2.   The Eight Rules of Safe Foodhandling: A Self-Assessment Test

Purpose: *After completing this activity, trainees will be able to assess foodhandling practices in their establishments, discover any weaknesses, and identify systems and procedures that could be put in place to improve foodhandling practices.*

Time: one hour

Directions: Ask trainees to honestly assess how well their establishments follow the Eight Rules of Safe Foodhandling listed at the end of Chapter 8.

Create a rating sheet with a list of the Eight Rules of Safe Foodhandling. Next to each rule, set up a rating scale from 1 to 4 with 1 being "perfect" and 4 being "poor." The sheet should look like this.

	Perfect	Good	Average	Poor
	1	2	3	4
Practice strict personal hygiene.	—	—	—	—

Allow ten or fifteen minutes for trainees to assess their establishment using the rating sheet. Then allow time for in-depth discussion of each rule. Ask any trainee indicating "average" or "poor" for one or more of the rules the following questions:

○ What food safety system and procedures do you currently have in place regarding the foodhandling rule?

○ What additional food safety systems and procedures could you put in place to improve foodhandling?

Solicit input from the group to help improve foodhandling practices at the establishment in question. Ask trainees scoring "good" or "perfect" on any rule if their establishment ever falls short of the rule, and if there is anything they could do to further improve current foodhandling practices.

Some trainees may be reluctant to report anything negative about their establishment, so you should ask for volunteers to answer some of these questions. The idea is to have a lively discussion about each of the rules—not to make anyone uncomfortable.

Remind trainees that while they may find their establishments to be safe in general, there is always room for improvement. Managers should take every opportunity to improve systems and procedures to minimize food contamination.

3. Front-of-the-House Food Safety Task Force

Purpose: *To involve service staff in an assessment of front-of-the-house foodhandling practices and to solicit recommendations for improving current practices.*

Time: several weeks

Directions: Develop a food safety task force made up of front-of-the-house personnel. Include at least one person from every front-of-the-house position in the task force. This should include servers, busers, cashiers, bartenders, hosts and hostesses, catering personnel, delivery drivers, shift supervisors, and managers. Plan to hold three or more meetings with your food safety task force.

First Meeting: Ask your food safety team to look for and document hazards and food safety violations in their particular areas of the establishment. Schedule another meeting to hear their reports.

Second Meeting: Allow time for each person to report on the hazards or safety violations found in his or her work area. Encourage and allow time for group feedback. The next assignment for the task force is to research possible solutions to problem areas. Suggestions for possible solutions can come from fellow employees and other establishments.

Third Meeting: Receive recommendations from members of the task force. Allow time for group feedback. Discuss how these recommended changes can be implemented. Remember to reward your task force for their work.

Follow-Up: Implement the solutions, evaluate, monitor, and modify where necessary.

4. Food Safety Training Materials

Purpose: *To assess current food safety training practices for front-of-the-house personnel in your establishment.*

Time: several weeks

Directions: Most front-of-the-house personnel have received some orientation and training materials, but these materials might lack up-to-date information about safe foodhandling procedures. Consider taking the following steps:

○ Assess all current training materials (videos, manuals, checklists, etc.) for various front-of-the-house personnel. Are the appropriate food safety issues addressed?

○ What additional safety issues and standards need to be addressed? How can these be integrated into existing materials?

○ How can you effectively communicate to training personnel the need for additional training in food safety issues and standards? How can you be certain this information is reinforced during training?

○ Are employees properly tested about food safety issues? Is there follow-up after the evaluation?

If you insist that employees follow safe-foodhandling standards from the very beginning of their employment, and consistently monitor employee performance, employees will be more likely to form safe-foodhandling habits.

DISCUSSION QUESTIONS

1. What are some ways a customer might contaminate food in a self-service area?

2. What are some potential threats to food safety that could occur when serving food? How might they be prevented?

3. Why should raw and ready-to-eat food be separated in self-service areas?

4. How often should you check the temperature of food being held for service?

MULTIPLE-CHOICE STUDY QUESTIONS

1. Which of the following is an unsafe serving practice?
   A. Stacking plates of food before serving them to the customer
   B. Holding flatware by the handles when setting a table
   C. Serving soup with a long-handled ladle
   D. Holding glassware by the stem

2. Which of the following statements about serving utensils is *not* true?
   A. They should be cleaned and sanitized at least once every four hours during continuous use.
   B. They can be used to handle more than one food item at a time.
   C. They can be stored in the food with the handle extended above the rim of the container.
   D. They must be cleaned and sanitized after each task.

3. Which item *cannot* be re-served?
   A. Bread sticks wrapped in cellophane
   B. Condiment packets
   C. Plate garnishes
   D. A package of crackers

4. Which of the following is an acceptable serving practice at a self-service bar?

A. Holding hot, potentially hazardous food at 120°F (49°C)

B. Storing raw meat next to ready-to-eat food

C. Allowing customers to use the same plate for a return trip to the self-service bar

D. Allowing customers to re-use glassware for beverage refills

5. When replenishing food on a self service bar,

A. mix fresh food with the remaining food being replaced.

B. replenish large amounts of food at one time.

C. practice the FIFO method of food rotation.

D. lower the temperature of the holding unit.

6. If potentially hazardous cold food has not been held at 41°F (5°C) or lower it should be discarded after

A. four hours.   C. two hours.

B. three hour.   D. one hour.

7. Several potentially hazardous hot food items are held on your luncheon cafeteria line. How can you help ensure that these items are safe to eat?

A. Check the internal temperature of the hot food with a thermometer every two hours.

B. Discard the hot food after five hours if it has not been held at or above 140°F (60°C).

C. Add fresh, hot food to the food already being held on the line to keep it hot.

D. Reheat food that has fallen below 140°F (60°C) by turning up the temperature of the hot-holding units.

8. Hot, potentially hazardous food should be held at an internal temperature of

A. 140°F (60°C) or higher.

B. 130°F (54°C) or higher.

C. 120°F (49°C) or higher.

D. 110°F (43°C) or higher.

For answers, please turn to Appendix A.

ADDITIONAL RESOURCES

Books and Periodicals

Equipped for food safety. 1998. *Best Practices.* 2 (4):6-10.

Fairbrook, P. 1994. The trouble with steam tables. *Food Management.* 29 (3):48.

Lachney, A. M. 1997. *The HACCP cookbook and manual,* 2nd ed. Washington, DC: Nutrition Development Systems.

The right way to…calibrate equipment. 1997. *Best Practices.* 1 (3):12.

The right way to…chill foods. 1998. *Best Practices.* 2 (2):12.

The right way to…prepare beef. 1998. *Best Practices.* 2 (3):12-13.

The right way to…prepare eggs. 1998. *Best Practices.* 2 (2):10-11.

The right way to…prepare fruits and vegetables. 1998. *Best Practices.* 2 (4):12-13.

The right way to…seafood safety. 1999. *Best Practices.* 3 (1):16.

Secrets of self-inspection. 1998. *Best Practices.* 2 (2):6-9.

Townsend, R. 1990. Proper holding patterns maintain food quality. *Restaurants & Institutions.* 100 (9):125.

Web Sites

Atkins Temptec

http://www.atkinstemptec.com

Atkins is a manufacturer of digital thermometers and probes for the restaurant and foodservice industry. This Web site provides information on how thermometers work, how to get the most out of your thermometer, and how to hold food safely during service. The site also provides an interesting demonstration on how food color is a poor indicator of internal temperature. Consult this site for useful tips related to thermometers.

Bunzl Distribution, Inc.

http://www.bunzldistribution.com

Bunzl Distribution is a supplier of disposable paper and plastic packaging supplies, as well as specialty items for service and packaging for the restaurant and foodservice industry.

Cooper Instrument Corporation

http://www.cooperinstrument.com

Cooper Instrument Corporation is a supplier of temperature, time, and humidity instruments for a variety of global markets. This Web site contains information on how to calibrate thermometers properly and maintain food temperatures during cooking, holding, cooling, etc.

Daydots International

http://www.daydots.com

Makers and distributors of many sanitation and food safety tools for restaurants and foodservice establishments most of which can be ordered online. Their site also contains articles and information related to food safety.

FDA-Center for Food Safety and Applied Nutrition (CFSAN)

http://www.cfsan.fda.gov

As the center within the FDA responsible for food safety and nutrition, CFSAN promotes and protects public health by researching and implementing guidelines, policies, and standards to ensure that food is safe, nutritious, wholesome, and properly labeled. This Web site provides a wealth of information on food safety and sanitation, including corresponding guidelines, policies, and standards.

FDA Food Code

http://vm.cfsan.fda.gov/~dms/foodcode.html

As the basis for many local sanitation codes, as well as the basis for the information in this textbook, the FDA Food Code, available at this Web address, is a useful resource for information relating to food safety for the restaurant and foodservice industry.

FoodHandler Inc.

http://www.foodhandler.com

FoodHandler Inc. is a manufacturer of foodservice safety equipment, such as gloves, aprons, and food-storage systems. The site contains information on products, as well as issues related to food safety.

National Automatic Merchandising Association (NAMA)

http://www.vending.org

NAMA is the national trade association of the merchandising vending and contract-foodservices management business. This Web site provides up-to-date information on all issues concerning this segment of the foodservice industry, as well as provides a resource for conferences and education for those working in this segment.

National Restaurant Association

http://www.restaurant.org

The National Restaurant Association is the leading business association for the restaurant industry. Together with the National Restaurant Association Educational Foundation, the Association's mission is to represent, educate, and promote the rapidly growing restaurant and foodservice industry. This Web site should be your starting place for all issues and concerns related to your restaurant. This Web site has it all, from tips for running your establishment to vital data on your customers' spending habits.

Chapter 9
Principles of a HACCP System

Table of Contents

TEST YOUR FOOD SAFETY KNOWLEDGE

1. **True or False:** A critical control point (CCP) is the last point before service where you can intervene to prevent, eliminate, or reduce the growth of microorganisms in food. *(See page 9-9.)*

2. **True or False:** Cooking chicken to a minimum internal temperature of 165°F (74°C) for fifteen seconds would be an appropriate critical limit if the cooking step is a CCP for chicken. *(See page 9-9.)*

3. **True or False:** The first step in developing a HACCP system is to identify all critical control points in the flow of food in your establishment. *(See page 9-6.)*

4. **True or False:** An establishment's HACCP plan must be reevaluated if there is a menu change. *(See page 9-12.)*

5. **True or False:** Receiving is a critical control point for chicken that will be served with pasta. *(See page 9-9.)*

For answers, please turn to Appendix A.

Learning Objectives

After completing this chapter, you should be able to

○ Identify the flow of food through the establishment.

○ Identify the importance of prerequisite programs for a Hazard Analysis Critical Control Point (HACCP) system.

○ Define HACCP.

○ Identify the HACCP principles for food safety.

○ Identify critical control points (CCPs) for various foods and processes.

Key Terms

Hazard Analysis Critical Control Point (HACCP)
HACCP plan
Prerequisite programs
Hazard analysis
Control point (CP)

Critical control point (CCP)
Critical limit
Monitoring
Corrective action
Verification

Previous chapters focused on the flow of food through a typical establishment. In these chapters, you learned how to receive, store, prepare, cook, cool, reheat, and serve food.

In this chapter, we'll discuss a system that will enable you to consistently serve safe food by identifying and controlling possible hazards throughout the flow of food. This system is called **Hazard Analysis Critical Control Point (HACCP).** A HACCP (pronounced has-sip) system is a dynamic process that uses a combination of proper foodhandling procedures, hazard and risk analysis, monitoring techniques, and record keeping to help ensure that the food you serve is safe. Because HACCP is dynamic, it allows you to continuously update and improve your food safety system as situations change.

The focus of this chapter is to provide you with a basic understanding of HACCP and its application to the restaurant and foodservice industry; however, if you want to develop and implement a HACCP plan, you will require additional information. A **HACCP plan** is a written document, based on HACCP principles, which describes the procedures a particular establishment will follow. Refer to the additional resources listed at the end of this chapter to help you develop your plan.

WHAT IS HACCP?

HACCP was developed by the Pillsbury Company in the early 1960s for the National Aeronautics and Space Administration (NASA). The system was designed to make the food served to astronauts as safe as possible. Since a foodborne illness on a space flight could be disastrous, the system had to ensure consistent food safety. The HACCP system is based on the idea that if significant biological, chemical, or physical hazards are identified at specific points within the flow of food, they can be prevented, eliminated, or reduced to safe levels.

HACCP was so beneficial that it was adopted by many segments of the food industry, from growers to manufacturers, and from distributors to operators. Over the years, the system has been refined and improved, and it is now accepted worldwide. The National Restaurant Association and the FDA recommend that all restaurants and foodservice establishments, no matter how large or small, develop and implement a HACCP system, which helps you to do the following:

○ Identify the food and procedures that are most likely to cause foodborne illness

○ Develop procedures that will reduce the risk of a foodborne-illness outbreak

○ Monitor procedures to keep food safe

○ Verify that the food you serve is consistently safe

Prerequisite Programs

To be effective, a HACCP program must have the commitment of management and employees, and it must be built on a solid foundation of prerequisite programs.

Prerequisite programs, also called standard operating procedures (SOPs), support your HACCP plan and are the basic operating conditions for producing safe food. They protect it from contamination, minimize microbial growth, and ensure the proper functioning of equipment. *(See Exhibit 9a).* They may include the following:

○ Proper personal hygiene practices

○ Proper facility-design practices

○ Supplier selection and specification programs

○ Proper cleaning and sanitation programs

○ Appropriate equipment-maintenance programs

Even the best HACCP plan can be undermined by ineffective prerequisite programs. For example, poor handwashing practices, improper cooling procedures, and faulty equipment that won't maintain proper temperatures can compromise food safety.

> **HACCP Principle**
>
> Prerequisite programs support your HACCP plan and are the basic operating conditions for producing safe food.

Proper personal hygiene practices

Proper facility-design practices

Supplier selection and specification programs

Proper cleaning and sanitation programs

Appropriate equipment-maintenance programs

Courtesy of Hobart Corporation

Exhibit 9a — **A HACCP System Must Be Built on a Solid Foundation of Prerequisite Programs**

DEVELOPING A HACCP PLAN

A HACCP plan is a written document describing the procedures an establishment will follow. It is developed using the HACCP principles and is specific to the facility and its menu, customers, equipment, processes, and operations.

The HACCP plan for a product prepared in one facility will be different from the HACCP plan for the same product prepared in another facility. For example, a pot of soup prepared in a deli may be purchased in cans and simply heated to the proper temperature, soup prepared in a restaurant might be made entirely from fresh ingredients on the premises, and soup prepared in a hospital may be made from a combination of *sous vide* and fresh ingredients. In each of these establishments, the soup will be prepared using a customized HACCP plan.

While generic HACCP plans can serve as useful guides, each facility must develop a HACCP plan that addresses its own unique conditions.

HACCP Principles

The plan you develop will be based on the seven basic HACCP principles outlined by the National Advisory Committee on Microbiological Criteria for Foods. Principles One, Two, and Three help you design your system. Principles Four and Five help you implement it. Principles Six and Seven help you maintain the system and verify its effectiveness. Each principle builds upon the information gained from the previous principle. For the plan to be complete, you must consider all seven principles in order. *(See Exhibit 9b.)*

HACCP Principle

A HACCP plan is developed using HACCP principles and is specific to the facility and its menu, equipment, processes, and operations.

HACCP Principle

Each HACCP principle builds upon the information gained from the previous principle.

Exhibit 9b: HACCP Principles

Principle One: Conduct a Hazard Analysis

Identify and assess potential hazards in the food you serve by taking a look at the flow of food in your establishment and determining where food safety hazards are likely to occur.

Principle Two: Determine Critical Control Points (CCPs)

Find the point in the flow of food that is essential to prevent, eliminate, or reduce a food safety hazard. If it is the last point before service where the hazard can be controlled, then it is a CCP.

Principle Three: Establish Critical Limits

For each CCP, establish minimum and maximum limits that must be met to prevent, eliminate, or reduce the identified hazard to an acceptable limit.

Principle Four: Establish Monitoring Procedures

Once critical limits have been established, determine ways to check them to make sure they are met. Determine who will monitor them and how often.

Principle Five: Identify Corrective Actions

Identify steps that must be taken when a critical limit has not been met. These steps should be determined in advance.

Principle Six: Verify that the System Works

Determine if the plan is working as intended. Plan to evaluate on a regular basis your monitoring charts, records, how you performed your hazard analysis, etc., and determine if your plan needs modification.

Principle Seven: Establish Procedures for Record Keeping and Documentation

Maintain records obtained while performing monitoring activities, whenever a corrective action is taken, when equipment is validated (checked to make sure it is in good working condition), and when working with suppliers. Also keep all documentation created while you were developing the plan.

Exhibit 9b

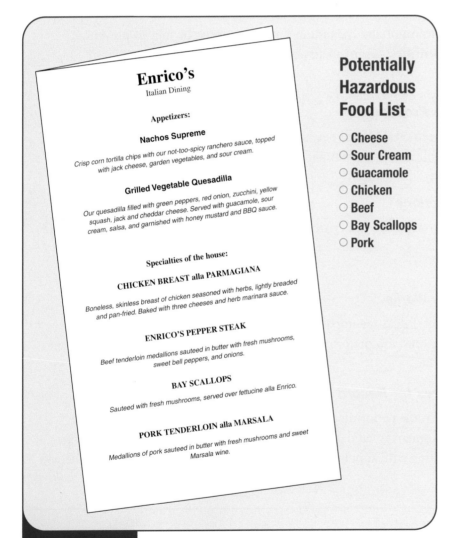

HACCP Principle

The hazard analysis is key to determining where hazards may occur in the flow of food if care is not taken to prevent or control them.

Principle One: Conduct a Hazard Analysis

Hazard analysis is the process of identifying and evaluating a potential hazard associated with a food, in order to decide what must be addressed in a HACCP plan.

The hazard analysis is key to determining where hazards may occur in the flow of food if care is not taken to prevent or control them. To perform a hazard analysis in your establishment, you need to consider the ingredients in your menu items, your equipment and processes, and your employees and customers. It is very important to look closely at each of these areas to make sure you identify all potential hazards. This will determine the success of your plan. Since most foodborne illnesses are caused by biological hazards, we will concentrate more on these. However, you also should be aware of any potential chemical or physical hazards as well.

Here are some key steps to follow when performing a hazard analysis.

○ **Identify potential food hazards.** Identify any food that may become contaminated if handled incorrectly at any step in its flow through the establishment, or that may allow the growth of harmful microorganisms. Because potentially hazardous food is known to support the rapid growth of these microorganisms, it is especially important to identify this food and determine how it is handled in your establishment. Make a list of these food items.

At Enrico's Italian Restaurant *(see Exhibit 9c),* their list of potentially hazardous food includes cheese, sour cream, guacamole, chicken, beef, scallops, and pork.

Enrico's
Italian Dining

Appetizers:

Nachos Supreme

Crisp corn tortilla chips with our not-too-spicy ranchero sauce, topped with jack cheese, garden vegetables, and sour cream.

Grilled Vegetable Quesadilla

Our quesadilla filled with green peppers, red onion, zucchini, yellow squash, jack and cheddar cheese. Served with guacamole, sour cream, salsa, and garnished with honey mustard and BBQ sauce.

Specialties of the house:

CHICKEN BREAST alla PARMAGIANA

Boneless, skinless breast of chicken seasoned with herbs, lightly breaded and pan-fried. Baked with three cheeses and herb marinara sauce.

ENRICO'S PEPPER STEAK

Beef tenderloin medallions sauteed in butter with fresh mushrooms, sweet bell peppers, and onions.

BAY SCALLOPS

Sauteed with fresh mushrooms, served over fettucine alla Enrico.

PORK TENDERLOIN alla MARSALA

Medallions of pork sauteed in butter with fresh mushrooms and sweet Marsala wine.

Potentially Hazardous Food List

○ **Cheese**
○ **Sour Cream**
○ **Guacamole**
○ **Chicken**
○ **Beef**
○ **Bay Scallops**
○ **Pork**

Exhibit 9c **Identifying Potentially Hazardous Food on a Menu**

○ **Determine where hazards can occur in the flow of food.** For each item on your list of potentially hazardous food, identify each step in its flow through the establishment. This might include receiving, storage, preparation, and cooking; holding and service; and cooling and reheating. For each step, identify any potentially adverse conditions the food might be exposed to, and write them down. You should also consider how your operation's equipment might affect the food.

At Enrico's, the chicken breast in *Chicken Breast alla Parmagiana* was identified as a potentially hazardous food. The steps in the chicken's flow through the restaurant include receiving, storage, preparation, cooking, and service. There are potentially adverse conditions at each step. During preparation, those conditions include thawing the chicken at room temperature (time-temperature abuse) and preparing the chicken on the same cutting board used to prepare other food (cross-contamination). *(See Exhibit 9d.)*

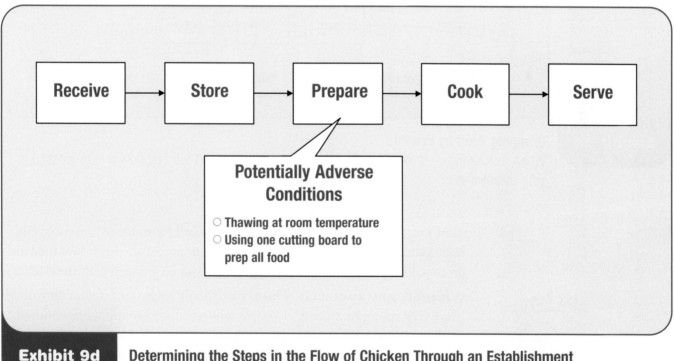

Exhibit 9d	**Determining the Steps in the Flow of Chicken Through an Establishment**
	The flow of chicken consists of five steps in this establishment. Potentially adverse conditions have been identified for the preparation step in this example, but must be identified for all steps in the food's flow through the establishment.

○ **Group food by how it is processed in your establishment.**
(See Exhibit 9e.) The most common groups include:

● Food prepared and served without cooking (salads, raw oysters, cheeses, and sandwich meats). Hazards involve contamination by employees, equipment, or other food.

● Food prepared and cooked for immediate service (hamburgers, scrambled eggs, and hot sandwiches). Hazards involve improper cooking, which may not eliminate the biological hazard.

● Food prepared, cooked, held, cooled, reheated, and served (chili, soups, and sauces). Hazards may occur at many points.

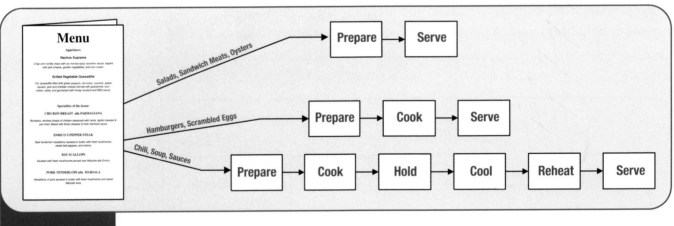

Exhibit 9e   **Grouping Food by Process**
When developing your HACCP plan, you may choose to group food by how it is processed in your establishment.

At Enrico's, the chicken is prepared and cooked for immediate service. The beef, scallops, and pork found on Enrico's list of potentially hazardous food are processed in the same manner and, therefore, can be grouped with the chicken.

○ **Identify your customers.** This is particularly important if your customers are very young or elderly, or people who are ill or immunocompromised. Hospitals, nursing homes, day-care centers, and schools need to be extra careful when preparing food because the groups they serve are more susceptible to foodborne pathogens.

Principle Two: Determine Critical Control Points

Once you've identified all potential food hazards, and the step or steps at which they occur in your establishment, the next task is determining where you can intervene to control them. A **control point (CP)** is any step in the

HACCP Principle

A control point (CP) is any step in the flow of food where a physical, chemical, or biological hazard can be controlled.

food's flow where a physical, chemical, or biological hazard can be controlled. While all control points are important in preventing a hazard from occurring, some are critical. To assess whether a control point is critical, you need to determine if it is the last step where you can intervene to prevent, eliminate, or reduce the growth of microorganisms before the food is served to customers. If it is, the step is called a **critical control point (CCP).**

The raw chicken breasts at Enrico's might have been received with the biological hazards *Salmonella* spp. or *Campylobacter* spp. While care is needed during preparation to prevent cross-contamination, proper cooking is *essential* to prevent, eliminate, or reduce these hazards. Therefore, preparation is a control point, while cooking is a critical control point for this menu item. *(See Exhibit 9f.)*

Typically, steps such as cooking, cooling, or holding food are identified as CCPs. However, these may not be the CCPs for all foods or all processes in your establishment. Each situation must be evaluated on an individual basis. Different establishments may prepare a similar food quite differently and will, therefore, identify different critical control points for that food.

Principle Three: Establish Critical Limits

Once you've identified the CCPs for potential hazards in the flow of food, you need to establish **critical limits**—minimum and maximum limits that the CCP must meet in order to prevent, eliminate, or reduce a hazard to an acceptable limit. For example, at Enrico's, cooking has been identified as a CCP for the chicken in the *Chicken Breast alla Parmagiana.* They've discovered that the critical limit is cooking it to a minimum internal temperature of 165°F (74°C) for fifteen seconds. *(See Exhibit 9g on next page.)* Enrico's ensures that the critical limit is met by placing the chicken in a convection oven set to 350°F (177°C) and cooking it for forty-five minutes.

> **HACCP Principle**
>
> A critical control point (CCP) is the last step where you can intervene to prevent, eliminate, or reduce the growth of microorganisms before the food is served to customers.

> **HACCP Principle**
>
> Cooking, cooling, or holding food are typically identified as CCPs. However, these may not be the CCPs for all foods or all processes in your establishment. Each situation must be evaluated on an individual basis.

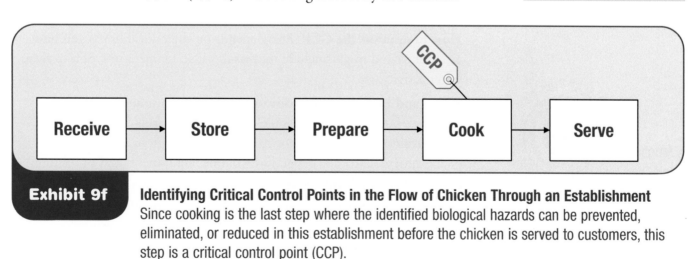

Exhibit 9f	**Identifying Critical Control Points in the Flow of Chicken Through an Establishment**

Since cooking is the last step where the identified biological hazards can be prevented, eliminated, or reduced in this establishment before the chicken is served to customers, this step is a critical control point (CCP).

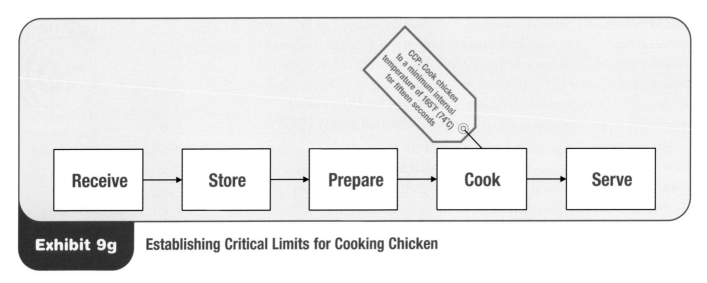

CCP: Cook chicken to a minimum internal temperature of 165°F (74°C) for fifteen seconds

Receive → Store → Prepare → Cook → Serve

Exhibit 9g   Establishing Critical Limits for Cooking Chicken

When establishing critical limits, keep in mind that they must be:

○ Measurable (such as a time or a temperature)

○ Based on scientific data, food regulations (such as the FDA Food Code), and expert advice

○ Appropriate for the food and equipment when prepared under normal conditions, and specific to your establishment

○ Clear and easy to follow

Principle Four: Establish Monitoring Procedures

Monitoring lets you know that critical limits are being met and that you are doing things right. By establishing procedures for monitoring your critical control points and consistently following them, you will be alerted to food safety problems needing correction in the flow of food.

To develop a successful monitoring program, you need to focus on each CCP and establish clear directions that specify the following:

○ **How to monitor the CCP.** This depends on the critical limits you have established and might include measuring time, temperature, pH, oxygen, water activity, etc.

○ **When and how often to monitor the CCP.** Continuous monitoring is preferred, but not always possible. Regular monitoring intervals should be determined based on the normal working conditions in your establishment, and they depend on volume, equipment, and procedures.

○ **Who will monitor the CCP.** Assign responsibility to a specific employee or position, and make sure that person is trained properly.

HACCP Principle

Critical limits are minimum and maximum limits the CCP must meet in order to prevent, eliminate, or reduce a hazard to an acceptable limit.

HACCP Principle

Monitoring lets you know critical limits are being met.

○ **Equipment, materials, or tools needed to monitor the CCP.** Provide employees with the proper tools to make the task as easy as possible.

At Enrico's, they have determined that cooking is a CCP for the chicken breasts, and that the critical limit is 165°F (74°C) for fifteen seconds. They will ensure that the critical limit is met by inserting a clean, sanitized, and calibrated thermometer into the thickest part of several chicken breasts. They will take at least two readings in different locations in the breast, and the result will be recorded in a temperature log. *(See Exhibit 9h.)*

Principle Five: Identify Corrective Actions

Because you have selected a particular point or step in the flow of food as critical to ensure food safety, specific and immediate actions must be established to correct errors as they occur. **Corrective actions** are predetermined steps taken when food doesn't meet a critical limit.

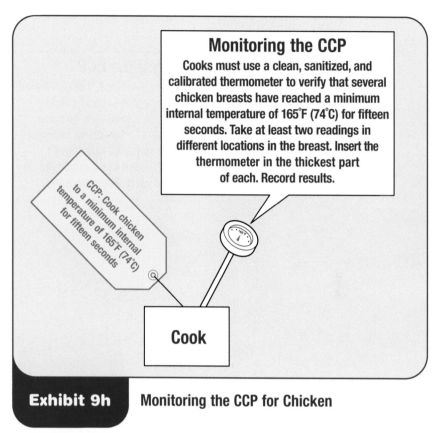

Monitoring the CCP

Cooks must use a clean, sanitized, and calibrated thermometer to verify that several chicken breasts have reached a minimum internal temperature of 165°F (74°C) for fifteen seconds. Take at least two readings in different locations in the breast. Insert the thermometer in the thickest part of each. Record results.

CCP: Cook chicken to a minimum internal temperature of 165°F (74°C) for fifteen seconds

Cook

Exhibit 9h   Monitoring the CCP for Chicken

Remember, this will be the last opportunity you have to ensure the safety of the food served. A corrective action might be as simple as continuing to cook the food to the required minimum internal temperature. Other actions might include throwing away food after a specified amount of time or rejecting a shipment that is not received at the temperature you specified. When developing corrective actions, be sure they are specific, stating exactly what should be done to correct the situation. Employees should also know who is responsible for taking the corrective action, how to perform the action, and where and how to record the action taken.

At Enrico's, if cooks check the internal temperature of a chicken breast and it hasn't reached the critical limit of 165°F (74°C) for fifteen seconds, they are instructed to keep cooking it until it does. This is the corrective action, which is recorded in the temperature log. *(See Exhibit 9i on the next page.)*

HACCP Principle

Corrective actions are predetermined steps taken when food doesn't meet a critical limit.

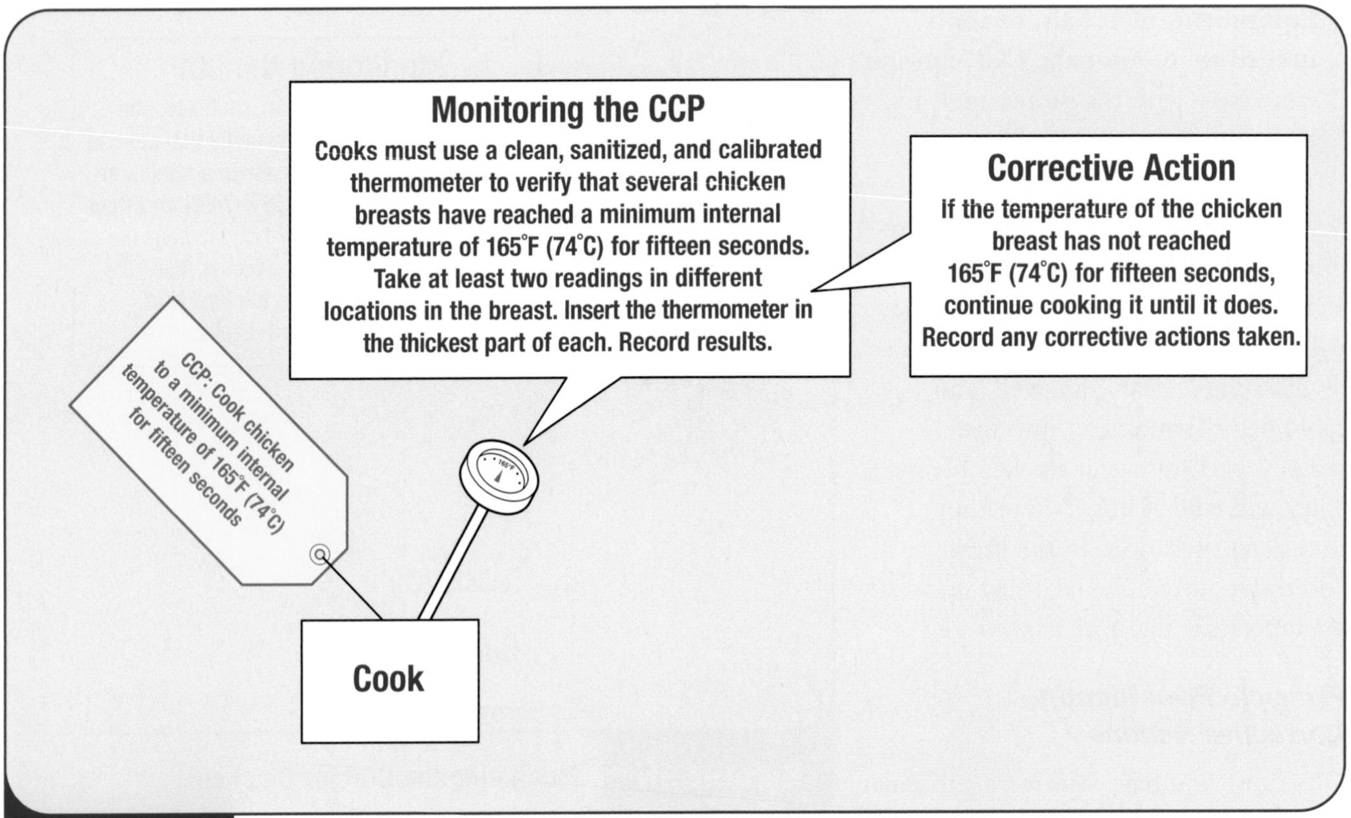

Monitoring the CCP
Cooks must use a clean, sanitized, and calibrated thermometer to verify that several chicken breasts have reached a minimum internal temperature of 165°F (74°C) for fifteen seconds. Take at least two readings in different locations in the breast. Insert the thermometer in the thickest part of each. Record results.

Corrective Action
If the temperature of the chicken breast has not reached 165°F (74°C) for fifteen seconds, continue cooking it until it does. Record any corrective actions taken.

CCP: Cook chicken to a minimum internal temperature of 165°F (74°C) for fifteen seconds

Cook

Exhibit 9i   Corrective Action for Cooking Chicken

Principle Six: Verify that the System Works

After you have developed your HACCP system, you need to confirm that it works according to the plan. This is called **verification.**

During this step verify that:

○ CCPs and critical limits selected are appropriate

○ Monitoring alerts you to hazards

○ Corrective actions are adequate to prevent foodborne illness

○ Employees are following established procedures

Verification of your plan should be performed on a regular basis, but if you notice that critical limits are frequently not being met, or if you receive a foodborne-illness complaint, you need to reevaluate your plan. You also need to reevaluate your plan if there are any changes in your menu, equipment, processes, suppliers, or product.

At Enrico's, the HACCP plan was reevaluated six months after implementation. The reevaluation revealed that chicken routinely failed to meet the set critical limit. Upon reevaluating the cooking process, it was

HACCP Principle

Verification confirms the system you developed works according to plan.

discovered that Enrico's vendor had started supplying a slightly larger chicken breast. This caused the chicken to be undercooked, given the equipment and established cooking parameters. Enrico's HACCP plan was revised by adjusting the cooking process to account for the larger chicken breast.

Principle Seven: Establish Procedures for Record Keeping and Documentation

HACCP Principle

Proper records allow you to document that you are continuously preparing and serving safe food.

Recording how food is handled as it flows through the establishment is important to the success of a HACCP system. Keeping proper records allows you to document that you are continuously preparing and serving safe food. Detailed records also serve as the basis for modifying your procedures if any food safety problems should occur.

Develop procedures specifying who should perform the documentation, as well as when and how it should be performed.

At Enrico's, records include time-temperature logs, procedures for checking temperatures, SOPs, calibration records, corrective actions, monitoring schedules, product specs, etc.

Training

Training is critical to the success of a HACCP plan. HACCP works best when it is integrated into each employee's job description and duties. Awareness of the HACCP plan depends on your role in the establishment.

To provide effective training, consider the following:

○ Help management understand the importance of food safety and the benefits of HACCP

○ Train staff (management and employees) to perform specific tasks required by the HACCP plan

A good training program should:

- Explain the importance of what the staff is learning
- Demonstrate steps and procedures
- Let employees practice
- Give feedback on employee performance
- Review materials
- Test employees on their knowledge
- Retrain employees, if needed
- Evaluate employees on job performance, as well as their food safety practices.
- Encourage employee involvement regarding food safety issues

SUMMARY

Hazard Analysis Critical Control Point (HACCP) is a food safety system designed to keep food safe throughout its flow in an establishment. HACCP is based on the idea that if biological, chemical, or physical hazards are identified at specific points within a food's flow through the operation, then the hazards can be prevented, eliminated, or reduced to safe levels. A successful HACCP system uses a combination of hazard and risk analysis, proper foodhandling procedures, monitoring techniques, and record keeping to keep food safe.

HACCP must be built on a solid foundation of prerequisite programs. These programs protect your food from contamination, minimize microbial growth, and ensure the proper functioning of equipment. They include programs for proper personal hygiene, cleaning and sanitation, and facility design, as well as equipment maintenance and the selection of suppliers.

While a generic HACCP plan can serve as a useful guide, each facility must develop a plan addressing its own unique conditions. The plan should be specific to the facility, its menu, customers, equipment, processes, and operations. An effective HACCP plan will be based on the following seven basic HACCP principles.

- **Principle One: Conduct a hazard analysis.** Identify and assess potential hazards in the foods you serve.

- **Principle Two: Determine critical control points (CCPs).** Find those points in the flow of food essential to preventing, eliminating, or reducing a food safety hazard. If this is the last point at which this hazard can be controlled before the food is served, then it is a CCP.

- **Principle Three: Establish critical limits.** For each CCP, establish minimum and maximum limits that must be met to prevent, eliminate, or reduce the hazard to an acceptable limit.

- **Principle Four: Establish monitoring procedures.** Once limits have been established, determine ways for checking them to make sure they are met. Decide who should check them and how often.

- **Principle Five: Identify corrective actions.** Determine what you will do if the critical limit is not met.

- **Principle Six: Verify that the system works.** Determine if the plan is working as intended.

- **Principle Seven: Establish procedures for record keeping and documentation.** Record how food is handled as it flows through the establishment. Documentation will include time-temperature logs, calibration records, corrective actions, etc.

Training is critical to making a HACCP plan successful. HACCP works best when it is integrated into each employee's job description and duties.

A CASE IN POINT

Case Study

Jason, the manager at Cal's Catering, received several calls from customers complaining they had contracted a foodborne illness after attending a picnic catered by Cal's the previous weekend. Concerned, Jason reviewed his menu for the event and noticed the only potentially hazardous food served was barbecued rib sandwiches. The ribs were received, stored, prepared, and cooked at the facility. Then the sandwiches were assembled, covered, and placed into insulated containers.

What steps in the barbecue sandwiches' flow are critical control points (CCPs)? What critical limits should have been established for each CCP? What records should be kept?

TRAINING TIPS

1. Control Point (CP) or Critical Control Point (CCP)?

Purpose: *After completing this activity, trainees should be able to differentiate between CPs and CCPs.*

Time: 20 minutes

Directions: Create a list of ten food items, along with a specific step or process related to the preparation and service of that particular item. Give a copy of the list to each trainee. Ask them to determine whether the step or process for each food in the list represents a CP or CCP. Here are some examples (with answers in bold):

Item	Step or Process
Fresh chicken	Receive chicken at 41°F (5°C) or lower. **(CP)**
Fresh ground beef	Discard ground beef that has been in the temperature danger zone for more than four hours. **(CP)**
Fresh pork	Cook pork to a minimum internal temperature of 145°F (63°C) for fifteen seconds. **(CCP)**
Iceberg lettuce	Wash lettuce prior to making salads. **(CP)**
Chili	Hold cooked chili for service at 140°F (60°C) or higher. **(CCP)**
Clam chowder	Cool cooked clam chowder from 140°F (60°C) to 70°F (21°C) within two hours and to 41°F (5°C) or lower within an additional four hours. **(CCP)**

As a group, discuss each food item. Discuss rationales for each trainee's position on CPs and CCPs. If possible, reach an agreement about each item.

2. The Seven Principles Pop Quiz

Purpose: *After completing this activity, trainees should be able to explain the seven HACCP principles and give an example of each.*

Time: 30 minutes

Directions: After discussing Chapter 9, give trainees a blank sheet of paper and ask them to write down, in order, the seven principles for setting up a HACCP system. They should explain what each principle means and give a specific example.

Allow fifteen minutes for this quiz. Then review the seven principles and provide trainees with a completed, one-page synopsis of the seven principles to use as a study aid. *(See page 9-5.)*

3. Developing HACCP Systems—A Breakout Exercise

Purpose: *After completing this activity, trainees should be able to assess hazards and identify CCPs for various menu items.*

Time: one hour

Directions: Make a list of four to eight different menu items. Identify the method of preparation for each item and the type of establishment that might serve it. Print handouts with one menu item per page. Examples may include:

Menu Item	Method of Preparation	Establishment
Chili	Prepare, cook, hold, cool, reheat	Full service
Potato Salad	Prepare, cook, cool, hold	Vending machine
Baked Chicken	Prepare, cook, hold, serve	Catered event
Cream Pie	Prepare, serve	Retail

When designing this list, include different types of food (meats, fish, dairy, vegetables, etc.). Include potentially hazardous food or items that contain them. Also, vary the methods of preparation (cook, cool, and reheat; prepare and serve; hot-hold), and assign these menu items to different types of establishments (quick service, full service, catering, vending, retail, and so on). When developing this list, keep your audience in mind. If possible, the list should reflect the types of food and the types of establishments that trainees represent.

Divide trainees into four to eight teams, with two to four members each. Assign a captain for each team. (Choose team captains who have some foodservice experience.) Assign each team a menu item. If the captain comes from a specific type of establishment, assign that team a menu item produced in a similar establishment.

Give each team a handout with the seven HACCP principles listed on it. Ask each team to develop a HACCP plan for their assigned menu item within the specified type of foodservice operation. It is not as important that each group develop a complete plan as it is that they properly assess the hazards associated with their menu item and identify the CCPs.

This exercise will last about an hour. Allow twenty minutes for the teams to develop the outline for the HACCP plan, and twenty minutes for the teams to present their plans to the group. You should use the remaining time for questions and comments.

4. Developing and Implementing a HACCP Plan

Purpose: *To provide some practical tips to help organize the development and implementation of a HACCP plan.*

Time: several months

Directions: Developing a HACCP plan can take a fair amount of time and effort, so it may be more effective if it is developed and implemented in several stages. Here are a few suggestions.

Stage 1. First, make sure your establishment has strong prerequisite programs in place. These include programs for personal hygiene, cleaning and sanitation, equipment maintenance, supplier selection, and so on. Strong prerequisite programs are the foundation of a strong food safety system, and are essential to the success of any HACCP plan.

Stage 2. Build a HACCP team. Ideally, the team should represent people from different job positions who handle some aspect of food production and service. For example, the team might include an employee responsible for ordering, receiving, and storing foods; a chef and a prep cook; a server; and a manager, among others. Team members should have food safety knowledge and strong foodservice experience. They should also be respected by their coworkers and committed to the program.

Stage 3. The members of the team should meet to discuss the food served in your establishment, your clientele, how food is handled, and the prerequisite programs currently in place.

Stage 4. After the initial discussion, the team will address the seven HACCP principles. This stage should probably be broken down into several meetings. Keep in mind that all seven principles must be performed in order, because each principle builds on the previous one. The seven principles can be combined into three groups to aid in development: Principles One, Two, and Three help you design your system; Principles Four and Five help you implement it; and Principles Six and Seven help you maintain it and verify its effectiveness. The team must begin by conducting a hazard analysis. A brainstorming session may help the team identify a list of potential hazards likely to occur in the establishment's flow of food (Principle One). Each team member brings to the table potential hazards specific to their area in the establishment. The team should review ingredients used in menu items, processes for preparing the items, equipment used to process it, and how the product will be stored and distributed.

After the list has been completed, the team must then determine which potential hazards will be addressed in the HACCP plan. The choice should be based on the severity of the hazard and the likelihood that it will occur in the establishment. These are the CCPs that must be identified by the team (Principle Two). Once the CCPs are identified, the team should brainstorm methods for controlling them (Principle Three). The controls will be critical to preventing, eliminating, or reducing the hazard to an acceptable limit. If necessary, the team should look to regulatory standards and guidelines, or to experts to help set the critical limits.

Stage 5. Next, the team must establish procedures for monitoring the CCPs (Principle Four), and corrective actions to take if the critical limit is not met. (Principle Five). Make sure these procedures are specific so the employee responsible for performing the monitoring or corrective action knows exactly what is expected. Training is the key to a successful implementation.

Stage 6. A HACCP plan should be implemented gradually over a period of time in order to be successful. Furthermore, you must frequently revisit and reassess your HACCP plan, particularly when changes in your establishment take place (Principle Six). Changes that might affect your plan could include employee turnover, a change in menu items and products, changes in equipment, changes in company standards, or new local or state laws.

Be sure your team establishes sound record-keeping procedures. Records will keep an accurate account of the who, what, when, where, and why for monitoring and corrective actions, and will also aid in verifying your plan (Principle Seven).

DISCUSSION QUESTIONS

1. What is HACCP?

2. What are the seven basic HACCP principles?

3. What are three types of hazards? What would be an example of each?

4. What is the difference between a control point and a critical control point?

5. What is a critical limit? Give an example of one.

MULTIPLE-CHOICE STUDY QUESTIONS

1. Checking the internal temperature of a roast with a bimetallic stemmed thermometer is an example of which HACCP principle?

    A. Verification   C. Record keeping
    B. Monitoring   D. Hazard analysis

2. You discard a pot of beef noodle soup that has been held at 120°F (49°C) for five hours. This is an example of:

    A. Monitoring   C. Hazard analysis
    B. Corrective action   D. Verification

3. Your restaurant plans to add linguini with meat sauce to the menu. The meat sauce will be cooked to 155°F (68°C) for fifteen seconds before being mixed with the pasta. Should receiving be a critical control point (CCP) for the meat?

    A. Yes, receiving is always a CCP for meat.
    B. No, meat is not potentially hazardous.
    C. No, because the meat will be cooked thoroughly.
    D. Yes, because the meat will be refrigerated prior to cooking.

4. All of the following are methods of monitoring a CCP *except*

   A. analyzing a food's flow through the establishment.
   B. observing the amount of time food remains in the temperature danger zone.
   C. checking a food's pH.
   D. taking the temperature of food during the cooking process.

5. Which of the following is *not* a corrective action?

   A. Continuing to cook a hamburger until it reaches a minimum internal temperature of 155°F (68°C) for 15 seconds
   B. Discarding cooked chicken that has been held at 120°F (49°C) for five hours
   C. Sanitizing a prep counter before starting a new task
   D. Rejecting a shipment of oysters received at 55°F (13°C)

6. Which statement best describes the purpose of verification in a HACCP plan?

   A. To determine if the CCPs and critical limits that have been chosen are appropriate
   B. To determine whether monitoring is alerting you to hazards
   C. To determine if corrective actions are adequate to prevent foodborne illness
   D. All of the above

7. Which of the following records would *not* be useful to your HACCP plan?

   A. Time and temperature logs
   B. Thermometer calibration records
   C. Corrective action logs
   D. Workplace accident records

8. Which of the following prerequisite programs should be included in your HACCP plan?

   A. A personal hygiene program
   B. An incentive program
   C. Workplace accident prevention program
   D. None of the above

9. Which of the following situations could sabotage even the best HACCP plan at an establishment?

A. Improper procedures for cooling food
B. Faulty equipment that won't maintain temperatures
C. Ineffective handwashing practices
D. All of the above

10. The purpose of a HACCP system is to

A. identify and control possible hazards throughout the flow of food.
B. identify the proper methods for receiving food.
C. keep the establishment pest free.
D. identify faulty equipment within the establishment.

11. Which of the following steps is likely to be a CCP for a steak cooked for immediate service?

A. Receiving
B. Storage
C. Preparation
D. Cooking

12. A chef took the temperature of a baine of minestrone soup being held in a hot-holding unit for service. The temperature of the soup was 120°F (49°C), which he recorded in a temperature log. The chef reheated the soup to 165°F (74°C) for fifteen seconds and placed it in a different holding unit. Which of the following actions was a corrective action?

A. Taking the temperature of the soup
B. Reheating the soup
C. Recording the temperature of the soup in the temperature log
D. Placing the soup in a different holding unit

13. Which of the following statements is an example of a critical limit?

A. Cook ground beef to 155°F (68°C) for fifteen seconds
B. Store ground beef at an internal temperature 41°F (5°C)
C. Reject ground beef if it is received at a temperature higher than 41°F (5°C)
D. All of the above

14. Your deli serves cold sandwiches in a grab-and-go display. Which step in the preparation of these sandwiches may be a CCP?

   A. Storage   C. Reheating
   B. Cooking   D. Cooling

15. Noting food temperatures in a temperature log is an example of which HACCP principle?

   A. Verification   C. Record keeping
   B. Monitoring   D. Hazard analysis

For answers, please turn to Appendix A.

ADDITIONAL RESOURCES

Books and Periodicals

Bertagnoli, L. A clean start. 1996. *Restaurants & Institutions.* 106 (22):90.

Bolton, L. 1997. Food safety I.Q.: Don't wait for the health inspector! *Restaurant Hospitality.* 81 (2):85.

Cichy, R. F. 1993. *Sanitation management.* East Lansing, MI: Educational Institute of the American Hotel & Motel Association.

Durocher, J. 1996. A clean sweep. *Restaurant Business.* 95 (4):202.

Durocher, J. 1997. Germ warfare. *Restaurant Business.* 96 (22):123.

Equipped for food safety. 1998. *Best Practices.* 2 (4):6-10.

Frable, F., Jr. 1997. 10 common errors in kitchen planning and ways to avoid them. *Nation's Restaurant News.* 31 (6):22.

Hertneky, P.B. 1996. You and your health inspector. *Restaurant Hospitality.* 80 (6):57.

ICMSF 1988. *Microorganisms in foods 4: Application of the Hazard Analysis Critical Control Point (HACCP) system to ensure microbiological safety and quality.* Oxford: Blackwell Scientific Publications.

Kratt, D. T. 1997. Running a clean operation. *Nightclub & Bar.* 13(11):79.

Marriott, N. G. 1994. *Principles of food sanitation.* New York: Chapman & Hall.

National Restaurant Association Educational Foundation. 1998. *A Practical Approach to HACCP: Coursebook.* Chicago: National Restaurant Association Educational Foundation.

National Restaurant Association. 1998. *Sanitation survival kit.* Washington, DC: National Restaurant Association Educational Foundation.

The right way to...organize workspaces. 1998. *Best Practices.* 2(3):15-16.

Sanson, M. 1996. A blueprint for safety: Restaurant kitchen design. *Restaurant Hospitality.* 80(8):65.

Secrets of self-inspection. 1998. *Best Practices.* 2(2):6-9.

Stevenson, K. E. and D. T. Bernard, eds. 1995. *Establishing Hazard Analysis Critical Control Point Programs: A workshop manual.* Washington, DC: The Food Processors Institute.

Vin Quetua, V. 1997. Scratching the surface. *Restaurants & Institutions.* 107(20):196.

Walczak, D. 1997. The sanitation imperative: Keep people from getting sick in your restaurant. *Cornell Quarterly.* 38(2):68.

Web Sites

FDA Bad Bug Book

http://vm.cfsan.fda.gov/~mow/intro.html

Produced by the FDA's Center for Food Safety and Nutrition (FDA CFSAN), this online handbook provides basic facts regarding pathogenic foodborne microorganisms and natural toxins. It brings together information from the FDA, the CDC, the USDA Food Safety Inspection Service, and the National Institutes of Health.

FDA-Center for Food Safety and Applied Nutrition (CFSAN)

http://www.cfsan.fda.gov

As the center within the FDA responsible for food safety and nutrition, CFSAN promotes and protects public health by researching and implementing guidelines, policies, and standards to ensure that food is safe, nutritious, wholesome, and properly labeled. This Web site provides a wealth of information on food safety and sanitation, including corresponding guidelines, policies, and standards.

FDA Food Code

http://vm.cfsan.fda.gov/~dms/foodcode.html

As the basis for many local sanitation codes, as well as the information in this textbook, the FDA Food Code, available at this Web address, is a useful resource for all food safety-related information for the restaurant and foodservice industry.

FDA Seafood Information and Resources

http://vm.cfsan.fda.gov/seafood1.html

The FDA operates an oversight compliance program for fishery products under which responsibility for the products' safety, wholesomeness, identity, and economic integrity rests with the processor or importer, who must comply with regulations under the Federal Food, Drug and Cosmetic (FD&C) Act. This Web site houses information on the seafood program, foodborne pathogens and contaminants associated with seafood, and HACCP compliance.

International HACCP Alliance

http://haccpalliance.org

The International HACCP Alliance was formed to assist the meat and poultry industry in preparing for mandatory HACCP. The Alliance promotes public health and safety by facilitating uniform development and implementation of HACCP. This Web site houses documents relating to food safety, HACCP, and its implementation.

National Advisory Committee for the Microbiological Criteria for Foods (HACCP Principles and Application Guidelines)

http://www-seafood.ucdavis.edu/Guidelines/nacmcf.htm

The guidelines found on this Web page are intended to facilitate the development and implementation of effective HACCP plans as appropriate to each segment of the food industry.

USDA-FDA HACCP Training Programs

http://www.nal.usda.gov/fnic/foodborne/haccp/index.shtml

The database is a joint project of the Food and Drug Administration, Food Safety and Inspection Service, Cooperative State Research, Education and Extension Service, and the National Agricultural Library. This database provides up-to-date listings of HACCP training programs and HACCP resource materials.

USDA-FSIS

http://www.fsis.usda.gov

FSIS is the public health agency of the U.S. Department of Agriculture responsible for ensuring that the nation's commercial supply of meat, poultry, and egg products is safe, wholesome, and correctly labeled and packaged. This Web site contains a wealth of information on food safety relating to meat, poultry, and eggs.

UNIT 3

CLEAN AND SANITARY FACILITIES AND EQUIPMENT

Chapter 10
Sanitary Facilities and Equipment

Knowledge

TEST YOUR FOOD SAFETY KNOWLEDGE

1. **True or False:** A hose attached to a utility sink faucet and left sitting in a bucket of dirty water could contaminate the water supply. *(See page 10-19.)*

2. **True or False:** It is acceptable to use nonpotable water for scrubbing pots and pans. *(See page 10-16.)*

3. **True or False:** The food-contact surfaces of equipment can be made from copper or galvanized steel. *(See page 10-9.)*

4. **True or False:** Any clean sink may be used for scrubbing baking potatoes. *(See page 10-8.)*

5. **True or False:** Grease on an establishment's ceiling can be a sign of inadequate ventilation. *(See page 10-21.)*

For answers, please turn to Appendix A.

Table of Contents

Learning Objectives

After completing this chapter, you should be able to

○ Respond to an interruption in the internal water supply.

○ Respond to wastewater overflows.

○ Identify methods for preventing backflow.

○ Identify potable water sources.

○ Identify uses of nonpotable water.

○ Handle waste properly.

○ Maintain restrooms properly.

○ Identify the requirements of a handwashing station.

○ Position equipment and facilities to make sanitation easier.

Key Terms

Americans with Disabilities Act (ADA)
Porosity
Resiliency
Coving
Service sink
NSF International

Underwriters Laboratories (UL)
Blast chiller
Tumble chiller
Cantilever-mounted equipment
Potable water

Booster heater
Cross-connection
Backflow
Flood rim
Vacuum breaker
Air gap
Foot-candle
Pulper

Many breakdowns in sanitation are caused by facilities and equipment that are simply too difficult to keep clean. Sanitary facilities and equipment are basic parts of a well-designed HACCP program. In this chapter, you will find a wide range of information on various equipment and facility-related issues key to keeping an establishment safe.

DESIGNING A SANITARY ESTABLISHMENT

When designing or remodeling a facility, consider how the building and the equipment in each area will be kept clean. Those areas not cleaned properly can allow microorganisms to remain, which can cause serious problems for the food coming in contact with them. Facilities should be arranged so contact with contaminated sources—such as garbage or dirty tableware, utensils, and equipment—is unlikely to occur.

This chapter focuses on four topics related to the sanitary layout and design of equipment and facilities:

○ Arrangement and design of equipment and fixtures to comply with sanitation standards

○ Material selection for walls, floors, and ceilings that will make cleaning these surfaces easier

○ Design of utilities to prevent contamination and to make cleaning easier

○ Proper solid waste management to avoid contaminating food and attracting pests

The Plan Review

A sanitary foodservice layout and design begins in the planning stage. Prior to starting construction, consult local regulations. Many jurisdictions require approval of layout and design plans by the health department or local regulatory agency prior to new construction or extensive remodeling. These plans should include proposed layout, mechanical plans (and specifications for utilities), type of construction materials to be used, and the types or models of proposed equipment. Specifications for utilities, plumbing, and ventilation will probably be required.

Some local jurisdictions will require the approval of design plans by building and zoning departments. In addition, the **Americans with Disabilities Act (ADA)** requires reasonable accommodation for access to the building by both patrons and employees with disabilities. These guidelines can be found in the ADA accessibility guide (ADAAG).

Even if local laws do not require it, layout and design plans should be reviewed by local or state regulatory agencies. In addition to ensuring compliance with sanitation requirements, such reviews can save time and money. This is true for remodeling, as well as for new construction.

To assist establishments, regulatory agencies might provide information they consider necessary for good sanitation. Ask for guides on how to submit plans and specifications.

Once construction is completed, the establishment applies for a permit to operate. Before granting the permit, the regulatory agency might conduct an inspection prior to opening to make certain all design and installation requirements have been met. After the establishment has passed the inspection and obtains a certificate of operation, it can open for business.

Materials for Interior Construction

Materials used during construction must be selected with several factors in mind. Sound-absorbent surfaces that also resist absorption of grease and moisture and reflect light will probably create an environment acceptable to your regulatory agency.

However, the most important consideration when selecting construction materials is how easy the establishment will be to clean and maintain.

Flooring

Flooring materials in kitchen and service areas should meet requirements for health and safety, strength and durability, and appearance. The floor surfaces should be easy to clean, wear-resistant, slip-resistant, and nonporous. Flooring should be kept in good repair and replaced if damaged or worn.

One of the most important factors to consider when selecting floor covering for an establishment is the **porosity** of the material. Porosity is the extent to which a floor covering can become saturated by liquids. When liquids are absorbed, the flooring can be damaged and microorganisms can grow. It might also cause a potential slip-and-fall situation. The FDA Food Code recommends the use of nonabsorbent flooring in food-preparation areas, walk-in refrigerators, warewashing areas, restrooms, and other areas subject to moisture, flushing, or spray cleaning.

Nonporous, Resilient Flooring

In most areas of the establishment, nonporous, resilient flooring is the best choice. **Resiliency** means a material has the ability to react to a shock without breaking or cracking.

Nonporous, resilient materials are relatively inexpensive and are easy to clean and maintain. They are rated for light, moderate, and heavy traffic, as well as for resistance to grease and alkalis. Some are easily damaged by cigarette burns

Key Point

Even if local laws do not require it, layout and design plans should be reviewed by the local regulatory agency.

Key Point

Select nonabsorbent flooring materials for food-preparation areas, walk-in refrigerators, warewashing areas, and restrooms.

or sharp objects, but they are also easy to repair or replace. They tend to be slippery when wet. Rubber tile and light- and medium-weight vinyl are good considerations. Vinyl tile is not recommended in dining rooms or public areas because it requires a high level of maintenance, such as waxing and frequent machine buffing. Vinyl tile is a practical choice in employee dressing rooms and break rooms, and foodservice offices. See *Exhibit 10a* for characteristics and recommended uses of nonporous, resilient flooring.

Hard-Surface Flooring

Hard-surface flooring is also commonly used in establishments. It includes quarry tile, ceramic tile, brick, terrazzo, marble, and hardwood. These materials are nonporous, but are not resilient.

Hard-surface floors are very durable, but may crack or chip if heavy objects are dropped on them. In addition, breakable objects dropped on hard-surface flooring will be more likely to shatter than those dropped on resilient flooring. These surfaces do not absorb sound and are somewhat difficult to clean compared to resilient surfaces. Some types, such as marble, are slippery. Quarry and ceramic tile are excellent for use in public restrooms or high-soil areas, but unglazed tiles should be selected for these areas due to their slip-resistant qualities. In general, hard-surface flooring is very heavy and more expensive to install and maintain than resilient flooring. See *Exhibit 10b* for characteristics and recommended uses of hard-surface flooring.

Carpeting

Carpeting is popular in dining rooms, because it absorbs sound. However, it is not recommended in high-soil areas, such as waitstaff service areas, tray

Exhibit 10a: Characteristics of Resilient Flooring

Material	Where to Use	Durability	Advantages	Disadvantages
Rubber Tile	Kitchens; restrooms	Less durable and less resistant to grease and alkalis	Nonslip; resilient	Use in moderate traffic areas
Vinyl Sheet	Offices; kitchens; corridors	Less resistant to grease and alkalis	Very resilient	Use only in light or moderate traffic areas
Vinyl Tile	Offices; employee restrooms	Wears out quickly with high traffic	Very resilient	Requires waxing and machine buffing

Exhibit 10a

Exhibit 10b: Hard-Surface Flooring				
Material	**Where to Use**	**Durability**	**Advantages**	**Disadvantages**
Marble; Terrazzo	Public corridors; dining rooms; public restrooms	Wear resistant	Nonporous; good appearance	Nonresilient; expensive; requires special care, such as buffing and polishing; heavy and difficult to install
Quarry Tile	Kitchen; dishwashing areas; receiving areas; offices; restrooms; dining rooms; service areas	Wear resistant	Nonporous	Nonresilient; heavy; relatively expensive; slippery when wet unless an abrasive is added
Wood	Offices; dining rooms	Durable in lower-traffic areas	Good appearance and sound absorption	Requires frequent polishing and periodic refinishing to maintain surface qualities
Acrylic Wood (plastic absorbed into wood)	Offices; dining rooms	Highly abrasion resistant	Less vulnerable to stains, scratches, chemical damage; high resistance to bacterial growth	Much like resilient flooring

Exhibit 10b

and dish drop-off areas, beverage stations, and major traffic aisles. Carpet can be maintained by simple vacuum cleaning. Areas prone to heavy traffic and moisture will require routine cleaning. Where sanitation, soiling, moisture, and fire safety are concerns, special carpet can be purchased.

Special Flooring Needs

Each area of an establishment has its own particular flooring needs. Nonslip surfaces should be used in traffic areas. In fact, nonslip surfaces are best for the entire kitchen, since slips and falls are a potential hazard. Rubber mats are allowed for safety reasons in areas where standing water may occur, such as the dish room. Rubber mats should be picked up and cleaned separately when scrubbing floors.

Coving is required in establishments using resilient or hard-surface flooring materials. **Coving** is a curved, sealed edge placed between the floor and the wall to eliminate sharp corners or gaps that would be impossible to clean. *(See Exhibit 10c.)* The coving tile or strip must adhere tightly to the wall to eliminate hiding places for insects and to prevent moisture from deteriorating the wall.

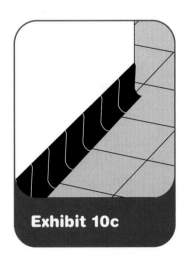

Exhibit 10c

Coving
Coving is a curved, sealed edge placed between the floor and the wall to eliminate sharp corners or gaps.

Finishes for Interior Walls and Ceilings

Interior finishes are the materials used on the surface of walls, partitions, or ceilings of an establishment. As with flooring, the most important criteria when choosing interior finishes are ease of cleaning and porosity.

When selecting finishes for walls and ceilings, consider location. One type of material might be suitable in one area, but may be a poor choice for another. Walls and ceilings in food-preparation areas must be light in color to distribute light and to make it easier to spot soil when cleaning. They should be kept in good repair, free of cracks, holes, or peeling paint. The best wall finish in cooking areas is ceramic tile; however, it must be monitored for grout loss and regrouted whenever necessary. Stainless steel is used occasionally because of its durability and resistance to moisture. The most common ceiling materials are acoustic tile, painted drywall, painted plaster, or exposed concrete.

The support structures for walls and ceilings (studs, joists, and rafters) and pipes should not be exposed unless they are finished and sealed for cleaning. Flexible materials such as paper, vinyl, and thin wood veneers are often used for walls and ceilings. Vinyl wall coverings are used in many areas of an establishment, because they are attractive, relatively inexpensive, easy to clean, and durable. Vinyl wall coverings are rated for flammability by testing agencies. Plaster or cinder-block walls sealed and painted with soil-resistant and easy-to-wash glossy paints are appropriate for dry areas of the facility.

CONSIDERATIONS FOR OTHER AREAS OF THE FACILITY

Dry Storage

Dry storerooms should be constructed of easy-to-clean materials that allow good air circulation. *(See Exhibit 10d.)* Shelving, table tops, and bins for dry ingredients should be made of corrosion-resistant metal or food-grade plastic.

Any windows in the storeroom should have frosted glass or shades. Direct sunlight can increase the temperature of the room and affect food quality.

Steam pipes, water lines, and other conduits have no place in a

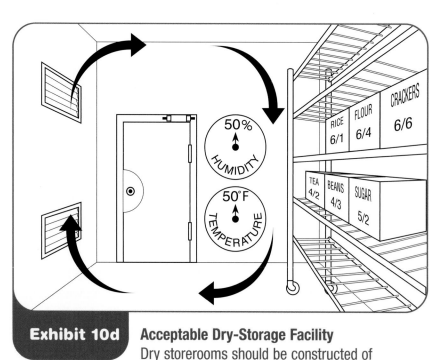

Exhibit 10d   Acceptable Dry-Storage Facility
Dry storerooms should be constructed of easy-to-clean materials that allow good air circulation.

well-designed storeroom. Dripping condensation or leaks in overhead pipes can promote microbial growth in such normally stable items as crackers, flour, and baking powder. Leaking overhead sewer lines can be a source of contamination for any food. Hot water heaters or steam pipes can increase the temperature of the storeroom to levels that will allow foodborne pathogens to grow.

Dry food is especially susceptible to attack by insects and rodents. Cracks and crevices in floors or walls should be filled. Doors leading to the exterior of the building should be self-closing. Screens for windows and doors should be sixteen mesh to the inch, without holes or tears.

Restrooms and Handwashing Stations

Local building and health codes usually specify how many sinks, stalls, toilets, and urinals are required in an establishment. It is best if separate restrooms are provided for employees and customers. If this is not possible, the establishment must be designed so patrons do not pass through food-preparation areas to reach the restroom, since they could contaminate food or food-contact surfaces.

Restrooms should be convenient and sanitary and have a fully equipped handwashing station, as well as self-closing doors. They must be adequately stocked with toilet paper, and trash receptacles must be provided if disposable paper towels are used. Covered waste containers must be provided in women's restrooms for the disposal of sanitary supplies.

Handwashing Stations

Handwashing stations must be conveniently located so employees will be encouraged to wash their hands often. They are required in food-preparation areas, service areas, warewashing areas, and restrooms. These stations must be operable and must be stocked and maintained. A handwashing station must be equipped with the following items: (See Exhibit 10e.)

Key Point

Handwashing stations must be conveniently located in food-preparation areas, service areas, warewashing areas, and restrooms.

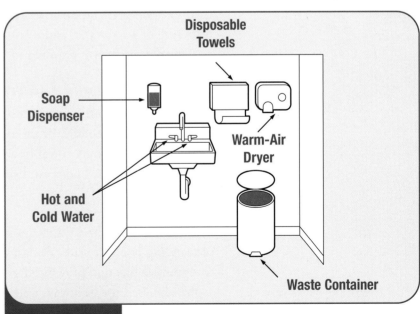

| Exhibit 10e | Acceptable Handwashing Station |

A handwashing station must be equipped with hot and cold running water, soap, a means to dry hands, and a waste container (if disposable towels are used).

- **Hot and cold running water.** Hot and cold water should be supplied through a mixing valve or combination faucet at a temperature of at least 100°F (38°C).
- **Soap.** The soap can be liquid, bar, or powder. Liquid soap is generally preferred, and some local codes require it.
- **A means to dry hands.** Most local codes require establishments to supply disposable paper towels in handwashing stations. Continuous-cloth towel systems, if allowed, should be used only if the unit is working properly and the towel rolls are checked and changed regularly. Installing at least one warm-air dryer in a handwashing station may provide an alternate method for drying hands if paper towels run out. The use of common cloth towels is not permitted, because they can transmit contaminants from one person's hands to another.
- **A waste container.** Waste containers are required if disposable paper towels are provided.

Handwashing stations must also be equipped with signage indicating employees are required to wash hands before returning to work.

Sinks

Each sink in an establishment must be used for its intended purpose only. Handwashing sinks are used for handwashing. Food-preparation sinks are used for food preparation. **Service sinks** are used for cleaning mops and disposing of wastewater, and must be kept separate. At least one service sink or curbed drain area is required in an establishment to dispose of soiled water.

Dressing Rooms and Lockers

Dressing rooms are not required. If available, they must not be used for food preparation, storage, or utensil washing. Lockers, if available, should be located in a separate room or a room where contamination of food, equipment, utensils, linens, and single-service items will not occur.

Premises

The parking lot and walkways should be kept free of litter and graded so standing pools of water do not form. In addition, they must be surfaced to minimize dirt and blowing dust. It is recommended that concrete and asphalt be used for walkways and parking lots. Gravel, while acceptable, is not recommended.

Patron traffic through the food-preparation area is prohibited, although guided tours are allowed. The premises may not be used for living or sleeping quarters.

SANITATION STANDARDS FOR EQUIPMENT

The task of choosing equipment designed for sanitation has been simplified by organizations such as **NSF International** and **Underwriters Laboratories (UL).** NSF International develops and publishes standards for sanitary equipment design. They also assess and certify that equipment has met these standards. The presence of the NSF mark on foodservice equipment means it has been evaluated, tested, and certified by NSF International as meeting international commercial food equipment standards. UL similarly provides sanitation classification listings for equipment found in compliance with NSF International standards. UL also lists products complying with their own published environmental and public health standards. Restaurant and foodservice managers should look for the NSF International mark or the UL EPH product mark on commercial foodservice equipment. *Exhibit 10f* shows examples of the NSF International mark and the UL EPH product marks.

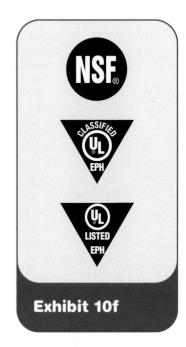

Exhibit 10f

NSF and UL EPH Marks
Look for the NSF International mark and UL EPH product marks on sanitary equipment.

Reprinted by permission of NSF International, Ann Arbor, MI, and Underwriters Laboratories, Inc., Northbrook, IL

NSF International Standards

The restaurant and foodservice manager should know the general features on which NSF International bases its standards:

○ **Equipment must be easy to clean.** All food-contact surfaces must be easy to access in order to ensure thorough cleaning.

○ **All food-contact surfaces must be smooth, nontoxic, nonabsorbent, corrosion resistant, and stable, and must not cause changes in the color, odor, or taste of food.** Food-contact surfaces should not react in any way with food products or cleaning compounds. Examples of materials that may not be used for food-contact surfaces include galvanized steel, lead, and copper.

○ **Internal corners and edges exposed to food must be rounded off (coved).** Solder and caulking are not acceptable rounding materials. External corners and angles must be sealed and finished smooth.

○ **All food-contact surfaces must be smooth and free of pits, crevices, ledges, inside threads and shoulders, bolts, and rivet heads.** Surfaces must be easy to clean and corrosion resistant.

○ **Coating materials must be nontoxic, cleanable, and must resist cracking and chipping.** They must not be used on food-contact surfaces.

○ **Equipment must be easy to disassemble to encourage frequent, thorough cleaning.**

○ **Only commercial foodservice equipment should be used in establishments.** Household equipment is not built to withstand the heavy use found in establishments and is not subject to NSF International standards.

Although all equipment used in an establishment must meet standards such as those set by NSF International, certain equipment requires particular attention.

Warewashing Machines

Warewashing machines vary widely by size, style, and method of sanitizing. High-temperature machines sanitize with extremely hot water; chemical-sanitizing machines use a chemical solution. Some states require the local regulatory agency's approval before installing a chemical warewashing system.

The size and type of machine you choose depends upon the nature and volume of the items to be cleaned and the required turnaround time for clean tableware and utensils. Because a warewashing machine is a big investment, the manager needs to carefully match the machine to the establishment's needs.

The following types of warewashing machines are common in restaurant and foodservice establishments:

○ **Single-tank, stationary-rack, with doors.** This machine holds a stationary rack of tableware and utensils. Items are washed by detergent and water from below and, sometimes, from above the rack. The wash cycle is followed by a hot water or chemical-sanitizer final rinse.

○ **Conveyor machine.** With this machine, a conveyor moves racks of items through the various cycles of washing, rinsing, and sanitizing. The machine may have a single tank or multiple tanks.

○ **Carousel or circular conveyor machine.** This multiple-tank machine moves tableware and utensils on a peg-type conveyor or in racks. Some models have an automatic stop after items go through the final rinse cycle. In other models, items must be removed after the final rinse, or they will continue to travel through the machine.

○ **Flight-type.** This is a high-capacity, multiple-tank machine with a peg-type conveyor. It may also have a built-in dryer. It is commonly used in institutions and very large establishments.

○ **Batch-type, dump.** This stationary-rack machine combines the wash and rinse cycles in a single tank. Each cycle is timed, and the machine automatically dispenses both the detergent and the sanitizing chemical or hot water. Wash and rinse water are drained after each cycle.

○ **Recirculating, door-type, non-dump.** This stationary-rack machine is not completely drained of water between cycles. The wash water is diluted with fresh water and reused from cycle to cycle.

Consider the following general guidelines regarding the installation and use of warewashing machines.

○ Water pipes to the warewashing machine should be as short as possible to prevent the loss of heat from hot water entering the machine.

○ The machine must be raised at least six inches off the floor to permit easy cleaning underneath.

○ Materials used in warewashing machines should be able to withstand wear, including the action of detergents and sanitizers.

○ Information should be posted on or near the machine regarding proper water temperature, conveyer speed, water pressure, and chemical concentration.

○ The machine's thermometer should be located so it is readable. The thermometer should have a scale in increments no greater than 2°F (1°C).

Clean-in-Place Equipment

Some equipment is designed to be cleaned and sanitized by having a detergent solution, hot-water rinse, and sanitizing solution passed through it. Certain soft-serve ice cream and frozen yogurt dispensers are cleaned and sanitized this way. These machines must be constructed so the cleaning and sanitizing solution remains within the fixed system of tubes and pipes for a predetermined amount of time. All food-contact surfaces must be reached by the solutions, but they must not leak into the rest of the machine.

Clean-in-place equipment must be self-draining. Some means of inspection must be provided to make sure the machine has been completely emptied of cleaning solution and has been thoroughly rinsed with fresh water. Manufacturers' instructions should be followed carefully.

Refrigerators and Freezers

There are several types of foodservice refrigerator and freezer units. The two most common types are walk-in and reach-in refrigerators and freezers. These units should be made of stainless steel or a combination of stainless steel and aluminum. The doors should be constructed to withstand heavy use and should close with a slight nudge. Door gaskets can be fixed in place, or removable for easy cleaning. A drain must be provided and maintained for disposal of condensation and defrost water. A properly plumbed, indirect drain can be used in the walk-in refrigerator. Excess condensation can be

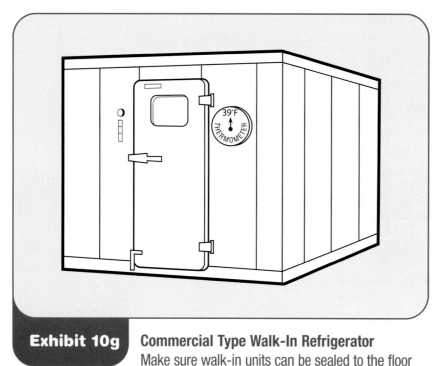

Exhibit 10g **Commercial Type Walk-In Refrigerator**
Make sure walk-in units can be sealed to the floor and wall.

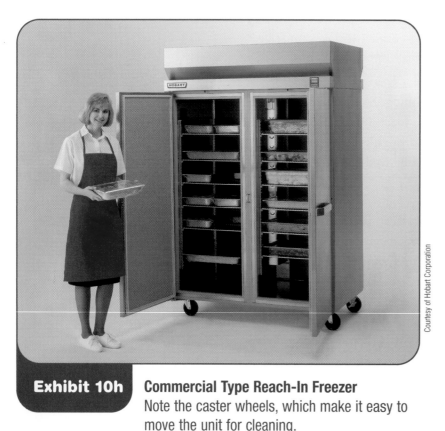

Courtesy of Hobart Corporation

Exhibit 10h **Commercial Type Reach-In Freezer**
Note the caster wheels, which make it easy to move the unit for cleaning.

minimized by maintaining a flush-fitting floor sweep (gasket) under the door.

Walk-in refrigerators and freezers with windows in the door may reduce unnecessary opening. Forced-air circulating fans are essential to help provide a quick recovery time so refrigerator and freezer temperatures remain at the appropriate level.

When purchasing a refrigerator or freezer unit, make sure it carries the NSF International mark or the UL EPH product mark or the equivalent. This will ensure that the unit is designed to protect food and simplify cleaning. In addition to the NSF International standards mentioned earlier in this chapter, consider these factors when purchasing a refrigerator or freezer unit:

○ **Choose a unit with adequate storage space.** An uncrowded unit can maintain required holding temperatures, is easier to clean, will prevent moisture buildup, and can minimize breakdowns.

○ **Make sure walk-in units can be sealed to the floor and wall.** *(See Exhibit 10g.)* They should offer no access to moisture or rodents. Flooring materials must be able to withstand heavy impact.

○ **Purchase reach-in refrigerator or freezer units with legs that elevate them six inches off the floor. Otherwise mount and seal them on a masonry base.** Casters making it easy to move the unit for cleaning are often preferred or required by local regulatory agencies. *(See Exhibit 10h.)*

○ **Make sure the unit meets the temperature requirements of the food you store.** Built-in thermometers should be easy to locate and read, and be accurate to within 3°F (2°C).

Blast Chillers and Tumble Chillers

Blast chillers are designed to move food through the temperature danger zone quickly. *(See Exhibit 10i.)* Many blast chillers are able to cool food from 140°F to 37°F (60°C to 3°C) within ninety minutes. Most units allow the operator to set target chill temperatures and monitor the temperature of food throughout the chill cycle. Once chilled to safe temperatures, the food then can be stored in conventional refrigerators or freezers.

Tumble chillers are also designed to cool food quickly. *(See Exhibit 10j.)* Prepackaged hot food is placed into a drum, which rotates inside a reservoir of chilled water. The tumbling action increases the effectiveness of the chilled water in cooling the food.

Cook-Chill Equipment

Some operations prepare food using a cook-chill system. By this method, food is partially cooked, rapidly chilled, and then held in refrigerated storage. When needed, the food simply is reheated. A cook-chill unit is an integrated piece of equipment capable of cooking, cooling, and reheating food.

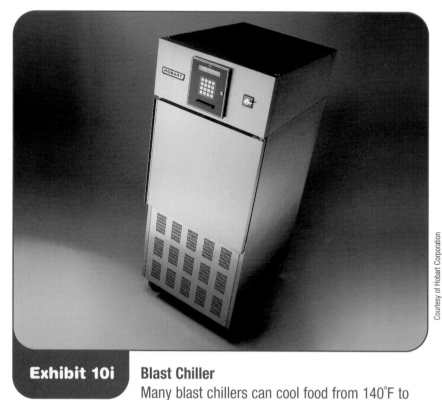

Courtesy of Hobart Corporation

Exhibit 10i　**Blast Chiller**
Many blast chillers can cool food from 140°F to 37°C (60°C to 3°C) within ninety minutes.

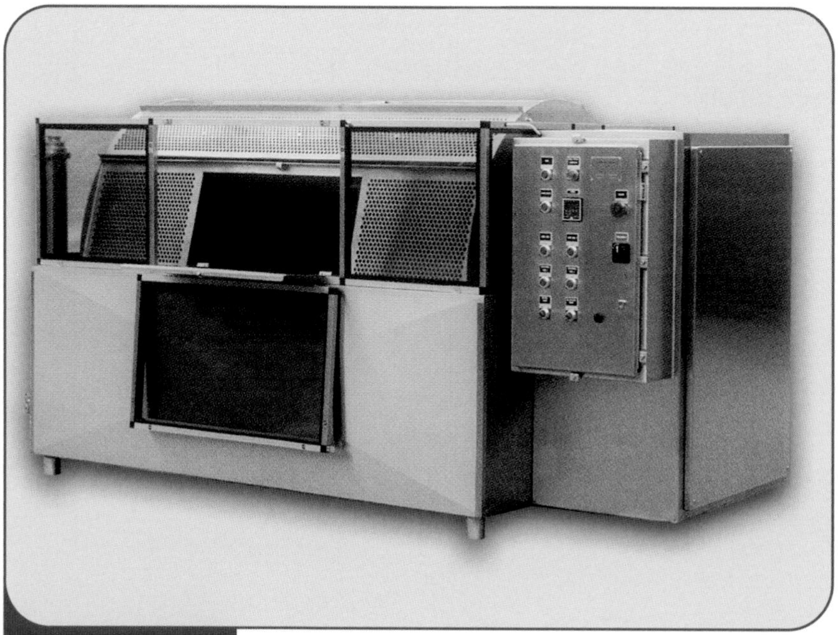

Courtesy of Chester-Jensen Company, Inc.

Exhibit 10j　**Tumble Chiller Unit**

Cutting Boards

Many jurisdictions allow the use of either wooden or synthetic cutting boards; however, some experts prefer synthetic boards, which can be cleaned and sanitized in a warewashing machine or by immersion in a three-compartment sink.

If wooden cutting boards and baker's tables are allowed by local codes, they must be made from a nonabsorbent hardwood, such as maple or oak. They must also be free of seams and cracks, be nontoxic, and must not transfer any odor or taste to food.

Separate cutting boards should be used for raw and ready-to-eat food in order to prevent cross-contamination. Cutting boards must be washed, rinsed, and sanitized between uses. Due to the high risk of cross-contamination, procedures for cleaning and sanitizing cutting boards should be included in your standard operating procedures.

Cross-Contamination

Synthetic cutting boards are generally preferred because they can be cleaned and sanitized in a warewashing machine.

CHOOSING AND INSTALLING KITCHEN EQUIPMENT

Well-designed kitchens make the job of keeping food safe easier. Generally, an efficient kitchen design is a more sanitary kitchen design.

Layout

A well-designed kitchen will address the following factors:

○ **Work flow.** A work flow must be established that will minimize the amount of time food spends in the temperature danger zone. It must also minimize the number of times food is handled. *(See Exhibit 10k.)* For example, storage areas should be located near the receiving area to prevent delays in storing food. Prep tables should be located near refrigerators and freezers for the same reason.

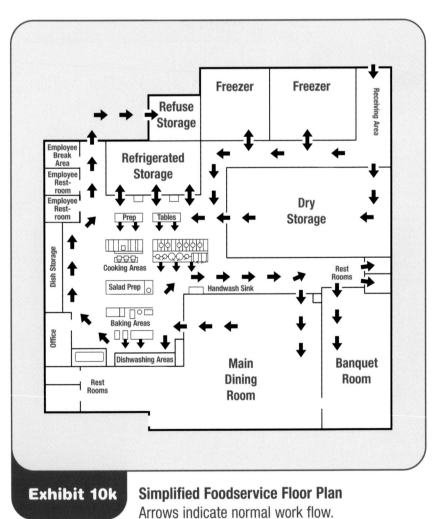

Exhibit 10k   **Simplified Foodservice Floor Plan**
Arrows indicate normal work flow.

○ **Contamination.** A good layout will minimize the risk of cross-contamination. Dirty equipment should not be placed where it will touch food or clean equipment. For example, it is not a good practice to place the soiled-utensil table next to the salad-preparation sink.

○ **Equipment accessibility.** A well-planned layout will ensure that equipment is accessible for cleaning. Hard-to-reach areas are less likely to be cleaned.

Equipment

When installing equipment, keep in mind that it should be easy for employees to clean both the equipment and surrounding floors, walls, and tabletops. Portable equipment makes it much simpler to do this.

When installing stationary equipment, it must be mounted on legs at least six inches off the floor or it must be sealed to a masonry base (allowing a toe space of four inches). *(See Exhibit 10l.)* How far the equipment is installed from the wall or from other equipment depends upon its size and the amount of surface to be cleaned. Follow specifications supplied by the manufacturer.

Stationary tabletop equipment should be mounted on legs, providing a minimum clearance of four inches between the base of the equipment and the tabletop. Alternatively, the equipment should be tiltable, or it should be sealed to the countertop with a nontoxic, food-grade sealant.

When equipment is sealed to the floor, wall, or counter, any crack or seam greater than 1/32" (1mm) must be filled with a nontoxic sealant to prevent food buildup or pests. Sealant should not be used to cover wide gaps resulting from faulty construction or repairs. These gaps should be properly repaired before equipment is installed.

> **Key Point**
>
> Install equipment so it is easy to clean both the equipment and surrounding areas.

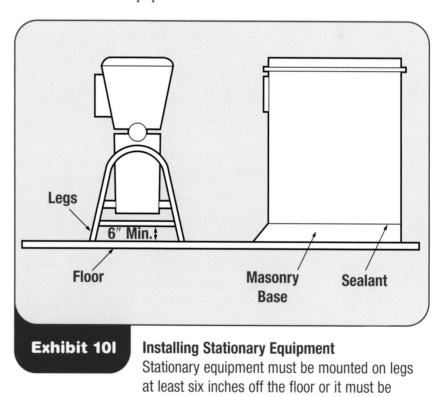

Exhibit 10l	**Installing Stationary Equipment**

Stationary equipment must be mounted on legs at least six inches off the floor or it must be sealed to a masonry base.

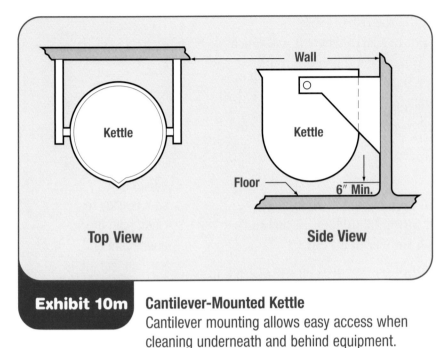

Exhibit 10m   **Cantilever-Mounted Kettle**
Cantilever mounting allows easy access when cleaning underneath and behind equipment.

Cantilever-Mounted Equipment

Attached to the wall with a bracket, **cantilever-mounted equipment** allows for easier cleaning of surfaces behind and underneath it. *(See Exhibit 10m.)*

UTILITIES

Establishments cannot operate without water and plumbing, electricity, gas, lighting, ventilation, sewage, and waste handling. Sanitary design of these utilities and services is important. Two basic goals must be met in their design: there must be enough utilities to meet the cleaning needs of the establishment, and the utilities themselves must not contribute to contamination.

Water Supply

Unsafe water can carry foodborne pathogens. Therefore, safe water is vital in every establishment. It is used as a beverage and in ice, as an ingredient in food, and for handwashing, cleaning, laundry, and showers, as well as for flushing in restrooms.

Water that is safe to drink is called **potable water.** Sources of potable water include approved public water mains, private water sources that are regularly maintained and tested, and bottled drinking water. Other sources include closed, portable water containers filled with potable water, on-premises water storage tanks, and properly maintained water-transport vehicles.

If your establishment uses a private water supply, such as a well, rather than an approved public source, you should check with your local regulatory agency for information on inspections, testing, and other requirements. Generally, nonpublic water systems should be tested at least annually and the report kept on file in the establishment.

The use of nonpotable water is extremely limited. If nonpotable water is allowed by local codes, the uses are generally limited to air conditioning, cooling equipment (nonfood), fire protection, and irrigation.

Water Emergencies

Occasionally an emergency occurs that causes the water supply to become unusable. Examples include natural disasters like floods, problems with a city water supply, or local problems with broken pipes. If there is a problem with the potable water supply, the establishment might have to be closed or an alternate source of water found. Anytime a water emergency occurs, water systems and equipment (e.g. ice machines) should be flushed and disinfected.

Establishments might not want to close during a water emergency due to the loss of revenue. Some establishments might *need* to stay open during the emergency, such as hospitals or nursing homes. In these cases, the regulatory agency might allow the establishment to continue to operate if certain precautions are followed. The following are issues an establishment must consider in order to continue serving food safely during a water emergency. Contact your local regulatory agency if you have any questions about the safety of a particular practice.

Water Used as a Beverage or Ingredient

If potable water must be obtained from an alternate source for use as a beverage or ingredient, there are several options, including buying bottled water and boiling water (the local regulatory authority should be contacted to determine the proper length of time for boiling). If you have been notified of an upcoming water emergency (water being shut off for repairs), you can store potable water in advance, using closed, food-grade water containers.

Ice

Most ice machines make new ice and drop it onto previously made ice. Ice made from contaminated water will, therefore, contaminate any previously made ice already in the bin. If the establishment knows in advance that the water supply might become unsafe, previously made ice can be stored before the emergency occurs. In an unexpected water emergency, ice can be purchased. If purchased, ice must be contained in single-use, food-grade plastic bags or wet-strength paper bags filled and sealed at the point of manufacture.

Cleaning

If a water emergency occurs, nonessential cleaning in the establishment should be minimized. When performing essential tasks, such as cleaning and sanitizing pots and pans, the water used should first be boiled. Establishments might consider using single-use items to eliminate the need for using a warewasher, which requires large amounts of water.

Handwashing

Warm, potable water is required for handwashing. Boiled water can be placed in large plastic containers that dispense water through a spout. Replenish the supply frequently to keep it warm.

Restrooms, Showers, and Laundries

Restrooms, showers, and laundries require large amounts of water. Managing them would be difficult without a potable water system. Portable toilets can be used, but it might not be possible to supply water to showers and laundry facilities. The local regulatory agency should be contacted to determine possible alternatives. Some codes allow the use of nonpotable water for flushing toilets and operating washing machines.

Hot Water

Providing a continuous supply of hot water can be a problem for many establishments serving the public. The FDA Food Code states that establishments must have enough hot water to meet peak demand. Water heaters should be evaluated regularly to make sure they can meet this demand. Consider how quickly the heater produces hot water, the size of the holding tanks, and the location of the heater in relation to sinks or warewashing machines.

Since most general-purpose water heaters will not heat water to temperatures required for hot-water sanitizing, a **booster heater** may be needed to maintain a water temperature of 180°F (82°C) for heat-sanitizing tableware and utensils. Many warewashing machines now come with booster heaters.

Plumbing

In almost every community in the United States, plumbing design is regulated by law, and with good reason. Improper plumbing design can have serious consequences. Improperly installed or poorly maintained plumbing that allows the mixing of potable and nonpotable water has been implicated in outbreaks of typhoid fever, dysentery, hepatitis A, Norwalk virus, and other gastrointestinal illnesses. Improperly installed water pipes can also lead to contamination from metals, such as toxic-metal poisoning from beverage dispensers, or contamination from chemicals used in the system, such as detergents, sanitizers, or drain cleaners.

Local plumbing codes vary widely regarding the types of connections permitted and the kind of protection required. If in doubt about the plumbing regulations governing your establishment, contact the local regulatory agency. Only licensed plumbers should install and maintain plumbing systems in an establishment.

Key Point

Only licensed plumbers should install and maintain plumbing systems in an establishment.

Cross-Connections

The greatest challenge to water safety comes from **cross-connections.** A cross-connection is a physical link through which contaminants from drains, sewers, or other wastewater sources can enter a potable water supply. Cross-connection is dangerous because it allows the possibility of **backflow,** or the unwanted reverse flow of contaminants through a cross-connection into a potable water system. Backflow can occur whenever the pressure in the potable water supply drops below the pressure of the contaminated supply. A running faucet located below the **flood rim** of a sink, or a running hose in a mop bucket, are examples of a cross-connection.

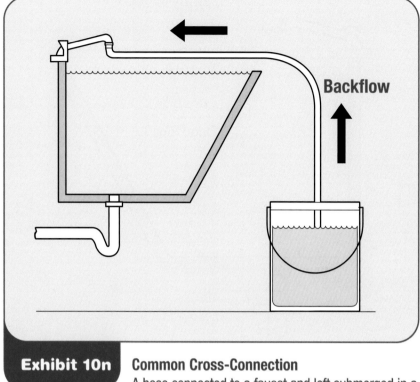

Backflow

Exhibit 10n

Common Cross-Connection
A hose connected to a faucet and left submerged in a mop bucket creates a dangerous cross-connection.

Exhibit 10n illustrates a situation in which an employee has attached a hose to the faucet of a utility sink to add hot water to a partially filled mop bucket. He leaves the nozzle of the hose submerged in the bucket of dirty water, accidentally creating a cross-connection. Because of heavy water usage somewhere else in the facility, the water pressure could drop low enough that contaminated water from the mop bucket would be drawn back through the hose and into the potable water supply.

To prevent cross-connections like this, do not attach a hose to a faucet unless a backflow-prevention device—such as a **vacuum breaker**—is attached. *(See Exhibit 10o.)* Threaded faucets and connections between two piping systems must have a vacuum breaker or other approved backflow-prevention device.

The only completely reliable method for preventing backflow is creating an **air gap.** An air gap is an air space used to separate a water supply outlet from any potentially contaminated source. A properly designed and installed sink typically has two air gaps to prevent backflow. The air space between the faucet and the flood rim of the sink is one. Another is located between the drain pipe of the sink and the floor drain of the establishment.

Exhibit 10o

Vacuum Breaker
A vacuum breaker should be installed on threaded faucets to prevent backflow.

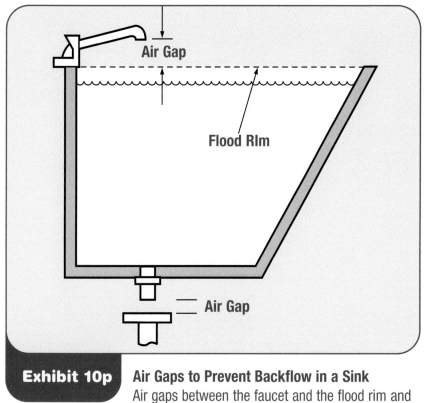

Exhibit 10p | **Air Gaps to Prevent Backflow in a Sink**
Air gaps between the faucet and the flood rim and between the drain pipe and floor drain of a sink prevent backflow.

Key Point

A backup of raw sewage on the floor is cause for immediate closure of the establishment, correction of the problem, and thorough cleaning.

(*See Exhibit 10p.*) Typically, the size of the air gap should be twice the diameter of the water supply outlet. For example, if a faucet has an opening with a diameter of one inch, the air gap between the faucet and the flood rim of the sink must be at least two inches.

Grease Condensation and Leaking Pipes

Grease condensation in pipes is another common problem in plumbing systems. Grease traps are often installed to prevent a buildup from creating a drain blockage. The trap must be cleaned and the grease removed periodically. If this is not done, or is not done properly, a backup of wastewater could lead to odor and contamination.

Overhead wastewater pipes or fire-safety sprinkler systems can leak and become a source of contamination. Even overhead lines carrying potable water can be a problem, since water can condense on the pipes and drip onto food. All piping should be serviced immediately when leaks occur.

Sewage

Sewage and wastewater are reservoirs of pathogens, soil, and chemicals. It is absolutely essential to prevent any possible contamination of food or food-contact surfaces by wastewater. The pipes of any nonpotable water system, such as those from toilets or sinks, should be clearly identified by a licensed plumber so they can be distinguished from pipes carrying potable water.

If there is a backup of wastewater, prompt action must be taken. The action taken depends on the type of backup. A backup of raw sewage on the floor is cause for immediate closure of the establishment, correction of the problem, and thorough cleaning. Other backups might not be as serious, but require an immediate correction of the problem, followed by a thorough clean-up.

Sufficient drainage must be provided to handle wastewater. Any area subjected to heavy water exposure should have its own floor drain. Wastewater from equipment and from the potable supply should be channeled into an open, accessible waste sink or floor drain. The drainage system should be designed to keep floors from being flooded.

Lighting

Building and health codes usually set minimum acceptable levels of lighting, typically based on the **foot-candle**—a unit of illumination one foot from a uniform source of light. Other units of measurement for light include lumens, luxes, and luminaires. Good lighting generally results in improved employee work habits, easier and more effective cleaning, and a safer work environment. Lighting requirements are different for various areas of the establishment. The following are some recommendations.

○ **Provide a minimum of fifty foot-candles of light (540 lux)** in food-preparation areas.

○ **Provide a minimum of twenty foot-candles of light (220 lux)** in handwashing or warewashing areas; at buffets and salad bars and where produce or packaged food is displayed for sale; inside certain equipment (e.g. a reach-in refrigerator); in utensil storage areas and wait stations; and in restrooms.

○ **Provide a minimum of ten foot-candles of light (110 lux)** inside walk-in refrigerator and freezer units, in dry-storage areas, and in the dining room for cleaning.

Overhead or ceiling lights above work stations should be positioned so employees do not cast shadows on the work surface. Using fluorescent lights helps minimize such shadows. Use shatter-resistant light bulbs and protective covers made of metal mesh or plastic to prevent broken glass from contaminating food or food-contact surfaces. Shields should also be provided for heat lamps.

Ventilation

Proper ventilation helps maintain an establishment's indoor air quality by removing steam, smoke, grease, and heat from the establishment. Adequate ventilation is particularly important in the food-preparation area, because it reduces the level of odors, gases, dirt, mold, humidity, grease, and fumes present in the air, all of which can contribute to contamination. Excess humidity can cause condensation on walls and ceilings, which may drip onto food. An accumulation of grease can cause fires. If ventilation is adequate, there will be little or no buildup of grease and condensation on walls and ceilings.

Mechanical ventilation must be used in areas for cooking, frying, and grilling. Ventilation must be designed so that hoods, fans, guards, and ductwork do not drip onto food or equipment. Hood filters or grease extractors must be tight-fitting and easy to remove, and should be cleaned on a regular basis. Thorough cleaning of the hood and ductwork should also be done periodically by a professional company.

Since so much air is moved through exhaust hoods, clean air must be taken in to replace it. This replacement air is called make-up air, which must be replaced without creating drafts. All outside air intakes must be screened to keep pests out.

In many areas of the United States, clean-air ordinances restrict the use of exhaust fans. Exhaust air containing food odors, smoke, and grease might have to be purified by filters or other devices. It is the establishment's responsibility to see that the ventilation system meets local regulations.

Solid Waste Management

Waste management is an important issue in establishments. There are many things foodservice managers can do to improve their waste management practices. The Environmental Protection Agency (EPA) has recommended three approaches to manage waste.

○ **Reduce the amount of waste produced.** Eliminate unnecessary packaging.
○ **Reuse when possible.** Reused containers must be cleaned and sanitized. *Never* reuse chemical containers as food containers.
○ **Recycle materials.** Store recyclables so they can't contaminate food or equipment or attract pests.

When these practices are followed, the amount of waste can be greatly reduced.

Garbage Disposal

Garbage is wet waste matter, usually containing food, that cannot be recycled. It attracts pests and has the potential to contaminate food, equipment, and utensils.

Garbage containers must be leak-proof, waterproof, pest-proof, easy to clean, and durable. They can be made of galvanized metal or an approved plastic. They must have tight-fitting lids and must be kept covered when not in use. Plastic bags and wet-strength paper bags may be used to line these containers.

Garbage should be removed from food-preparation areas as soon as possible. Frequent disposal prevents odor and pest problems. Garbage-storage areas, inside or outside, should be large enough to contain all garbage and must be located away from food-preparation and storage areas. When removing garbage, employees should not carry it above or across a food-preparation area.

All garbage containers should be cleaned frequently and thoroughly. Both the inside and the outside of containers must be cleaned. A cleaning area equipped with hot and cold water and a floor drain is recommended for indoor cleaning. It must be located so food being prepared or in storage will not be contaminated when garbage containers are being cleaned.

Food waste can also be disposed of through the use of in-drain garbage disposals, which reduce the amount of waste that goes into garbage containers. However, local regulations often limit their use in establishments. These disposals create huge amounts of food and water waste, which might overload local wastewater systems.

Exhibit 10q **Outdoor Trash Receptacles**
Receptacles and compactor systems should be located on or above a smooth surface of nonabsorbent material, such as concrete or machine-laid asphalt.

Pulpers or grinders are another garbage-disposal alternative. Pulpers grind food and some other types of waste (such as paper) into small parts that are flushed with water. The water is then removed so the processed solid waste weighs less and is more compact for easier disposal.

Outdoor trash receptacles should be kept covered (with their drain plugs in place) at all times, except during cleaning. Receptacles and compactor systems should be located on or above a smooth surface of nonabsorbent material such as concrete or machine-laid asphalt. The area must be kept clean. *(See Exhibit 10q.)*

SUMMARY

An establishment that is difficult to clean will not be cleaned well. Sanitation efforts will be more effective if the establishment is designed and equipped with ease of cleaning in mind. In most communities, plans for new construction or extensive remodeling are subject to review and approval by local regulatory agencies.

Sanitation can be built into the facility through the proper design and construction of floors, walls, and ceilings. The selection of materials should be based on ease of cleaning and durability, as well as appearance.

Separate restrooms should be provided for employees and patrons. If this is not possible, restrooms should be positioned so patrons don't pass through food-preparation areas to reach them. Restrooms should be sanitary and have a fully equipped handwashing station and self-closing doors. They should also be stocked adequately with toilet paper, and have trash receptacles.

Handwashing stations must be conveniently located, operable, and fully stocked and maintained. They are required in food-preparation areas, service areas, warewashing areas, and restrooms. They must be equipped with hot and cold running water, soap, a means to dry hands, and a waste container.

Equipment must meet the sanitation standards set by NSF International, UL, or the equivalent. It should be installed so that both the equipment and the area surrounding it can be cleaned easily. Stationary equipment must be mounted on legs at least six inches off of the floor, or it must be sealed to a masonry base. Stationary tabletop equipment should be mounted on legs with a clearance of four inches between the equipment and the tabletop, or it should be sealed to the tabletop. Any gap between a piece of equipment and the floor, wall, or tabletop greater than 1/32″ (1mm) should be filled with a nontoxic sealant to prevent food buildup and pests.

Utilities must be designed so they meet the establishment's cleaning needs, and they must not contribute to contamination. Potable water—water that is safe to drink—is vital in an establishment. Sources include public water mains, private water sources regularly maintained and tested, and bottled drinking water. The use of nonpotable water is usually limited to fire protection, irrigation, and air conditioning. In a water emergency, an establishment might be allowed to remain open if certain precautions are followed. These could include boiling or purchasing water, storing potable water and ice in advance, and boiling water for handwashing and essential tasks. The local regulatory agency should be contacted any time there is a question about the safety of a particular practice.

Improperly installed or maintained plumbing can have serious consequences. Only licensed plumbers should install and maintain plumbing systems. The greatest challenge to water safety comes from cross-connections—a physical link through which contaminants from drains, sewers, and other wastewater sources can flow back into the potable-water supply. Vacuum breakers and air gaps can be used to prevent backflow. It is essential to prevent wastewater from contaminating food and food-contact surfaces. A backup of raw sewage is cause for immediate closure of the establishment, correction of the problem, and thorough cleaning.

Good lighting in an establishment generally results in improved employee work habits, easier and more effective cleaning, and a safer work environment. Use shatter-resistant bulbs and protective covers to prevent broken glass from contaminating food or food-contact surfaces. Proper ventilation improves the indoor air quality of the establishment by removing smoke, grease, steam, and heat. If ventilation is adequate, there will be little or no buildup of grease and condensation on walls and ceilings. Ventilation must be designed so hoods, fans, guards, and ductwork do not drip onto food or equipment. Hood filters and grease extractors must be cleaned regularly.

Garbage containers must be leak-proof, waterproof, pest-proof, easy to clean, and durable. They must have tight-fitting lids and must be kept covered when not in use. All garbage containers should be cleaned frequently and thoroughly both inside and out. Garbage should be removed from food-preparation areas as soon as possible, and must not be carried above or across a food-preparation area.

A CASE IN POINT

Case Study

Several people became ill shortly after drinking beverages at the bar in a local restaurant. They all complained that their iced drinks had an odd taste. At the time of the incident, the glasswasher in the bar had been out of service.

When interviewed, the manager explained that the glasswasher was functional, but that the unit could not be used because the large volume of water discharged after each wash load was worsening a recent drain-blockage problem. He mentioned there had been intermittent backups in the plumbing during the previous week, and a large pool of water was found under the glasswasher. Drain cleaners had been used repeatedly with no change in the blockage.

The icemaker shared piping with the glasswasher in the bar and the grease trap on the sink in the restaurant. The manager revealed he had installed the plumbing himself.

When ice cubes were removed from the icemaker, congealed grease and food debris were found on them. Grease and debris also covered the bottom of the ice bin.

Why do you think the people became ill? What should the manager do to correct the problem?

TRAINING TIPS

1. Facility Food Safety Pop Quiz

Purpose: *After completing this activity, trainees will be able to identify possible ways in which facilities and equipment can contaminate food if not properly designed, constructed, or installed.*

Time: 30 minutes

Directions: Design a quiz with the following categories at the top of a sheet of paper:

Area or Item   **Cause of Contamination**   **Prevention**

Divide trainees into teams of two, and tell them they are to list ways food might become contaminated by improperly designed, constructed, or installed facilities and equipment. Use the following example to show the teams how to organize their lists.

Area or Item	Cause of Contamination
Plumbing	○ Cross-connection
Prevention	○ Overhead sewage pipe dripping on food
○ Install backflow-prevention devices	
○ Create air gaps	○ Grease-trap backup
○ Repair pipes	
○ Clean grease traps weekly	

Limit the activity to five to ten minutes. When time expires, have each team give itself one point for each item listed under Area or Item, Cause of Contamination, and Prevention. Ask the teams to add up their points. Have the team with the most points read its list. Discuss any missed items.

2. Expect the Unexpected: Crisis Management Group Activity

Purpose: *After completing this activity, trainees will be able to identify food safety risks associated with various operational emergencies and develop action plans to address them.*

Time: one hour

Directions: List a number of possible crisis situations that could occur in an establishment related to topics discussed in Chapter 10.

○ **Plumbing:** Broken or stopped-up pipes or drains; cross-connections; backup of wastewater or raw sewage

○ **Water:** Water turned off for eight hours or longer; contamination of city or private water supply

○ **Electric:** Power outage for eight hours or longer

○ **Equipment failure:** Broken booster heater in a warewashing machine; freezer or refrigerator malfunction; hot-water heater breakdown

○ **Natural disaster:** Major flooding

○ **Waste management:** Strike by garbage workers

Divide the trainees into groups. Assign each group a specific crisis and give them this set of directions.

○ Determine specific food safety risks, if any, involved with the crisis.

○ Create an action plan to address the crisis.

○ Develop a plan to prevent an occurrence of that crisis.

Allow twenty minutes, then have a spokesperson from each group present their work. Solicit feedback and additional suggestions after each presentation. Allow time for trainees to talk about any personal experiences related to the crisis situations discussed. Conclude by asking trainees if their establishments have adequate crisis management plans in place and, if not, what steps they will take to help implement such plans.

3. Let's Design and Build a Restaurant

Purpose: *After completing this activity, trainees will be able to identify food safety issues that should be considered when designing, constructing, and equipping an establishment.*

Time: one hour

Directions: Divide trainees into five groups. Assign each group a restaurant or foodservice concept, complete with menu, hours of operation, and projected volume of business. Include the possibility of additional business from carryout and catering sales.

Next, assign each group one of the following topics discussed in Chapter 10:

○ Design and layout   ○ Utilities

○ Equipment selection   ○ Ventilation

○ General construction (floors, walls, ceiling)

Have the groups develop an action plan that includes food safety concerns and issues related to their topic. Allow ten minutes for this activity.

Ask each group to make a brief presentation of its action plan. Solicit group discussion. Conclude by reminding trainees of all the food safety issues related to these areas. Reiterate how a well-designed and efficiently operated establishment can contribute to a successful food safety program.

4. "Find the Seal" Search

Purpose: *To identify equipment in your operation that has met sanitation standards developed by NSF International and/or Underwriters Laboratory.*

Time: one week

Directions: Invite your kitchen staff to participate in a competition. Over the course of a week, have them search for NSF and/or UL marks on equipment throughout the establishment and record on paper where the label is located. You might limit the search to the receiving, storage, kitchen, and warewashing areas, or you might extend it to other areas of your operation.

At the end of the week, collect the lists, record the number of items found by each participant, and announce a winner at the next kitchen staff meeting. Offer the winner an appropriate reward. At the meeting, discuss with the staff the significance of NSF and UL, and explain why it is important to use equipment and utensils designated for commercial food service.

5. Develop a "Waste Management Task Force"

Purpose: *To involve staff members in an effort to improve the waste-handling program at your establishment. Note:* Involving your staff and allowing them to initiate approved changes in the existing program will create buy-in.

Time: several weeks

Directions: Solicit a team of employees to perform the following research and development assignment:

○ Assess the waste-handling methods currently used in the establishment. Consider all aspects: number, type, and conditions of garbage cans; location and condition of outside garbage bins; other waste-handling equipment; the recycling program; and so on.

○ Research and compare waste-handling methods in other operations similar to yours.

○ Research local health department codes and regulations regarding waste handling.

○ Develop recommendations for improved waste-handling techniques and systems, along with a list of expected benefits from the proposed improvements.

Give the team clear direction and a deadline, and keep in touch with its progress. When the report and recommendations are complete, schedule a formal meeting with the team in order to receive its proposals.

Authorize the team to initiate any approved changes. Provide funding if necessary. Monitor the results, offer suggestions, and solicit feedback from the team. Acknowledge the team's effort with an appropriate reward.

DISCUSSION QUESTIONS

1. What is one of the most important considerations when choosing flooring for food-preparation areas?

2. What action must be taken in the event of a backup of raw sewage in an establishment?

3. What can be done to prevent backflow in an establishment?

4. What can an establishment do to continue to serve safe food during an interruption in the water supply?

5. What are some potable-water sources for an establishment? If allowed by local codes, what can nonpotable water be used for?

6. What are the requirements of a handwashing station? In what areas of an establishment are handwashing stations required?

7. What are the requirements for installing stationary equipment?

MULTIPLE-CHOICE STUDY QUESTIONS

1. All of the following are proper responses to an interruption in the water supply *except:*

   A. Minimizing nonessential cleaning tasks
   B. Boiling drinking water
   C. Contacting the local regulatory agency about practices that should be followed
   D. Using nonpotable water for handwashing

2. An establishment should respond to a backup of raw sewage by

   A. closing.
   B. correcting the problem that caused the backup.
   C. thoroughly cleaning the establishment.
   D. All of the above

3. Which of the following *will not* prevent backflow?

   A. An air gap between the sink drain pipe and the floor drain
   B. The air space between the faucet and the flood rim of a sink
   C. A vacuum breaker
   D. A cross-connection

4. The most important consideration when selecting construction materials for an establishment is:

   A. Cost   C. Ease of cleaning
   B. Appearance   D. Durability

5. Local codes may allow nonpotable water to be used for all of the following *except:*

   A. Air conditioning   C. Handwashing
   B. Fire protection   D. Irrigation

6. Which of the following statements about garbage containers is true?

   A. They can be made of galvanized metal
   B. They can be lined with wet-strength paper bags
   C. They must have tight-fitting lids
   D. All of the above

7. Which hand-drying method is *not* allowed in a handwashing station?

   A. Warm-air dryer

   B. Single-use paper towels

   C. A common hand towel

   D. All these methods are allowed

8. Handwashing stations are required in

   A. service areas.

   B. food-preparation areas.

   C. warewashing areas.

   D. All of the above

9. Handwashing stations must be

   A. stocked with hand sanitizers.

   B. conveniently located to encourage handwashing.

   C. stocked with a common hand towel.

   D. supplied with water at least 170°F (77°C).

10. A good kitchen layout will

    A. minimize the number of times food is handled.

    B. minimize the risk of cross-contamination.

    C. ensure that equipment is accessible for cleaning.

    D. All of the above

11. Potable water is water that is

    A. drinkable.

    B. not drinkable.

    C. contaminated.

    D. highly acidic.

12. Wooden cutting boards, if allowed by the local jurisdiction, should

    A. be made from softwoods, such as pine.

    B. be made from hardwoods, such as oak.

    C. contain seams.

    D. All of the above

For answers, please turn to Appendix A.

ADDITIONAL RESOURCES

Books and Periodicals

Bjerklie, S. 1997. The super soaker. *Institutional Distributor.* 50(11):35.

Cichy, R. F. 1993. *Sanitation Management.* East Lansing, MI: Educational Institute of the American Hotel & Motel Association.

Durocher, J. 1996. A clean sweep. *Restaurant Business.* 95(4):202.

Durocher, J. 1995. Working with sanity. *Restaurant Business.* 94(9):154.

Equipped for food safety. 1998. *Best Practices.* 2(4):6-10.

Frable, F., Jr. 1998. Cleanability of equipment should be a factor in your purchasing decisions. *Nation's Restaurant News,* 32 (33):19.

Frable, F., Jr. 1997. 10 common errors in kitchen planning and ways to avoid them. *Nation's Restaurant News.* 31(6):22.

Kratt, D. T. 1997. Running a clean operation. *Nightclub & Bar.* 13(11):79.

Longree, K and G. Armbruster. 1996. *Quantity Food Sanitation.* New York: Wiley.

Marriott, N. G. 1994. *Principles of Food Sanitation.* New York: Chapman & Hall.

Riell, H. 1997. Beverage dispenser sanitation. *FoodService Director.* 10(12):166.

The right way to…calibrate equipment. 1997. *Best Practices.* 1(3):12.

The right way to…organize workspaces. 1998. *Best Practices.* 2(3):15-16.

Sanson, M. 1996. A blueprint for safety: Restaurant kitchen design. *Restaurant Hospitality.* 80(8):65.

Secrets of self-inspection. 1998. *Best Practices.* 2(2):6-9.

Troller, J. A. 1993. *Sanitation in Food Processing.* San Diego, CA:
    Academic Press.

Web Sites

American National Standards Institute (ANSI)

http://www.ansi.org

ANSI is a private, nonprofit organization administering and
coordinating the U.S. voluntary standardization and conformity
assessment system. Use this Web site to order ANSI/NSF standards
outlining the materials to be used in the construction of commercial
food equipment.

Americans with Disabilities Act Info Page

http://www.usdoj.gov/crt/ada/adahom1.htm

Department of Justice Americans with Disabilities Act (ADA)
publications and information, including ADA requirements, design
standards for new construction and alterations, status reports, featured
settlements and consent agreements, technical assistance materials,
and new or proposed regulations, are all found on this Web site.

FDA Food Code

http://vm.cfsan.fda.gov/~dms/foodcode.html

As the basis for many local sanitation codes, as well as the basis for
information in this textbook, the FDA Food Code, available at this
Web address, is a useful resource for information relating to food for
the restaurant and foodservice industry.

FDA Food Establishment Plan Review Guide 2000

http://vm.cfsan.fda.gov/~dms/prev-toc.html

The food establishment plan review document found at this site was
developed for the purpose of assisting both regulatory and industry
personnel in achieving greater uniformity in the plan review process.

National Restaurant Association

http://www.restaurant.org

The National Restaurant Association is the leading business association for the restaurant industry. Together with the National Restaurant Association Educational Foundation, the Association's mission is to represent, educate, and promote the rapidly growing restaurant and foodservice industry. This Web site should be your starting place for all issues and concerns related to your restaurant. This Web site has it all, from tips for running your establishment to vital data on your customers' spending habits.

NSF International

http://www.nsf.org

Use this Web site to find a wealth of information related to food standards and resources. NSF works closely with the food-equipment industry, regulators, and foodservice organizations to develop public health safety standards used by the restaurant and foodservice industry.

Occupational Safety and Health Administration (OSHA)

http://www.osha.gov

OSHA's mission is to prevent work-related injuries, illnesses, and deaths. This Web site provides information related to workplace-safety conferences, technical links, and regulation and compliance documents.

Underwriters Laboratories, Inc. (UL)

http://www.ul.com/eph

UL is an independent, not-for-profit product-safety testing and certification organization offering certification programs to demonstrate compliance with public health codes to manufacturers of foodservice equipment. UL classifies products in accordance with food-related NSF International standards. This Web site contains a foodservice-equipment and drinking water product directory, information on the indoor air quality program, ISO 14001, and other information related to the foodservice industry.

Chapter 11
Cleaning and Sanitizing

TEST YOUR FOOD SAFETY KNOWLEDGE

1. **True or False:** Chemicals can be stored in food-preparation areas if they are properly labeled. *(See page 11-16.)*

2. **True or False:** The temperature of the final sanitizing rinse in a high-temperature warewashing machine should be 140°F (60°C). *(See page 11-8.)*

3. **True or False:** Cleaning reduces the number of microorganisms on a surface to safe levels. *(See page 11-2.)*

4. **True or False:** Utensils cleaned and sanitized in a three-compartment sink should be dried with a clean towel. *(See page 11-11.)*

5. **True or False:** Tableware and utensils that have been cleaned and sanitized should be stored at least two inches off of the floor. *(See page 11-16.)*

For answers, please turn to Appendix A.

Learning Objectives

After completing this chapter, you should be able to

- Explain the difference between cleaning and sanitizing.
- Identify factors affecting the efficiency of sanitizers.
- Select appropriate cleaners and sanitizers, and safely store and handle them.
- Properly clean and sanitize items in a three-compartment sink.
- Identify proper machine warewashing techniques.
- Identify storage requirements for food-service chemicals.

Key Terms

Cleaning
Sanitizing
Cleaning agent
Detergent
Solvent cleaners
Abrasive cleaners
Acid cleaners
Heat sanitizing
Chemical sanitizing
Sanitizer
Chlorine
Iodine
Quaternary ammonium compounds (quats)
Hazard Communication Standard (HCS)
Master cleaning schedule

In Chapter 9, you learned that a HACCP plan depends on prerequisite programs. A cleaning and sanitation program is one of the most important of these.

If you don't maintain a high standard of cleanliness and sanitation, food can easily become contaminated. No matter how carefully you prepare and cook food, without a clean and sanitary environment, bacteria and viruses—such as those that cause salmonellosis and hepatitis A—can spread quickly to both cooked and uncooked food. Cleaning and sanitizing must be done carefully and correctly, however. If not used properly, cleaning and sanitizing chemicals can be just as harmful to customers and employees as the illnesses they help prevent.

CLEANING AND SANITIZING

It is important to understand the difference between **cleaning** and **sanitizing**. Cleaning is the process of removing food and other types of soil from a surface, such as a countertop or plate. Sanitizing is the process of reducing the number of microorganisms on that surface to safe levels. To be effective, cleaning and sanitizing must be a two-step process. Surfaces must *first* be cleaned and rinsed *before* being sanitized.

Key Point

Surfaces must *first* be cleaned and rinsed *before* being sanitized.

Everything in your operation must be kept clean; however, any surface that comes in contact with food must be cleaned *and* sanitized. All food-contact surfaces must be washed, rinsed, and sanitized:

○ After each use

○ Any time you begin working with another type of food

○ Any time you are interrupted during a task and the tools or items you have been working with may have been contaminated

○ At four-hour intervals, if the items are in constant use

Cleaning

Several factors affect the cleaning process. The table in *Exhibit 11a* lists these factors and provides a brief explanation of each.

Cleaning Agents

Cleaning agents are chemical compounds that remove food, soil, rust, stains, minerals, or other deposits. They must be stable, noncorrosive, and safe for employee use. Since cleaning agents have different cleaning properties, ask your supplier to help you select those that will best meet your needs.

Cleaning agents work best when used as directed. They can be ineffective and even dangerous if misused. Follow manufacturers' instructions carefully. Employees should never combine compounds or attempt to make up their

Exhibit 11a: Factors Affecting the Cleaning Process

Factor	Effect on Cleaning Process
Type of Soil	Certain types of soil require special cleaning methods.
Condition of Soil	The condition of the soil or stain affects how easily it can be removed. Dried or baked-on stains will be more difficult to remove than soft, fresh stains.
Water Hardness	Cleaning is more difficult in hard water because minerals react with the detergent, decreasing its effectiveness. Hard water can cause scale or lime deposits to build up on equipment, requiring the use of lime-removal cleaners.
Water Temperature	In general, the higher the water temperature, the better a detergent will dissolve and the more effective it will be in loosening dirt.
Cleaning Agent and Surface Being Cleaned	Different surfaces require different cleaning agents. Some cleaners work well in one situation, but might not work well or might even damage equipment when used in another.
Agitation or Pressure	Scouring or scrubbing a surface helps remove the outer layer of soil, allowing a cleaning agent to penetrate deeper.
Length of Treatment	The longer soil on a surface is exposed to a cleaning agent, the easier it is to remove.

Exhibit 11a

own cleaning agents. Combining ammonia and chlorine bleach, for example, produces chlorine gas, which can be fatal. Also, do not substitute one type of detergent for another unless the intended use is stated clearly on the label. Detergents used for warewashing machines, for example, can cause severe burns to the skin if used for manual warewashing.

Cleaning agents are divided into four categories: **detergents, solvent cleaners, abrasive cleaners,** and **acid cleaners.** Some categories may overlap. For example, most abrasive cleaners and some acid cleaners contain detergents. Some detergents also may contain solvents.

Detergents

There are different types of detergents for different types of cleaning jobs. All detergents contain surfactants (surface acting agents) that reduce surface tension between the soil and the surface, so the detergent can quickly penetrate and soften the soil. General-purpose detergents are mildly alkaline and are used to clean fresh soil from floors, walls, ceilings, prep surfaces, and most equipment and utensils. Heavy-duty detergents are highly alkaline and are used to remove wax, aged or dried soil, and baked-on grease. Warewashing detergents also are highly alkaline.

Solvent Cleaners

Solvent cleaners, often called degreasers, are alkaline detergents containing a grease-dissolving agent. These cleaners work well in areas where grease has been burned on, such as grill backsplashes, oven doors, and range hoods. Solvents are usually effective only at full strength, so they are costly to use on large areas.

Acid Cleaners

Acid cleaners are used on mineral deposits and other soils alkaline cleaners can't remove. These cleaners are often used to remove scale in warewashing machines and steam tables, as well as rust stains and tarnish on copper and brass. The type and strength of the acid varies with the cleaner's purpose. Follow the instructions carefully and use acid cleaners with caution.

Abrasive Cleaners

Abrasive cleaners contain a scouring agent like silica that helps scrub off hard-to-remove soil. These cleaners are often used on floors or to remove baked-on food in pots and pans. Use abrasives with caution since they can scratch surfaces.

Sanitizing

There are two methods used to sanitize surfaces: **heat sanitizing** and **chemical sanitizing.** Which you use depends on the application.

Heat Sanitizing

The higher the heat, the shorter the time required to kill microorganisms. The most common way to heat sanitize tableware, utensils, or equipment is to immerse or spray the items with hot water. Use a thermometer to check water temperature when heat sanitizing by immersion. To check water temperature in a high-temperature warewashing machine, attach a temperature-sensitive label or tape, or a high-temperature probe to items being run through the machine.

Chemical Sanitizing

Chemical sanitizers are widely used in establishments because they are effective, reasonably priced, and easy to use. They are regulated by federal and state environmental protection agencies (EPAs). Do not use a sanitizer on a food-contact surface unless it is EPA approved.

Chemical sanitizing is done in two ways: either by immersing a clean object in a specific concentration of sanitizing solution for a required period of time or by rinsing, swabbing, or spraying the object with a specific concentration of sanitizing solution.

The three most common types of **sanitizer** used in the restaurant and foodservice industry are **chlorine, iodine,** and **quaternary ammonium compounds (quats)**. The advantages and disadvantages of each type are listed in *Exhibit 11b.*

In some instances, detergent-sanitizer blends may be used to sanitize surfaces, but items still must be cleaned and rinsed first. Scented or oxygen bleaches are not acceptable as sanitizers for food-contact surfaces. Household bleaches are acceptable only if the labels indicate they are EPA registered.

Key Point

Do not use a sanitizer on a food-contact surface unless it is EPA approved.

Exhibit 11b: Advantages and Disadvantages of Different Sanitizers

Types	Advantages	Disadvantages
Chlorine	Most commonly used sanitizer Kills a wide range of microorganisms Leaves no film on surfaces Least expensive Effective in hard water	Soil quickly inactivates chlorine solutions Corrosive to some metals, such as stainless steel and aluminum, when used improperly Adversely affected by temperatures above 115°F (46°C)
Iodine	Effective at low concentrations Not as quickly inactivated by soil as chlorine	Less effective than chlorine Becomes corrosive to some metals at temperatures above 120°F (49°C) More expensive than chlorine May stain surfaces
Quaternary ammonium compounds (quats)	Not as quickly inactivated by soil as chlorine Remains active for a short period of time after it has dried Noncorrosive Nonirritating to skin Works in most temperature and pH ranges	Leaves a film on surfaces Does not kill certain types of microorganisms Hard water reduces effectiveness

Exhibit 11b

Key Point

The
effectiveness
of a chemical-sanitizing
solution depends upon
its contact time,
temperature, and
concentration.

Factors Influencing the Effectiveness of Sanitizers

Different factors influence the effectiveness of chemical sanitizers. The most critical include contact time, temperature, and concentration. *(See Exhibit 11c.)*

Contact Time

In order for a sanitizing solution to kill microorganisms, it must make contact with the object for a specific amount of time. Since minimum times may differ for each sanitizer, check with your supplier.

Temperature

Generally, sanitizers work best at temperatures between 55°F and 120°F (13°C and 49°C). Some may not be effective at temperatures lower than 55°F (13°C), while others may corrode metals or evaporate at temperatures higher than 120°F (49°C).

Concentration

Chemical sanitizers are mixed with water until the proper concentration—ratio of sanitizer to water—is reached. Concentration is measured using a sanitizer test kit and is expressed in parts per million (ppm). The test kit should be designed for the sanitizer you are using and is usually available from the manufacturer or your supplier. *(See Exhibit 11d.)*

Concentrations below those required in your jurisdiction or recommended by the manufacturer could fail to sanitize objects. Concentrations higher than recommended can be unsafe, leave an odor or bad taste on objects, and might corrode metals.

The concentration of a sanitizing solution must be checked frequently. This is important since a sanitizer is depleted during use. It can also become bound up by hard water, food particles, or detergent not adequately rinsed from a surface.

A sanitizing solution must be changed when it is visibly dirty or when the concentration of the sanitizer has dropped below the level required.

Exhibit 11c: General Guidelines for Chemical Sanitizers

	Chlorine	Iodine	Quats
Minimum Concentration			
○ For immersion	50 ppm	12.5-25 ppm	200 ppm
○ For spray cleaning	50 ppm	12.5-25 ppm	200 ppm
Temperature of Solution	55°F-115°F (13°C-46°C)	75°F (29°C) (leaves solution at 120°F [49°C])	75°F (24°C) or higher
Contact Time			
○ For immersion	7 seconds	30 seconds	30 seconds—some products require longer contact time—read label
○ For spray cleaning	Follow manufacturer's directions	Follow manufacturer's directions	Follow manufacturer's directions
pH (detergent residue raises pH, so rinse completely)	Must be below 8.0	Must be below 5.0	Most effective at 7.0, but varies with compound
Corrosiveness	Corrosive to some substances	Noncorrosive	Noncorrosive
Reaction to Organic Contaminants in Water	Quickly inactivated	Made less effective	Not easily affected
Reaction to Hard Water	Not affected	Not affected	Some compounds inactivated—hardness over 500 ppm is undesirable
Indication of Proper Strength	Test kit required	Amber color indicates presence. Use test kit to determine concentration.	Test kit required; follow directions closely

Exhibit 11c

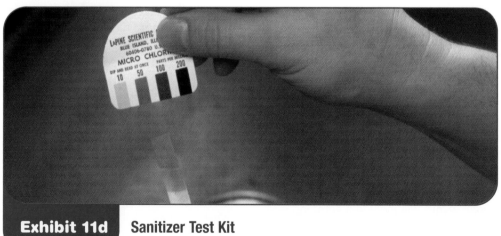

Exhibit 11d **Sanitizer Test Kit**
Use a test kit to check the concentration of a sanitizing solution.

MACHINE WAREWASHING

Most tableware, utensils, and even pots and pans can be cleaned and sanitized in warewashing machines. Warewashing machines sanitize by using either hot water or a chemical-sanitizing solution.

High-Temperature Machines

High-temperature machines rely on hot water to clean and sanitize. Water temperature is critical and may vary by model. The temperature of the final sanitizing rinse must be at least 180°F (82°C). For stationary-rack, single-temperature machines, the temperature of the final sanitizing rinse must be at least 165°F (74°C). Water that is too hot might vaporize before sanitizing items, or might bake food onto tableware and utensils, making it even harder to get them clean. You may need a booster heater to provide enough hot water to sanitize a high volume of tableware. Make sure your warewasher has a built-in thermometer to measure the temperature of water at the manifold, where it sprays into the tank.

Chemical-Sanitizing Machines

Warewashing machines that use chemical sanitizing often wash at much lower temperatures, but not lower than 120°F (49°C). Rinse-water temperature in these machines should be between 75°F and 120°F (24°C and 49°C) for the sanitizer to be effective. Items washed and rinsed at these lower temperatures may take longer to air dry, so you may need more room at the clean end of the machine and more tableware during peak periods.

The effectiveness of your warewashing program will depend on a number of factors:

○ A well-planned layout *(see Exhibit 11e)* in the warewashing area, with a scraping and soaking area and adequate space for both soiled and clean items

Exhibit 11e   **Warewashing Area Layout**
A well-planned layout will include a scraping and soaking area and adequate space for both soiled and clean items.

○ A sufficient water supply, especially hot water

○ A separate area for cleaning pots and pans

○ Devices that indicate water pressure and temperature of the wash and rinse cycles

○ Protected storage areas for clean tableware and utensils

○ Employees who are trained to operate and maintain the equipment and use the proper chemicals

All warewashing machines should be operated according to manufacturers' instructions. These instructions will typically be located on the machine. No matter what type of machine you use, there are some general procedures to follow to clean and sanitize tableware, utensils, and related items.

○ **Check the machine for cleanliness at least once a day, cleaning it as often as needed.** Fill tanks with clean water. Clear detergent trays and spray nozzles of food and foreign objects. Use an acid cleaner on the machine whenever necessary to remove mineral deposits caused by hard water.

○ **Make sure detergent and sanitizer dispensers are filled properly.**

○ **Scrape, rinse, or soak items before washing.** Presoak items containing dried-on food.

○ **Load warewasher racks correctly and use racks designed for the items being washed.** Make sure all surfaces are exposed to the spray action. Never overload racks.

○ **Check temperatures and pressure.** Follow manufacturers' recommendations.

○ **Check each rack as it comes out of the machine for soiled items.** Run dirty items through again until they are clean. Most items will need only one pass if the water temperature is correct and proper procedures are followed.

○ **Air dry all items.** Towels can recontaminate items.

○ **Keep your warewashing machine in good repair.**

Key Point

All ware-washing machines should be operated according to manufacturers' instructions.

MANUAL WAREWASHING

Establishments that do not have a warewashing machine may use a three-compartment sink to wash items (some local regulatory agencies allow the use of two-compartment sinks; others require four-compartment sinks). These sinks may also be used to wash larger items. A properly set up warewashing station includes:

○ An area for rinsing away food or for scraping food into garbage containers

○ Drain boards to hold both soiled and clean items

○ A thermometer to measure water temperature

○ A clock with a second hand, allowing employees to time how long items have been immersed in the sanitizing sink

Cleaning and Sanitizing in a Three-Compartment Sink

Before cleaning and sanitizing items in a three-compartment sink, each sink and all work surfaces must be cleaned and sanitized. Follow the steps listed below when cleaning and sanitizing tableware, utensils, and equipment. *(See Exhibit 11f.)*

Step 1: Rinse, scrape, or soak all items before washing.

Step 2: Wash items in the first sink in a detergent solution at least 110°F (43°C). Use a brush, cloth, or nylon scrub pad to loosen the remaining soil. Replace the detergent solution when the suds are gone or the water is dirty.

Step 3: Immerse or spray-rinse items in the second sink, using water at least 110°F (43°C). Remove all traces of food and detergent. If using the immersion method, replace the rinse water when it becomes cloudy or dirty.

Step 4: Immerse items in the third sink in hot water or a chemical-sanitizing solution. If hot-water immersion is used, the water must be at least 171°F (77°C) (some health codes require a temperature of 180°F) and the items must be immersed for thirty seconds. If chemical sanitizing is used, the sanitizer must be mixed at the proper concentration and the water temperature

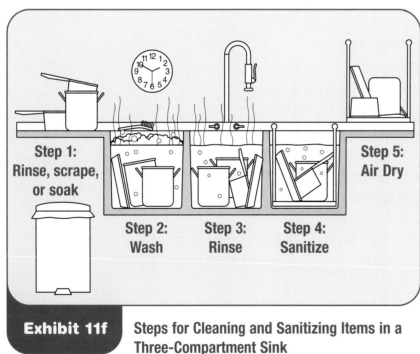

Exhibit 11f   **Steps for Cleaning and Sanitizing Items in a Three-Compartment Sink**

must be correct. Check the concentration of the sanitizing solution at regular intervals with a test kit.

Step 5: Air dry all items.

Wood surfaces—such as cutting boards, handles, and bakers' tables—need special care. After each use, scour them in a detergent solution with a stiff-bristle nylon brush, rinse them in clean water, and sanitize them. Do not soak wood surfaces in detergent or sanitizing solutions.

CLEANING AND SANITIZING EQUIPMENT

Because equipment must be kept clean and food-contact surfaces must be cleaned and sanitized, employees should be taught how to clean each type of equipment properly.

Clean-in-Place Equipment

Some pieces of equipment, such as soft-serve yogurt machines, are designed to have cleaning and sanitizing solutions pumped through them. Since many of them hold and dispense potentially hazardous food, they must be cleaned and sanitized every day unless otherwise indicated by the manufacturer.

Stationary Equipment

Equipment manufacturers will usually provide cleaning instructions. In general, follow these steps:

○ Turn off and unplug equipment before cleaning.

○ Remove food and soil from under and around the equipment.

○ Remove detachable parts and manually wash, rinse, and sanitize them, or run them through a warewasher if permitted. Allow them to air dry.

○ Wash and rinse fixed food-contact surfaces, then wipe or spray them with a chemical-sanitizing solution.

○ Keep cloths used for food-contact and nonfood-contact surfaces in separate, properly marked containers of sanitizing solution.

○ Air dry all parts, then reassemble according to directions. Tighten all parts and guards. Test equipment at recommended settings, then turn off.

○ Resanitize food-contact surfaces handled when putting the unit back together by wiping with a cloth that has been submerged in sanitizing solution.

Key Point

Clean-in-place equipment used to hold and dispense potentially hazardous food must be cleaned and sanitized every day unless otherwise indicated by the manufacturer.

Key Point

Cloths used for food-contact and nonfood-contact surfaces should be stored between use in separate, properly marked containers of sanitizing solution.

In some cases, you may be able to spray-clean fixed equipment. Check with the manufacturer. If allowed, spray each part with solution in the right concentration and let it sit for the recommended amount of time.

Refrigerated Units

As mentioned in Chapter 6, clean up spills in refrigerators and freezers immediately. These units should be thoroughly cleaned and sanitized regularly to remove soil, mold, and odors. When cleaning and sanitizing refrigeration units, follow these suggestions.

○ Clean before storing deliveries so less food has to be moved.

○ Move food to another unit before starting to clean.

○ Clean shelves regularly. Thoroughly clean walls, floors, door edges, and gaskets.

CLEANING THE KITCHEN

Kitchen floors, walls, shelves, ceilings, light fixtures, and drains are nonfood-contact surfaces, but they still require regular cleaning. Sanitizing these surfaces is not required, but might be a good practice. How often these surfaces are cleaned will depend on several factors, including the type of food served, surfaces to be cleaned, and rate of ventilation in the kitchen. Spills should be cleaned up immediately. Floors and walls around food-preparation and cooking areas should be cleaned at least daily—before or after each shift is preferable. Other nonfood-contact surfaces may need to be cleaned less often.

Floors

Floors are a safety hazard if they aren't cleaned regularly or rinsed properly. They can also be a source of cross-contamination since soil and spills can be tracked through the entire operation. To clean floors, follow these steps:

○ Mark the area being cleaned with signs or safety cones to prevent slips and falls.

○ Sweep the floor.

○ Use a deck or scrub brush and full-strength detergent on extra-soiled areas to remove grease and dirt.

○ Mop or pressure-spray the area, working from the walls toward the floor drain. Soak the mop in a bucket of detergent solution and wring it out. Clean a ten-foot by ten-foot area with both sides of the mop, using a figure-eight motion. Soak and wring out the mop, then clean the same area again.

○ Remove excess water with a damp mop or squeegee, working away from the walls and toward the floor drain.

○ Rinse thoroughly with clean water, using the same mopping procedure.

Walls and Shelves

Clean tile and stainless-steel surfaces by spraying or sponging with a detergent solution. Use a nylon scrub brush to clean dried-on soil or grease. Rinse with clean water. When spray-cleaning, take care not to damage walls. Protect food, equipment, and nearby supplies. Use a sponge or wet cloth to clean other wall surfaces, such as painted drywall.

Ceilings and Light Fixtures

Ceilings do not need to be cleaned as often as floors or walls. Check ceilings and light fixtures daily to ensure that cobwebs, dust, dirt, or condensation won't fall and contaminate food or food-contact surfaces below. Wipe and rinse ceilings and light fixtures with a sponge or cloth.

CLEANING THE PREMISES

Although the kitchen requires the most attention, all areas of your operation need to be kept clean. Areas such as restrooms, busing or serving stations, and tables and booths need to be kept clean and sanitary as well. Pay attention to the exterior of the facility, too. Dirty receiving docks and garbage areas can attract pests. A dirty parking lot or exterior is unsightly and can hurt business.

Tables

Tables, booths, and counters are nonfood-contact surfaces, since food is usually served on dishes. However, tables should always be kept clean. If you use table linens or butcher's paper, change them before seating new customers. If flatware is placed directly on the table surface, tables should also be sanitized.

○ Use a dry wiping cloth to clean crumbs and dry-food spills from tables. Replace the cloth whenever it becomes soiled or sticky, so it doesn't transfer soil and microorganisms to other surfaces.

○ Use a moist cloth to clean up other types of food spills. Keep moist cloths in a bucket of chemical-sanitizing solution. Replace both the sanitizing solution and wiping cloths as necessary. Check the concentration of the sanitizing solution often.

Serving Stations

Serving stations might be little more than a small space to store tableware and linens. In many operations, however, they are also used to dispense water, coffee, and other beverages, to prepare and serve bread and condiment baskets, and to serve desserts. Serving stations often contain a sink, a coffee brewer, beverage dispensers, ice makers or bins, and even small coolers to hold butter, cream, condiments, and more. To keep serving stations clean, follow these procedures:

○ Clean up spills immediately. Wipe or sweep up dry food with a dry cloth or broom. Clean wet spills with a damp cloth kept in sanitizing solution.

○ Wash, rinse, and sanitize sinks and countertops at least daily or after each shift. If work areas are used to prepare potentially hazardous food, such as cheesecake, clean at least every four hours.

○ Clean equipment daily or as often as recommended by the manufacturer. Items such as ice bins, beverage-dispensing nozzles and lines, and coffee grinders should be cleaned as often as needed to prevent dirt or mold from accumulating.

○ Wash, rinse, and sanitize bus tubs manually or in the warewashing machine at least daily or after every shift.

Public Restrooms

Unsanitary restrooms pose a danger to your customers and your business. Never underestimate cleaning needs in this area. Restrooms can quickly become visibly dirty and may harbor unpleasant odors and disease-causing microorganisms. Clean restrooms are important to customers. Many customers associate the cleanliness of the restroom with the establishment's commitment to food safety. When maintaining restrooms, consider the following suggestions.

○ Check the condition of public and employee restrooms regularly.

○ Restock soap, toilet paper, and towels before they run out.

○ Clean sinks, mirrors, walls, floors, counters, dispensers, toilets, urinals, and waste receptacles at least daily. Sanitize toilets and urinals at least once daily. Clean up accidents and spills as often as necessary.

○ Remove trash at least once daily, or as often as necessary.

Key Point

Many customers associate the cleanliness of the restroom with the establishment's commitment to food safety.

Exterior Premises

The exterior of the premises should be kept clean. Windows, walls, and fixtures should be cleaned on a regular basis, and should be included on the master cleaning schedule. Check the grounds at least daily, pick up trash, and sweep walkways. Clean garbage areas as often as necessary to prevent odors or trash from attracting pests.

TOOLS FOR CLEANING

Cleaning is easier when you have the right tools at hand. For example, worn out tools won't give you the pressure or friction needed for cleaning, and tools that are the wrong size will be ineffective.

Keep tools used for cleaning separate from those used for sanitizing, and tools used for food-contact surfaces separate from those used to clean nonfood-contact surfaces. Using color-coded tools is one way to accomplish this. Always use a separate set of tools for the restroom.

Key Point

Use a separate set of cleaning tools for the restroom.

Brushes

Brushes apply more effective pressure than wiping cloths, and the bristles loosen soil more easily. Worn brushes will not clean effectively and can be a source of contamination.

Brushes come in different shapes and sizes for each task. Lacquered wood or plastic brushes with synthetic bristles are preferred. They don't absorb moisture, are nonabrasive, and last longer. Use the right brush for the job.

Scouring Pads

Steel wool and other abrasives are sometimes used to clean heavily soiled pots and pans, equipment, or floors. However, metal scouring pads can break apart and leave residue on surfaces, which can later contaminate food. Nylon scouring pads can provide an alternative.

Mops and Brooms

Keep both light- and heavy-duty mops and brooms on hand. Mop heads can be all-cotton or synthetic blends. It makes sense to have a bucket and wringer for both the front and back of the facility. Both vertical and push-type brooms will come in handy.

Key Point

Store flatware and utensils with handles up so employees can pick them up without touching food-contact surfaces.

STORING UTENSILS, TABLEWARE, AND EQUIPMENT

Once tableware, utensils, and equipment are clean and sanitary, store them so they stay that way. It's equally important to ensure that cleaning tools and supplies are stored properly.

Tableware and Equipment

○ Store tableware and utensils at least six inches off the floor. Keep them covered or otherwise protected from dirt and condensation.

○ Clean and sanitize drawers and shelves before clean items are stored.

○ Clean and sanitize trays and carts used to carry clean tableware and utensils. Do this daily or as often as necessary.

○ Store glasses and cups upside down. Store flatware and utensils with handles up so employees can pick them up without touching food-contact surfaces.

○ Keep the food-contact surfaces of clean-in-place equipment covered until ready for use.

Cleaning Tools and Supplies

Cleaning tools and supplies should be cleaned and sanitized before being put away. Tools and chemicals should be stored in a locked area away from food and food-preparation areas. The area should be well-lighted so employees can identify chemicals easily. It should also be equipped with hooks for hanging mops, brooms, and other cleaning tools, a utility sink for filling buckets and cleaning tools, and a floor drain. *(See Exhibit 11g.)* Never use handwashing sinks, food-preparation sinks, or warewashing sinks to clean mops, brushes, or tools. When storing tools and supplies, consider the following suggestions:

○ Air dry wiping cloths overnight.

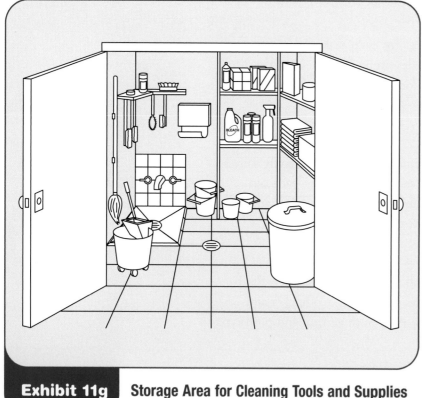

Exhibit 11g   **Storage Area for Cleaning Tools and Supplies**
Tools and chemicals should be stored in a locked area away from food and food-preparation areas.

○ Hang mops, brooms, and brushes on hooks to air dry. Do not leave brooms or brushes standing on their bristles.

○ Clean, rinse, and sanitize buckets. Let them air-dry, and store them with other tools.

USING HAZARDOUS MATERIALS

Chemicals are both useful and necessary to keep an establishment clean, sanitary, and pest free. Used properly, they may pose little threat to an employee's safety. Used improperly, they may become a health hazard that can cause injury.

Because of the potential danger of chemicals used in the workplace, the Occupational Safety and Health Administration (OSHA) requires employers to comply with their **Hazard Communication Standard (HCS).** This standard, also known as Right-to-Know or HAZCOM, requires employers to tell their employees about chemical hazards to which they might be exposed at the establishment. It also requires employers to train employees on how to use the chemicals they work with safely. Employers must comply with OSHA's HCS by developing a hazard communication program for their establishment.

A hazard-communication program must include the following components:

○ An inventory of hazardous chemicals used at the establishment

○ Chemical labeling procedures

○ Material Safety Data Sheets (MSDS)

○ Employee training

○ A written plan addressing the HCS

Inventory of Hazardous Chemicals

A hazardous chemical is any chemical that is a physical or health hazard to humans. Chemicals known to have acute or chronic health effects, or are explosive, flammable, or unstable, are considered hazardous. Virtually any chemical with toxic properties should be included in an establishment's HAZCOM program.

Take an inventory of the hazardous chemicals stored in your establishment. List the name of the chemical and where it is stored. Update the list when chemicals are added or no longer used.

Key Point

OSHA requires employers to tell employees about the chemical hazards they might be exposed to and to train employees to use the chemicals they work with safely.

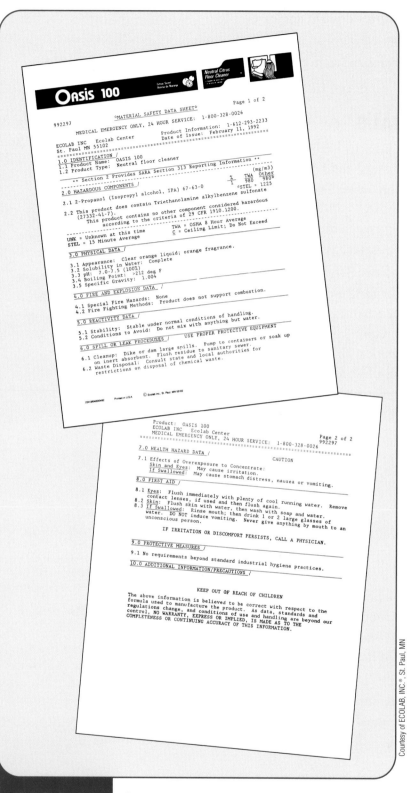

Courtesy of ECOLAB, INC.® St. Paul, MN

Exhibit 11h   **Sample Material Safety Data Sheets**

Labeling Procedures

OSHA requires chemical manufacturers to clearly label the outside of containers with the chemical name, manufacturer's name and address, and possible hazards. When receiving chemicals, only accept containers with proper labels, and make sure they remain readable and attached to the container.

If a chemical is transferred from a manufacturer's container to another container, the new container's label must contain the following information:

- Chemical name
- Manufacturer's name and address
- Potential hazards of the chemical

Material Safety Data Sheets (MSDS)

OSHA requires chemical suppliers and manufacturers to provide Material Safety Data Sheets *(see Exhibit 11h)* for each hazardous chemical at your establishment. These sheets are a part of employees' right to know about the hazardous chemicals they work with. MSDS contain the following information about the chemical:

- Information about safe use and handling
- Physical, health, fire, and reactivity hazards
- Precautions
- Appropriate personal protective equipment (PPE) to wear when using the chemical
- First-aid information and steps to take in an emergency

○ Manufacturer's name, address, and phone number

○ Preparation date of MSDS

○ Hazardous ingredients and identity information

MSDS should be collected in a binder and stored in a location accessible to all employees while on the job.

Training

OSHA requires that every employee who might be exposed to hazardous chemicals during normal or regular working conditions be informed of the hazards and trained to use chemicals properly. Employees should receive this training annually, and new employees must receive it when first assigned to a new department or area. The following topics should be covered during training:

○ Existence and requirements of the HCS

○ How the HCS is implemented in the workplace

○ Operations and processes in which hazardous chemicals are used

○ Inventory of chemicals in your establishment

○ Location of MSDS

○ How to read MSDS and product labels

○ Physical and health hazards of all chemicals used

○ Specific procedures adopted to provide protection, such as work practices and the use of engineering controls

○ Use of PPE, and steps to prevent or reduce exposure to chemicals

○ Safety and emergency procedures

○ Information on the normal use of chemicals

A Written Plan

OSHA requires employers to develop a written plan describing how they will meet the requirements of the HCS in their establishment. The following items should be included in your written plan:

○ List of hazardous chemicals stored on the premises, and their amounts

○ Purchasing specifications for chemicals

○ Procedures for receiving and storing chemicals

○ Labeling requirements in your establishment

○ Procedures for accessing MSDS

○ List of PPE

○ Employee training procedures

Key Point

MSDS should be collected in a binder and stored in a location accessible to all employees while on the job.

Key Point

OSHA requires employers to develop a written plan that describes how they will meet the requirements of the Hazard Communication Standard (HCS) in their establishments.

○ Reporting and record-keeping procedures

○ How the employer will inform employees of the hazards of nonroutine tasks

IMPLEMENTING A CLEANING PROGRAM

A clean and sanitary establishment is a prerequisite for an effective HACCP-based food safety program. As with HACCP, an effective cleaning program takes commitment from management and the involvement of employees. The following are some basic steps for designing and implementing a cleaning program.

Identify Cleaning Needs

○ **Identify all surfaces, tools, and equipment in the facility that need cleaning.** Walk through every area of the facility.

○ **Look at the way cleaning is done currently.** Get input from employees. Ask them how and why they clean a certain way. Find out which procedures can be improved.

○ **Estimate the time and skills needed for each task.** Some jobs may be done more efficiently by two or more people. Others might require an outside contractor. Determine how often things need to be cleaned.

Create a Master Cleaning Schedule

Take information gathered while identifying your cleaning needs and develop a **master cleaning schedule.** *(See Exhibit 11i.)* Organize the schedule by area and list the items that need to be cleaned and how often. Provide a brief description of how to do the job. Assign responsibility by job title. The schedule should include the following:

○ **What should be cleaned.** Arrange the schedule in a logical way so nothing is left out. List all cleaning jobs in one area, or list jobs in the order they should be performed. Keep the schedule flexible enough so you can make changes if needed.

○ **Who should clean it.** Assign each task to a specific individual. In general, employees should clean their own areas. Rotate other cleaning tasks to distribute them fairly.

○ **When it should be cleaned.** Employees should clean as they go, and clean and sanitize at the end of their shifts. Schedule major cleaning when food won't be contaminated or service interrupted—usually after closing. Schedule work shifts to allow enough time. Employees rushing to clean before their shifts end may cut corners.

Key Point

As with HACCP, an effective cleaning program takes commitment from management and the involvement of employees.

Key Point

When developing a master cleaning schedule for equipment and the facility, consider what should be cleaned, when, how, and by whom.

Exhibit 11i: Sample Cleaning Schedule for a Food-Preparation Area

Item	What	When	Use	Who
Floors	Wipe up spills	Immediately	Cloth mop and bucket, broom and dustpan	Busers
	Damp mop	Once per shift, between rushes	Mop, bucket, safety signs	
	Scrub	Daily, at closing	Brushes, squeegee bucket, detergent, safety signs	
	Strip, reseal	Every 6 months	Check written procedure	
Walls and ceilings	Wipe up splashes	As soon as possible	Clean cloth, detergent	Dishwashing staff
	Wash walls	Food-prep and cooking areas: daily		
		All other areas: first of month		
Worktables	Clean and sanitize tops	Between uses and at end of day	See cleaning procedure for each table	Prep cooks
	Empty, clean, and sanitize drawers	Weekly	See cleaning procedure for table	

Exhibit 11i

○ **How it should be cleaned.** Provide clearly written procedures for cleaning. Lead employees through the process step by step. Always follow manufacturer's instructions when cleaning equipment. Specify cleaning tools and chemicals by name. Post cleaning instructions near the item to be cleaned. *(See Exhibit 11j.)*

Exhibit 11j: Sample Cleaning Procedure: Cleaning a Food Slicer

Step	Process	Notes
1	Turn off slicer and unplug	
2	Set blade control to zero	
3	Remove meat carriage	
4	Remove the back blade guard	
5	Remove the top blade guard	
6	Scrub meat carriage and guards in pot-and-pan sink	Use detergent solution and a scrub brush
7	Rinse parts	Immerse in clean water at 171°F (77°C) for 30 seconds. Use an S-hook to remove parts from water
8	Allow parts to air dry on a clean and sanitary surface	

Exhibit 11j

Choose Cleaning Materials

When selecting cleaning tools for your establishment, consider the following:

○ **Select tools and cleaning agents according to the needs identified on the master cleaning schedule.** Talk to suppliers for suggestions on which tools and supplies are appropriate for your operation. Make sure your supplies match the needs listed on the master schedule.

○ **Replace worn out tools.** Equipment that is too worn or soiled may not clean or sanitize surfaces properly.

○ **Provide employees with the right protective gear.** Make sure there's an adequate supply of rubber gloves, aprons, goggles, and other supplies.

Training Employees

Training is critical to the success of the cleaning program. Employees must understand what you want them to be able to do, how well, and under what conditions. To ensure the success of the program, follow these guidelines:

○ **Schedule a kick-off meeting to introduce the program to employees.** Explain the reason behind it. Stress how important cleanliness is to food safety. If people understand why they are supposed to do something, they're more likely to do it.

○ **Schedule enough time for proper training.** Work with small groups or conduct training by area. Show employees how to clean equipment and surfaces in each area.

○ **Provide plenty of motivation.** Reward employees for any job well done. Create small incentives for individuals or teams, such as "clean team of the month" awards. Tie performance to specific measurements or goals, such as achieving high marks during health code inspections.

Monitor the Program

Once you have implemented the cleaning program, you must monitor it to make sure it is working. This includes:

○ **Supervising daily cleaning routines.**

○ **Monitoring the daily completion of all cleaning tasks against the master cleaning schedule.**

○ **Reviewing the master schedule every time there is a change in menu, procedures, or equipment.** Make sure the cleaning program addresses any changes.

Key Point

Once a cleaning program has been established, it must be monitored. Supervise cleaning routines, conduct spot inspections, and review the master schedule often.

○ **Requesting employee input on the program during staff meetings.**
Ask employees if they need additional equipment, supplies, manpower, time, or training to get cleaning jobs done. Find out if they have suggestions for improving the program.

○ **Conducting spot inspections.**

SUMMARY

All the work that goes into a HACCP-based food safety system is undermined if you don't keep your utensils, equipment, and facility clean and sanitary. Cleaning is the process of removing food and other types of soil from a surface. Sanitizing is the process of reducing the number of harmful microorganisms on a clean surface to safe levels. You must clean and rinse a surface before it can be sanitized effectively. Surfaces can be sanitized with hot water or with a chemical-sanitizing solution.

All surfaces should be cleaned on a regular basis. Food-contact surfaces must be cleaned and then sanitized after every use, any time a task is interrupted, and at four hour intervals if items are in constant use.

Warewashing machines can be used to clean, rinse, and sanitize most tableware and utensils. Follow manufacturers' instructions and make sure your machine is clean and in good working condition. Check the temperature and pressure of wash and rinse cycles daily.

Items that are too large to be placed into a warewashing machine can be cleaned and sanitized manually. This may be done in a three-compartment sink or, if the items are stationary, by cleaning and then spraying them with a sanitizing solution. Items cleaned in a three-compartment sink should be presoaked or scraped clean, washed in a detergent solution, rinsed in clean water, and sanitized in either hot water or in a chemical sanitizing solution for a predetermined amount of time. All items should then be air dried.

Clean all nonfood-contact surfaces regularly. Areas such as public restrooms, floors, shelves, and floor drains should be cleaned daily (or more often as needed) and sanitized when appropriate. Ceilings, walls, and fixtures, as well as exterior areas, such as docks, garbage containers, driveways, and parking lots, can be cleaned less frequently.

Keep all cleaning tools clean and sanitary. Cleaning tools and supplies should be stored in a well-lighted, locked room separate from food storage and preparation areas. Make sure chemicals are clearly labeled. Keep MSDS on hand, inform employees of potential hazards, and train them how to handle and use chemicals correctly.

Develop and implement a cleaning program. Identify cleaning needs by walking through the operation and talking to employees. Create a master cleaning schedule listing all cleaning tasks, as well as when and how they are to be cleaned. Assign responsibility for each task by job title. Enlist employee support by including their input in the program's design and rewarding good performance. Explain to employees the important relationship between cleaning and sanitizing and food safety.

Monitor the cleaning program to keep it effective. Supervise cleaning procedures. Check completion of each job against the master schedule. Adjust cleaning and sanitizing procedures when there is a change in menu, equipment, or procedures.

A CASE IN POINT I

Case Study

It was only 9:05 A.M., but already, Tim, the day shift manager, could tell it was going to be a bad day. While the executives from American Widget munched unenthusiastically on complimentary doughnuts, they threw annoyed looks at Tim and the buser, who were cleaning the banquet room American Widget had reserved for 9:00 A.M. Tim had opened the room at 8:50 A.M. and found, to his dismay, the remains of the annual banquet of the Pine Valley Martial Arts Club still strewn about the room.

Later, Tim sat down with his cleaning schedule and tried to figure out what had gone wrong. The banquet had been scheduled to end at 11:30 P.M. the previous night. The buser was scheduled to clean the room at midnight. But a note from Norman, the night shift manager, told Tim the banquet had been a wild one and the last guest had left long after the 1:00 A.M. closing time. The buser had punched out at 12:30 A.M., as he always did. Tim sighed.

What do you think went wrong? How can Tim prevent this from happening again? How should any cleaning-policy changes be introduced?

A CASE IN POINT II

Case Study

The kitchen employees in the university cafeteria had scraped and rinsed every piece of tableware, placed them in racks, and fed them into the high-temperature warewashing machine. When the wash and rinse cycles were complete, one of the employees noticed that the items were spotted. The thermometer registering the final rinse temperature for the sanitizing cycle had a reading of 140°F (60°C), rather than the required 180°F (82°C) indicated on the manufacturer's label. The employee went to inform the manager, who called for service.

What alternative methods can be used to clean and sanitize the tableware since the machine is not functioning properly? How often should the temperature in the warewashing machine be checked?

TRAINING TIPS

1. Tools of the Trade

Purpose: *After completing this activity, trainees should be able to identify different tools for cleaning and sanitizing, as well as describe the specific use or uses of each tool.* Note: This activity should help trainees evaluate the adequacy of their establishment's current inventory of tools.

Time: 20 minutes

Directions: Divide trainees into teams of two or three. Each team has the same challenge.

- List as many different tools for cleaning and sanitizing an establishment as possible—for example, chemicals, buckets, spray bottles, brooms, brushes, scrapers, gloves, mops, sanitizers.
- List the specific use or uses for each item.

Allow five minutes for this activity, and then ask to see all of the lists or have each team read its list to the group. Give a prize (or a round of applause) to the team listing the most tools and the uses for each.

On a flipchart, build a master inventory of cleaning tools and supplies from the lists. Add additional items not listed by the teams. Ask trainees if their establishments have all of the tools and supplies necessary to conduct an effective cleaning and sanitizing program. If not, ask them to identify additional items needed. Remind trainees that food safety depends on proper cleaning and sanitizing. Stress that no establishment can maintain proper cleaning and sanitizing without the proper tools and supplies.

2. The Clean Team Competition

Purpose: *After completing this activity, trainees should be able to write step-by-step cleaning and sanitizing procedures for kitchen areas and equipment.*

Time: 30 minutes

Directions: Divide trainees into groups of two to four people, and ask each group to write out the procedures for cleaning and sanitizing a specific piece of equipment in the kitchen. Pick a piece of equipment common to most trainees in the group.

Have the groups list the specific cleaning and sanitizing agents they would use, specific tools, and the step-by-step procedures for cleaning and sanitizing the equipment.

After five or ten minutes, have the teams describe their cleaning and sanitizing procedures. Compare their work, and award a prize to the team whose procedures were the clearest, most thorough, and most accurate.

3. Creating a Cleaning Program

Purpose: *To enlist the help of employees from different areas of the establishment to develop a master cleaning schedule.*

Time: several weeks

Directions: Enlist a "clean team" of employees from different positions in your establishment. Work closely with the team to design and implement an effective, workable, comprehensive master cleaning schedule. Assign team members to work with you on the following tasks:

○ **Identifying cleaning needs.** Walk through the establishment with team members and identify tools, equipment, and surfaces that need to be cleaned.

○ **Creating a master cleaning schedule.** Identify what should be cleaned, how it should be cleaned, when it should be cleaned, and by whom.

○ **Selecting cleaning tools and supplies.**

Once the master cleaning schedule has been created, schedule a kick-off meeting with all staff to introduce the program. Empower "clean team" members to train their coworkers in the new cleaning and sanitizing procedures in their department. Give team members an appropriate reward for their hard work. Schedule periodic meetings with the "clean team" to monitor the progress of the cleaning program. Modify the program if necessary.

4. How To HAZCOM

Purpose: *To enlist the help of employees in assessing the effectiveness of the establishment's Hazard Communication Program and to modify the program if any deficiencies are found.*

Time: several weeks

Directions: Enlist a "safety team" of employees from different positions in your establishment. Work with the team to assess the establishment's Hazard Communication Program. Focus on the following areas:

○ **Hazardous chemical inventory.** Is there a written inventory indicating the chemicals by name and the places they are stored? Is the list current?

○ **Labeling procedures.** Are chemical containers labeled with the chemical's name, the manufacturer's name and address, and the potential hazards of the chemical?

○ **MSDS.** Are there MSDS for each chemical stored at the establishment? Are these sheets accessible to all employees at all times while on the job?

○ **Training.** Are employees aware of all chemicals' health hazards, and have they been trained to use them properly? Do employees know where to find MSDS, and do they know how to read them? Do they know what personal protective equipment (PPE) is available, and can they use it properly?

○ **Written HAZCOM plan.** Does the establishment have a written plan describing how it will meet the requirements of the Hazard Communication Standard (HCS)?

You may assign different "safety team" members to assess the current status in each of the above areas. Each team member should report his or her findings to the group so an action plan can be created to correct any deficiencies. The "safety team" should meet frequently (perhaps monthly) to assess how well the establishment is meeting the requirements of the HAZCOM program. Team members should receive an appropriate reward for their efforts.

DISCUSSION QUESTIONS

1. When should food-contact surfaces be cleaned and sanitized?

2. Why should different cleaning agents not be mixed together?

3. What are the steps that should be taken (in order) when cleaning and sanitizing items in a three-compartment sink?

4. How should clean and sanitized tableware, utensils, and equipment be stored?

5. What must be included in a master cleaning schedule?

6. Where should MSDS be stored?

MULTIPLE-CHOICE STUDY QUESTIONS

1. Which factor influences the effectiveness of a chemical sanitizer?
   A. The amount of time the sanitizer is in contact with the item
   B. The temperature of the sanitizing solution
   C. The concentration of the sanitizer in the solution
   D. All of the above

2. You want to make a spray solution for use in sanitizing food-contact surfaces in the establishment. What should you do to ensure that you have made a proper sanitizing solution?
   A. Compare the color of the solution to another solution of known strength
   B. Try out the solution on a food-contact surface
   C. Test the solution with a sanitizer test kit
   D. Use very hot water when making the solution

3. What is the proper procedure for sanitizing a table that has been used to prepare food?
   A. Spray it with a strong sanitizing solution, then wipe it dry
   B. Wash it with a detergent, rinse it, then wipe it with a sanitizing solution
   C. Wash it with a detergent, then wipe it dry
   D. Wipe it with a dry cloth, then wipe it with a sanitizing solution

4. Which of the following items need to be both cleaned and sanitized?
   A. Floors   C. Cutting boards
   B. Walls   D. Ceilings

5. What is the best cleaner to use to remove mineral deposits from a warewashing machine or steam table?
   A. Detergent   C. Solvent cleaner
   B. Acid cleaner   D. Abrasive cleaner

6. Which one of the following steps is incorrect for cleaning and sanitizing a standing food mixer?

   A. Clean the food and dirt from around the base of the mixer

   B. Remove the detachable parts and wash them in the warewashing machine

   C. Wash and rinse nondetachable food-contact surfaces. Wipe them with a sanitizing solution

   D. Dry the detachable parts with a clean cloth and reassemble the machine

7. Your warewashing machine is not working and you must use your three-compartment sink to clean and sanitize tableware. What is the first thing you must do?

   A. Fill the first sink with hot water and detergent

   B. Clean and sanitize the sinks and drainboards

   C. Prepare the sanitizing solution in the third sink

   D. Gather clean towels for drying items

8. Which of the following is an *improper* method for storing clean and sanitized tableware and equipment?

   A. Storing glasses and cups upside down

   B. Storing tableware six inches off the floor

   C. Storing flatware in containers with the handles down

   D. Storing utensils in a covered container until needed

9. What is the purpose of Material Safety Data Sheets (MSDS)?

   A. To protect customers from potentially dangerous food substances

   B. To provide information regarding the proper use of potentially dangerous tools in the kitchen

   C. To keep employees informed of the hazards associated with the chemicals they work with

   D. To detail the specifications of proper construction materials for establishments

For answers, please see Appendix A.

ADDITIONAL RESOURCES

Books and Periodicals

Hernandez, J. 1998. Keeping it clean. *Restaurant Hospitality.* 82 (9):118.

Marriott, N. G. 1994. *Principles of food sanitation.* New York: Chapman & Hall.

National Restaurant Association. 1998. *Sanitation survival kit.* Washington, DC: National Restaurant Association.

Web Sites

Chlorine Chemistry Council®

http://www.c3.org

The Chlorine Chemistry Council, comprised of chlorine and chlorinated product manufacturers, is a business council of the American Chemistry Council. At this Web site, you will find the history of chlorine, its many uses at home, in the restaurant, and in the foodservice industry, and chlorine in the news.

Colgate-Palmolive Company

http://www.colpalipd.com

A manufacturer of powerful, performance-tested cleaners and sanitizers, Colgate-Palmolive provides a Web site that contains useful information on its foodservice products and on food safety issues.

ECOLAB, INC.

http://www.ecolab.com

An excellent source of information on housekeeping and sanitation supplies for the restaurant and foodservice industry from the world's leading sanitation product supplier.

FDA Food Code

http://vm.cfsan.fda.gov/~dms/foodcode.html

As the basis for many local sanitation codes, as well as the basis for information in this textbook, the FDA Food Code, available at this Web address, is a useful resource for information relating to food safety for the restaurant and foodservice industry.

Johnson Wax Professional

http://www.jwp.com

Johnson Wax Professional is a global marketer of hygiene and appearance products and solutions for most every type of sanitation, cleaning and maintenance task—from floor care, carpet care, and restroom and laundry care to foodservice, food safety, and pest elimination. Information on setting up food safety systems tailored specifically to your type of operation, as well as product information, can be found on this Web site.

KatchAll/San Jamar

http://www.sanjamar.com

KatchAll is an industry leader in innovative food safety products for the restaurant and foodservice industry. The Web site contains information on all KatchAll products that control time and temperature and help avoid cross-contamination.

MSDSonline

http://www.msdsonline.com

MSDSonline is a provider of electronic solutions for the outbound and inbound management of Material Safety Data Sheets (MSDS). From this site your employees can access MSDS for all chemicals used in our operation.

National Restaurant Association

http://www.restaurant.org

The National Restaurant Association is the leading business association for the restaurant industry. Together with the National Restaurant Association Educational Foundation, the Association's mission is to represent, educate, and promote the rapidly growing restaurant and foodservice industry. This Web site should be your starting place for all issues and concerns related to your restaurant. This Web site has it all, from tips for running your establishment to vital data on your customers' spending habits.

Occupational Safety and Health Administration (OSHA)

http://www.osha.gov

OSHA's mission is to prevent work-related injuries, illnesses, and deaths. This Web site provides information related to workplace-safety conferences, technical links, and regulation and compliance documents.

Procter & Gamble

http://www.pg.com

Procter & Gamble produces cleaning and food and beverage products for the restaurant and foodservice industry. This Web site provides information on its products.

Reckitt Benckiser Professional

http://www.reckittprofessional.com

The Web site of this producer of household products and over-the-counter pharmaceutical brands offers product and brand information.

Chapter 12
Integrated Pest Management

Knowledge

TEST YOUR FOOD SAFETY KNOWLEDGE

1. **True or False:** Once pests enter the building, it is very difficult to eliminate them. *(See page 12-2.)*

2. **True or False:** Prevention is the most critical aspect of integrated pest management (IPM). *(See page 12-2.)*

3. **True or False:** Glue traps should be used to control roaches. *(See page 12-6.)*

4. **True or False:** Your establishment should have corresponding Material Safety Data Sheets (MSDS) on hand any time pesticides are used. *(See page 12-15.)*

5. **True or False:** Pesticides can be stored in food-storage areas if they are labeled properly and closed tightly. *(See page 12-16.)*

Key Terms

Integrated pest management (IPM)
Pest control operator (PCO)
Infestation
Air curtains
Pesticide
Residual sprays
Contact sprays
Glue boards

Learning Objectives

After completing this chapter, you should be able to

○ Develop an integrated pest management (IPM) program.

○ Identify ways to prevent pests from entering the facility.

○ Detect signs of infestation and identify the pests involved.

○ Evaluate and select a pest control operator (PCO).

○ Work with a PCO to choose the best method to control pests in your facility.

Pests such as insects and rodents can pose serious problems for restaurants and foodservice establishments. Not only are they unsightly to customers, they also damage food, supplies, and facilities. The greatest danger from pests comes from their ability to spread diseases, including foodborne illnesses.

THE INTEGRATED PEST MANAGEMENT (IPM) PROGRAM

Once pests have come into the facility in large numbers—an **infestation**—they can be very difficult to eliminate. Developing and implementing an **integrated pest management (IPM)** program is the key. An IPM program uses *prevention* measures to keep pests from entering the establishment and *control* measures to eliminate any pests that do infest it.

For your IPM program to be successful, it is best to work closely with a licensed **pest control operator (PCO).** These professionals use safe, up-to-date methods to prevent and control pests. Prevention is critical in pest control. If you wait until there is evidence of pests in your establishment, they may already be there in large numbers—a major infestation.

There are three basic rules of an IPM program:

○ Deny pests access to the establishment

○ Deny pests food, water, and a hiding or nesting place

○ Work with a licensed PCO to eliminate pests that do enter

Deny Pests Access to the Establishment

Pests can enter an establishment in one of two ways. They either are brought inside with deliveries, or they enter through openings in the building itself. To prevent pests from entering your establishment in deliveries, start by using reputable suppliers. Check all deliveries before they enter your facility. Refuse any shipment in which you find pests or signs of infestation, such as egg cases and body parts (legs, wings, etc.).

Holes, cracks, and other openings in the exterior of a building allow pests to get inside and hide, but there are many other ways pests can enter an establishment.

Doors, Windows, and Vents

○ **Screen all windows and vents with at least sixteen mesh per square inch screening.** Anything larger might let in mosquitoes or flies. Check screens regularly, and clean and replace them as needed.

Key Point

Check all deliveries before they enter the establishment, and refuse any shipment in which you find pests or signs of infestation.

○ **Install self-closing devices or door sweeps on all doors.** Repair gaps and cracks in doorframes and thresholds. Use weather stripping on the bottoms of doors with no threshold.

○ **Install air curtains (also called air doors or fly fans) above or alongside doors.** These devices blow a steady stream of air across the entryway, creating an air shield around open doors. Insects avoid air curtains.

○ **Keep drive-through windows closed when not in use.**

○ **Keep all exterior openings closed tightly.**

Pipes

Mice, rats, and insects such as cockroaches use pipes as highways through a facility.

○ **Use concrete to fill holes or sheet metal to cover openings around pipes.** *(See Exhibit 12a.)*

○ **Install screens over ventilation pipes and ducts on the roof.**

○ **Cover floor drains with hinged grates.** Rats are good swimmers and can enter buildings through drainpipes.

Floors and Walls

Rodents often burrow into buildings through decaying masonry or cracks in building foundations. They move through floors and walls the same way.

○ **Seal all cracks in floors and walls.** Use a permanent sealant recommended by your PCO or local health department.

○ **Properly seal spaces or cracks where stationary equipment is fitted to the floor.** Use an approved sealant or concrete, depending on the size of the spaces.

○ **Paint a white stripe around the edge of your storeroom six inches from the walls.** This stripe will remind employees to store supplies six inches from the walls. It will also help you monitor the area for signs of infestation, since hairs, tracks, and droppings will show up better against the white stripe.

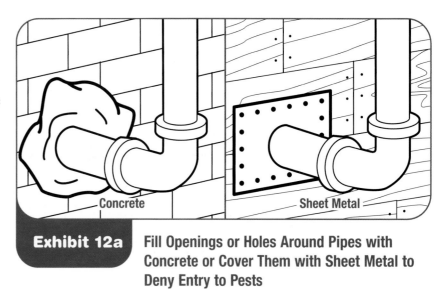

Exhibit 12a Fill Openings or Holes Around Pipes with Concrete or Cover Them with Sheet Metal to Deny Entry to Pests

Concrete — *Sheet Metal*

Key Point

Garbage should be stored in clean, tightly covered containers that are in good condition. Dispose of garbage quickly so it will not attract pests.

Deny Food and Shelter

Pests are usually attracted to damp, dark, dirty places. A clean and sanitary establishment offers them little in the way of food and shelter. The stray pest that might get in cannot thrive or multiply in a clean kitchen. Besides adhering to your master cleaning schedule, follow these additional guidelines:

○ **Dispose of garbage quickly and correctly.** Garbage attracts pests and provides them a breeding ground. Keep garbage containers clean, in good condition, and tightly covered in all areas (indoors and outdoors). Clean up spills around garbage containers immediately. Wash and rinse containers regularly.

○ **Store recyclables in clean, pest-proof containers, as far away from your building as local regulations allow.** Bottles, cans, paper, and packaging material provide shelter and food for pests.

○ **Store all food and supplies properly and as quickly as possible.**

- Keep all food and supplies at least six inches off the floor and six inches away from the walls.

- When possible, keep humidity at 50 percent or lower. Low humidity helps prevent roach eggs from hatching. Use dehumidifiers as well as ventilation that forces air outside the building.

- Refrigerate food such as powdered milk, cocoa, and nuts after opening. Most insects that might be attracted to this food become inactive at temperatures below 41°F (5°C).

- Follow FIFO, so pests don't have time to settle into products and breed.

○ **Clean your establishment thoroughly.** Careful cleaning eliminates the food supply, destroys insect eggs, and reduces the number of places pests can safely take shelter.

- Clean up food and beverage spills immediately, including crumbs and scraps.

- Clean toilets and restrooms as often as necessary.

- Train employees to keep lockers and break areas clean. Food and dirty clothes should not be kept in or around lockers. Break rooms should be cleaned properly after use.

- Keep cleaning tools and supplies clean and dry. Store wet mops on hooks, rather than on the floor since roaches love to hide in them.

- Empty water from buckets to keep from attracting rodents.

Key Point

Clean your establishment thoroughly. Careful cleaning eliminates a pest's food supply, destroys insect eggs, and reduces the number of places pests can take shelter.

Grounds and Outdoor Dining Areas

The popularity of outdoor dining presents different concerns. Birds, flies, bees, and wasps can be annoying and dangerous to the health of your customers. The key to protecting customers lies in minimizing these outdoor pests by denying them food and shelter. (*See Exhibit 12b.*)

○ **Mow the grass, pull weeds, get rid of standing water, and pick up litter.**

○ **Cover all outdoor garbage containers.**

○ **Remove dirty dishes and uneaten food from tables, cleaning them as quickly as possible.**

○ **Do not allow employees or customers to feed birds or wildlife on the grounds.**

○ **Locate electronic insect eliminators ("zappers") away from food, customers, employees, serving areas, and combustible material.**

○ **Call your PCO to remove hives and nests.**

Exhibit 12b **Minimize Pests in Outdoor Dining Areas** Cover garbage containers, remove dirty dishes, and clean up spills to deny food to pests.

IDENTIFYING PESTS

Pests may still get into your establishment even if you take careful preventive measures. Pests are good hitchhikers, hiding in delivery boxes and even coming in on employees' clothing or personal belongings. Learn how to spot signs of pests and identify what kind they are. If possible, record the time, date, and location of any pest sighting (or evidence of pests) and report this to your PCO. Early detection gives your PCO a chance to start treatment as soon as possible.

Cockroaches

Roaches often carry disease-causing microorganisms such as *Salmonella* spp., fungi, parasite eggs, and viruses. Research shows that many people are allergic to residue left by roaches on food and surfaces. Roaches reproduce quickly and can adapt to some pesticides, making it difficult to control them.

There are several different types of roaches. Most live and breed in dark, warm, moist, hard-to-clean places. You will typically find them:

○ Behind refrigerators, freezers, and stoves

○ In sink and floor drains

○ In spaces around hot-water pipes

○ Inside equipment, often near motors and other electrical devices

○ Under shelf liners and wallpaper

○ Underneath rubber mats

○ In delivery bags and boxes

○ Behind unsealed coving (especially rubber-based)

Roaches generally feed in the dark. If you see a cockroach in daylight, you may have a major infestation. Only the weakest roaches come out in daylight.

If you suspect you have a roach problem, check for these signs:

○ A strong oily odor

○ Droppings (feces), which look like grains of black pepper

○ Capsule-shaped egg cases that are brown, dark red, or black and may appear leathery, smooth, or shiny

You may have problems with more than one type of roach. Glue traps—containers with sticky glue on the bottom (*see Exhibit 12c*)—should be used to find out what type of roaches might be present. Work with your PCO to place the traps where roaches typically can be found. If possible, place them on the floor in the corner where two walls meet. Check the traps after twenty-four hours and show them to your PCO. The type and stage (nymph or adult) of roaches present will determine the type of treatment needed. Common types of roaches include American, Brown-banded, German, and Oriental. (*See Exhibit 12d.*)

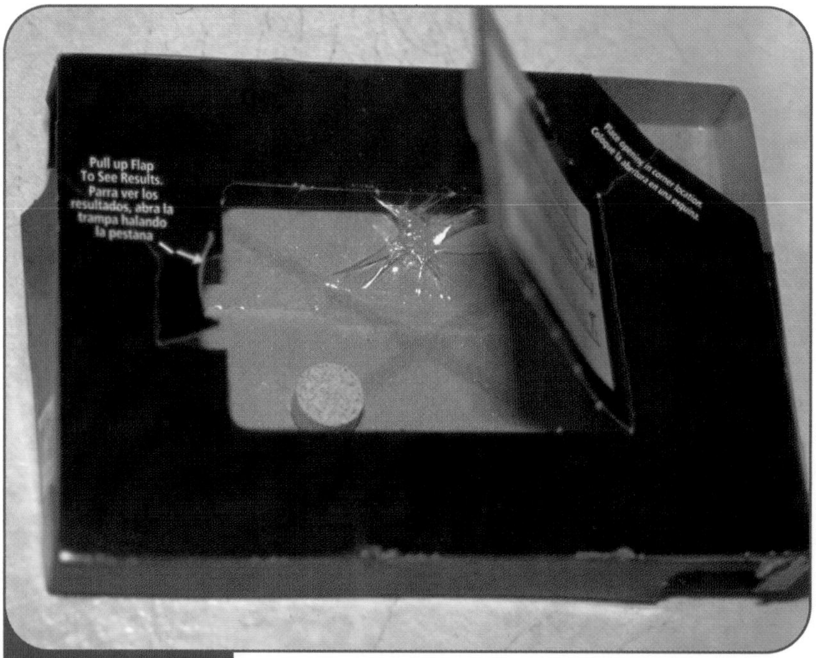

Courtesy of the National Pest Management Association

Exhibit 12c **Roach Glue Trap**
Glue traps can be placed in areas where roaches typically are found to monitor their presence.

| American | Brown-banded |
| German | Oriental |

Courtesy of Orkin Commercial

Exhibit 12d Common Cockroaches Found in Restaurants and Foodservice Establishments

Flies

The common housefly is also a great threat to human health. Because they feed on garbage and animal waste, flies transmit foodborne illnesses such as typhoid fever and dysentery. They also spread pathogens such as *Streptococcus* and *Staphylococcus,* which can stick to their feet, hair, and mouths. Their feces and vomitus—swarming with bacteria—contaminate food. Flies have no teeth, so they only can eat liquids or dissolved foods. They inject a long spike and vomit into solid food, let it dissolve, and then eat it.

In general, houseflies have the following characteristics:

○ They prefer calm air and the edges of objects, such as rims of garbage cans.

○ To find food and lay their eggs, they are drawn to odors of decay, garbage, and animal waste.

○ In warm weather, they reproduce rapidly (eggs can hatch in as few as thirty hours). For their eggs to hatch, they need warm, moist, decayed material located out of the sun. Eggs hatch into maggots, which can grow into adult flies in six days.

Other types of flies can be just as bothersome. Fruit flies are somewhat less harmful because they are primarily attracted to spoiled fruit, not animal waste. They can still transmit diseases, however. "Biting" flies, such as deer flies and horse flies, can be an additional nuisance to outdoor diners.

Other Insects

○ **Beetles, weevils, and moths.** Flour-moth larvae and beetles are usually found in dry storage areas. Look for insect bodies, wings, or webs, as well as clumped-together food and holes in food and packaging. To prevent these pests, cover food tightly, keep storage areas clean and sanitary, and practice FIFO.

○ **Ants.** Ants often nest in walls and floors near stoves and hot-water pipes. They are drawn to grease and sweet food. Clean up all food scraps and spills to keep them out.

○ **Termites and carpenter ants.** These insects can cause great structural damage, boring into wood and weakening walls, floors, and ceilings. They are rarely visible. Carpenter ants nest in wood, but forage for food elsewhere. Look for signs of sawdust that has fallen from the ceiling. Call your PCO at once if you see signs they are present.

○ **Spiders.** Look for webs in corners and around fixtures. Spiders can be controlled by checking for and removing webs daily. The bites of some spiders—such as the black widow and brown recluse—are poisonous and can cause illness, but are rarely fatal. If you see either type, call your PCO.

○ **Bees, wasps, and hornets.** Bees, wasps, and hornets can sting outdoor diners. Some people are allergic to their venom and can go into shock, or even die, from just one or two stings. In general, they are drawn to sweet food. Here are some suggestions for handling them:

  • Bees make hives in hollow trees, under eaves, or in other protected places. They usually will not bother people unless their hives are threatened. If you find a hive, call your PCO to remove it.

  • Wasps and hornets often build nests under eaves. Call your PCO to remove them.

  • Yellow jackets usually nest in the ground and are drawn to proteins in early summer and sweets in late summer. They aggressively defend their nests, so call your PCO to remove them.

○ **Mosquitoes and gnats.** Mosquitoes carry diseases such as malaria, typhoid, yellow fever, and encephalitis. These members of the fly family feed on the blood of animals and humans. Mosquitoes are drawn to heat and are most active at dawn and dusk. Gnats are annoying to outdoor diners. To minimize their presence, keep all outdoor areas free of stagnant water and move lights away from the building.

○ **Deer ticks.** Ticks can carry Lyme disease and cause other serious illnesses. Minimize risk by keeping the grass short around dining areas.

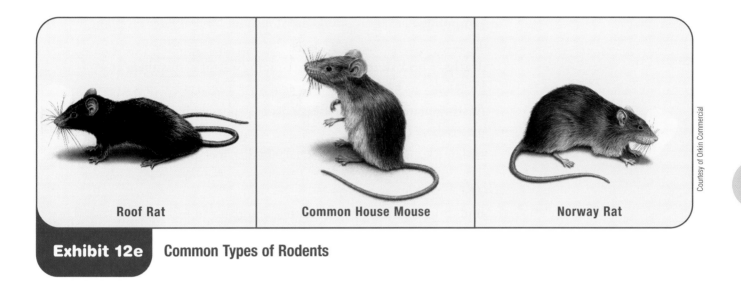

| Roof Rat | Common House Mouse | Norway Rat |

Courtesy of Orkin Commercial

Exhibit 12e Common Types of Rodents

Rodents

Rodents are a serious health hazard. They eat and ruin food, damage property, and can spread disease. Most rodents have a simple digestive system. They urinate and defecate as they move around a facility. Their waste can fall into food and can contaminate surfaces.

Rats and mice are the most common type of rodent. *(See Exhibit 12e.)* They hide during the day and search for food at night. Like other pests, they reproduce often. Typically, they do not travel far from their nests—rats travel only 100 to 150 feet, while mice travel only ten to thirty feet. Mice can squeeze through a hole the size of a dime to enter a facility, rats through quarter-sized holes. Rats can stretch to reach an item as high as eighteen inches, can jump three feet in the air, and can even climb straight up brick walls. They have very good senses of hearing, touch, and smell, and are smart enough to avoid poison bait and poorly laid traps. Effective control of rats and mice requires the knowledge and experience of professionals.

A building can be infested with both rats *and* mice at the same time. Look for these signs:

○ **Signs of gnawing.** Rats and mice gnaw to reach food and to wear down their teeth, which grow continuously. Rats' teeth are so strong they can gnaw through pipes, concrete, and wood. *(See Exhibit 12f.)*

Courtesy of the National Pest Management Association

Exhibit 12f Signs of a Rodent Infestation
Rats and mice gnaw to reach food and to wear down their teeth.

○ **Droppings.** Fresh droppings are shiny and black. Older droppings are gray.

○ **Tracks.** Check dusty surfaces by shining a light across them at a low angle.

○ **Nesting materials.** Mice use scraps of paper, cloth, hair, and other soft materials to build nests.

○ **Holes.** Rats nest in burrows, usually in dirt, rock piles, or along foundations.

Birds

Bird droppings carry fungi and bacteria that can make people sick. They also may carry mites and microorganisms that can cause encephalitis and other diseases. Birds can be drawn to crumbs and food scraps in outdoor dining areas. Keep these areas clean, and remove food from tables quickly. Post signs asking customers not to feed birds. If birds become a problem, call a PCO specializing in bird-control measures.

Other Animals

Though less common, other animals such as bats, raccoons, and squirrels can infest your building too. Building damage, bites, and the possible spread of rabies are the biggest concerns. Prevent them from getting inside. If there is evidence any of these pests have entered your establishment, work with a PCO or your local animal control department to remove them.

○ **Bats.** Bats will nest in high places that are warm, dark, and dry, often in eaves or attics. Bats are beneficial, since they eat insects, and in many places it is against the law to kill them. However, they will bite if cornered and can spread rabies. Established bat colonies can be hard to eliminate. At dusk, after bats have left to feed, block all crevices and entrances. Light the area for several days to keep them away until they find a new place to roost.

○ **Raccoons.** Raccoons have adapted to humans and urban growth. They feed at night, but it's not unusual to see them during the day. They prefer wooded areas, but females will nest in dark places, such as attics and chimneys, to give birth. Raccoons can contract rabies, and they will bite if cornered.

○ **Squirrels.** Like raccoons, squirrels will nest in attics if given the chance. Clear branches away from buildings and fill any holes or cracks around eaves and gutters.

WORKING WITH A PEST CONTROL OPERATOR (PCO)

Few pest problems are solved simply by spraying pesticides—chemical agents used to destroy pests. Although you can take most preventive measures yourself, most control measures should be carried out by professionals. Employ a licensed, certified PCO to handle pest control. Working as a team, you and the PCO can prevent and/or eliminate pests and keep them from coming back. There are many advantages to working with a professional PCO, who will:

○ Work with you to develop an integrated approach to pest management. This may include using a combination of chemical and nonchemical treatments to solve and prevent problems.

○ Stay up to date on new equipment and products.

○ Provide prompt service to address problems as they occur. Most contracts should provide regular visits plus immediate service when pests are spotted.

○ Take responsibility for keeping records on the use and safe handling of potentially hazardous chemicals, as well as all steps taken to prevent and control pests.

How to Choose a PCO

Hiring a PCO is like choosing any other supplier or service provider: You must do your homework. Use these guidelines to help make your decision.

○ **Talk to other foodservice managers.** Find out what PCO they use and what their experiences have been. When you call a PCO, ask for references and check them thoroughly. Make sure the PCO has experience working with restaurants and other foodservice establishments.

○ **Make sure the PCO is licensed or certified by your state, as required by federal law.** Certification means the PCO has passed a test on proper pesticide use and other control methods.

○ **Ask the PCO if they belong to any professional organizations.** Membership in the National Pest Management Association (NPMA) or state or local groups usually means the PCO has up-to-date information on IPM.

○ **Ask for proof of insurance.** Make sure the PCO has adequate coverage to protect you and your employees, customers, and facility.

○ **Weigh all factors, not just price.** Make sure the PCO you choose has the expertise and resources you need to provide the service promised. A low bid may be costly in the long run.

Key Point

Most pest control measures should be carried out by professional pest control operators (PCOs).

Key Point

When choosing a PCO, make sure he or she is licensed or certified, belongs to professional organizations, and is insured.

Key Point

Always require a contract from your PCO. Service contracts should spell out the work to be performed and what is expected from both you and your PCO.

Service Contract

Always require a written service contract from your PCO. Service contracts outline the work to be performed and what is expected from both you and the PCO. Read your contract carefully and have your lawyer review it, if possible. A contract should include these items:

○ A description of services to be provided, including an initial inspection, regular monitoring visits, follow-up visits, and emergency service

○ A warranty for work to be done

○ Legal liability of the PCO

○ The period of service

○ Your duties, including preventive measures and facility preparation before and after treatment

○ Records to be kept by the PCO, such as:

  • Pests sighted and trapped, species, location, and actions taken

  • All chemicals used and Material Safety Data Sheets (MSDS) for each

  • Building and maintenance problems noted and fixed

  • Maps or photos of the facility noting location of traps, bait, and problem spots

  • Schedule for checking and cleaning traps, replacing bait, and reapplying chemicals

  • Regular written summary reports from the PCO

TREATMENT

Effective treatment starts with a thorough inspection of your facility and grounds. Give the PCO complete access to the building and cooperate fully during the inspection.

○ Prepare employees to answer the PCO's questions.

○ Provide building plans and equipment layouts.

○ Point out possible trouble spots.

After the initial inspection, your PCO should outline—in writing— a treatment plan. In addition to price, the plan should contain the following information:

○ **Exactly what treatment will be used for each area or problem and the potential risks involved.**

Key Point

After the initial inspection, your PCO should outline a treatment plan in writing.

Courtesy of the National Pest Management Association.

○ **Dates and times of each treatment.** The federal government requires a PCO to give you enough advance warning to prepare the facility properly. Employees must not be on site during the treatment.

○ **Steps you can take to control pests.**

○ **Building defects that may cause problems for prevention and control measures.**

○ **Timing of follow-up visits.** The PCO should review how well treatment is working and suggest alternate treatments if pests reappear.

CONTROL MEASURES

PCOs can use a variety of pest-control methods that are environmentally sound and safe for establishments. They are trained to know which techniques will work best to control different types of pests in your area. Since new technologies are being developed all the time, the more you know about each of these methods, the better you can evaluate how well your PCO is doing.

Controlling Insects

There are several methods your PCO can use to control insects, depending on the type of insect and degree of infestation.

○ **Repellents.** Repellents are liquids, powders, or mists that keep insects away from an area, but do not kill them. Repellents are often used in hard-to-reach places, such as the spaces behind wallboards and plaster.

○ **Sprays.** Chemical **pesticide** sprays are used to control insects. Because you can buy them in any supermarket, they are easy to misuse and abuse. Let your PCO, not your employees, use these types of sprays to control insects. Letting employees apply pesticides is risky, and, applied improperly, they will be ineffective. Prepare the area to be sprayed by removing all food and food-contact surfaces. Cover equipment and food-contact surfaces that can't be moved. Wash, rinse, and sanitize food-contact surfaces after the area has been sprayed.

  ● **Residual sprays** leave a film of insecticide that insects absorb as they crawl across it. Used in cracks and crevices like those along baseboards, these sprays can be liquid or a dust, such as boric acid.

  ● **Contact sprays** kill insects on contact. They are usually used on groups of insects, such as clusters of roaches or a nest of ants.

○ **Bait.** Chemical bait sometimes is used to control roaches or ants. The bait contains an attractant, and, when insects eat it, the chemical kills them. The advantage of using baits is that kitchen areas don't need to be prepped and people can remain on site while they are set.

○ **Traps.** There are several types of traps.

- Light-only units simply entice insects to crawl inside where they find it hard to escape.

- Electronic insect eliminators or "zappers" use an electrically charged grid to kill insects attracted to the light.

- Other units use both light and chemical attractants to lure insects onto a glue board.

Wasp and hornet traps are designed to hold nectar, which attracts them. Once inside, they cannot escape. Most fly traps use a light source—usually UV light—to attract insects to the trap. Tests have shown that different insects respond to different types or ranges of UV light. Make sure your PCO is using the most current technology.

The placement of traps is important. Never place them above or near food-preparation areas or food-contact surfaces.

Controlling Rodents

Rats and mice tend to use the same routes through an establishment. Your PCO will choose the best method to eliminate these pests, which could include the following: *(See Exhibit 12g.)*

○ **Traps.** Traps are a safe, effective way to kill rats and mice. If the infestation is large, however, traps will take time. Work with your PCO to set traps near or in rodent runways. Check traps often and remove dead rodents carefully. If a trapped rodent is still alive, have your PCO remove it.

○ **Glue boards.** Glue boards kill mice. When these devices are placed in runways, they stick to the board and die in several hours—from exhaustion or lack of water or air. Check boards often and throw away any with trapped mice. Glue boards are not effective for controlling rats since these rodents are usually strong enough to escape.

○ **Bait.** Chemical bait should only be used by a PCO in areas where it cannot contaminate food or food-contact surfaces. It is usually placed in special covered, locked containers near rodent runways and possible entry points. Your PCO may change the baits and their locations often until they begin to work. Rats can easily detect chemical bait and often avoid it.

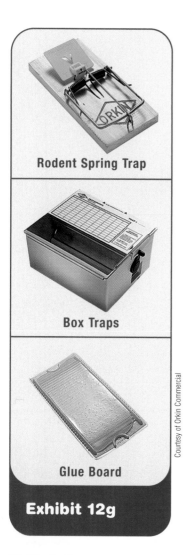

Rodent Spring Trap

Box Traps

Courtesy of Orkin Commercial

Glue Board

Exhibit 12g

Methods for Controlling Rodents
Several devices can be used to control rodents.

Controlling Birds

Birds can be a serious problem in outdoor dining areas. While there is no way to eliminate birds, your PCO can use several techniques to keep them from nesting and roosting on your building.

○ **Repellents.** Chemical pastes are sometimes used on gutters and ledges to repel birds. Paste must be used carefully so it doesn't fall into food or onto tables.

○ **Netting.** Fine-mesh wire netting is used to keep birds from roosting on statues and bas-relief carvings on buildings.

○ **Wires.** A PCO might string wires across the roof to prevent birds from roosting. The wires can be electrified to deliver a mild shock.

○ **Sound.** In some instances, birds can be frightened away by the sound of other birds in distress. Prerecorded tapes are played through loudspeakers near roosting sites, though birds might eventually ignore the sounds and return to roost.

○ **Balloons.** Birds sometimes can be frightened away with helium-filled, mylar balloons. Ask your PCO about this technique.

USING AND STORING PESTICIDES

While it may seem more cost-effective to purchase and apply pesticides yourself, there are many reasons not to do so.

○ Pesticides can be dangerous to employees, customers, and food.

○ Pests can develop resistance and an immunity to pesticides.

○ Each region has its own pest-control problems, and some control measures are more effective than others.

○ Pesticides are regulated by federal, state, and local laws, and some are not approved for use in restaurants and foodservice establishments.

Rely on your PCO to decide if and when pesticides should be used in your establishment. They are trained to determine the best pesticide for each pest, and how and where to apply it.

Pesticides are hazardous materials. Any time one is used or stored on the premises, you should have corresponding MSDS. To minimize the hazard to people, have your PCO use pesticides only when you are closed for business, and employees are not on site.

Your PCO should store and dispose of all pesticides used in your facility. If they are stored on the premises, follow these guidelines:

○ **Keep pesticides in their original containers.**

Key Point

Pesticides should only be applied by a PCO. They are trained to determine the best pesticide for each pest, and how and where to apply it.

Courtesy of the National Pest Management Association.

○ **Store pesticides in locked cabinets away from areas where food is stored or prepared.**

○ **Store aerosol or pressurized spray cans in a cool place.** Exposure to temperatures higher than 120°F (49°C) could cause them to explode.

○ **Check local regulations before disposing of pesticides.** Many are considered hazardous waste. Dispose of empty containers according to manufacturer's directions and local regulations.

SUMMARY

Pests can carry and spread a variety of diseases. Once they have infested a facility, it can be very difficult to eliminate them. Developing and implementing an integrated pest management (IPM) program is the key. An IPM program uses *prevention* measures to keep pests from entering the establishment and *control* measures to eliminate any pests that do get inside. To be successful, establishments must deny pests access to the facility and deny them food, water, and shelter. Finally, you must work with a licensed pest control operator (PCO) to eliminate pests that do enter.

Pests can be brought inside the establishment with deliveries, or they can enter through openings in the building itself. To prevent them from getting inside, check deliveries before they enter your facility and refuse any shipment in which you find pests or signs of infestation. Screen all windows and vents, install self-closing doors and air curtains, and keep exterior openings closed when not in use. Fill or cover holes around pipes, and seal cracks in floors and walls.

Pests are usually attracted to damp, dark, dirty places. A clean and sanitary establishment offers them little in the way of food and shelter. Stick to your master cleaning schedule. Dispose of garbage quickly and keep containers clean and tightly covered in all areas. Store recyclables as far from your building as allowed. Keep food and supplies at least six inches off the floor and six inches away from the walls. Follow FIFO, so pests don't have time to settle into products and breed. Protect outdoor diners by denying pests food and shelter in these areas as well. Mow grass, remove standing water, and pick up litter. Cover outdoor garbage containers. Remove dirty dishes and uneaten food from tables, and clean them as quickly as possible. Do not allow employees or customers to feed birds or wildlife on the grounds.

Understanding pests is the key to controlling them. Roaches live and breed in dark, warm, and moist places. Check for a strong oily odor and droppings, which look like grains of black pepper, and egg cases. Use glue traps to find out what types of roaches are present. Rodents also are a serious health hazard.

A building can be infested with both rats and mice at the same time. Look for droppings, signs of gnawing, tracks, nesting materials, and holes.

Although you can take most prevention measures yourself, most control measures should be carried out by professionals. Employ a licensed, certified PCO to handle pest control and require them to provide a written service contract outlining the work to be performed, as well as what is expected from both you and them. Working as a team, you can prevent and/or eliminate pests and keep them from coming back.

There are several methods your PCO can use to control insects, depending on the type of insect and degree of infestation. Control methods include repellents, sprays, chemical bait, and traps. Rodents can be controlled through the use of traps set near runways, glue boards (for mice), and chemical bait.

While it might seem cost-effective to apply pesticides yourself, there are many reasons for not doing so. Pesticides can be dangerous to your employees, customers, and food. Some pesticides are not approved for use in a restaurant or foodservice establishment, and pests can develop immunity to them. Rely on your PCO to decide if and when pesticides should be used in your establishment. They are trained to determine the best pesticide for each pest, and how and where to apply it.

Pesticides are hazardous materials. Any time pesticides are used or stored on the premises, you should have corresponding Material Safety Data Sheets (MSDS). To minimize the hazard to people, have your PCO use pesticides only when you are closed for business and your employees are not on site. Your PCO should store and dispose of all pesticides used in your facility. If they are stored on the premises, they should be kept in their original containers and stored in locked cabinets away from food-storage and food-preparation areas.

A CASE IN POINT

Case Study

Now We're Cooking is a restaurant located in the middle of town on the first floor of a landmark ninety-year-old building. The building is in a shopping area that includes several other restaurants. Fred, the manager, recently remodeled the restaurant. He chose materials that are easy to clean. All of the new foodservice equipment is designed for ease of cleaning as well. The cleaning procedures listed on Fred's master cleaning schedule are written out in detail and completed by the employees as scheduled, with frequent self-inspections. Spills are cleaned up immediately. FIFO is followed, and food is stored on metal racks, which keep it six inches off of the floors and away from the walls.

During a self-inspection two weeks after he finished remodeling, Fred finds live cockroaches behind the sinks, in the vegetable storage area, in the public restrooms, and near the garbage-storage area.

What factors might Fred have overlooked in his recent remodeling that could have led to this infestation? What should Fred do to eliminate the roaches?

TRAINING TIPS

1. *"Pests" versus "People"*

Purpose: *After completing this activity, trainees will be able to identify prevention measures to keep pests from entering a facility and describe control measures that can be taken to treat an infestation.*

Time: 30 minutes

Directions: Divide trainees into two groups: "Pests" and "People." Divide the "Pests" into four subgroups: roaches, flies, rats, and mice. Divide the "People" into two subgroups: foodservice managers and pest control operators (PCOs). Print handouts for each group with the following questions:

Pests

○ How can you get into a facility?

○ Once inside, where do you typically seek shelter?

○ What do you need to survive and reproduce?

○ What do you prefer to eat? Where and when do you eat?

Foodservice Managers

○ How can you prevent roaches, flies, and rodents from entering the facility?

○ What steps can you take to deny these pests food and shelter once they are inside?

Pest Control Operators (PCOs)

○ What methods can you use to control roaches, flies, and rodents once they are inside a facility?

Ask each group to appoint a spokesperson to present its answers. Discuss each group's answers and provide additional input. Remind the entire group that an integrated pest management (IPM) program, with vigilant prevention and control measures, is essential in the fight against pests.

2. Help Wanted: PCO!

Purpose: *After completing this activity, trainees will be able to identify prevention and control measures for pest infestations.*

Time: one hour

Directions: Ask the group for three volunteers. These volunteers are PCOs. The group will assume the role of a foodservice manager in charge of hiring a new PCO for their establishment.

While the three PCOs wait in another area, ask the group to develop a series of interview questions that focus on testing the PCOs' knowledge of pest prevention and control measures. Questions asked could include:

○ How can you determine if an establishment has a cockroach infestation?

○ How would you control a cockroach infestation?

○ How would you prevent a rodent infestation?

After interviewing each of the three candidates separately, have the group evaluate their responses and choose a PCO. Bring all three PCO candidates back into the room, announce the winner, and let a spokesperson for the group explain why a particular candidate was chosen.

3. Pest Prevention Inspection

Purpose: *To identify areas in your facility where pests may enter, and to determine steps that should be taken to prevent entry.*

Time: several days

Directions: Conduct a thorough inspection of your entire facility. If possible, ask your PCO to perform the inspection with you. Pay particular attention to the following areas:

- Doors, windows, and vents
- Floors and walls
- Pipes
- The building's foundation
- Receiving areas

Once possible entry points have been identified, a plan should be outlined for correcting these problems. While conducting the inspection, look for possible sources of food and shelter for potential pests. Check the way you currently clean the facility, store food and supplies, and store and dispose of garbage and recyclables. Identify ways in which you might improve your current practices.

4. Pest Patrol

Purpose: *To enlist the help of your employees in the early detection of a pest infestation in your establishment. This information can then be used by your PCO to isolate specific problems and problem areas quickly.*

Time: ongoing activity

Directions: Solicit volunteers from your establishment to become members of the Pest Patrol. Explain the signs of insect and rodent infestations and ask them to watch for and report any signs of pests.

If an employee observes signs of an infestation, the sighting should be recorded in a "Pest Log" posted in or near the manager's office. The log should be a record of all relevant information, including the type of pest sighted, when, where, and by whom.

DISCUSSION QUESTIONS

1. How can you prevent pests from entering an establishment?

2. What pests are associated with outdoor dining? How can you minimize the presence of outdoor pests?

3. How can you tell if your establishment has been infested with cockroaches or rodents?

4. Why is it important to be able to identify the type of pest that has infested an establishment?

5. What factors should be considered when choosing a pest control operator (PCO)?

MULTIPLE-CHOICE STUDY QUESTIONS

1. Which of the following is a sign you might have a problem with rodents?
   A. You find capsule-shaped egg cases
   B. You find scraps of paper and cloth gathered in the corner of a drawer
   C. You see droppings that look like black grains of pepper
   D. You smell a strong, oily odor

2. Cockroaches typically are found in places that are
   A. cold, dry, and light.
   B. cold, moist, and dark.
   C. warm, moist, and dark.
   D. warm, dry, and light.

3. Which of the following is a sign you might have a problem with cockroaches?

A. You find small holes burrowed through the storeroom wall

B. You find droppings that look like black grains of pepper underneath a refrigeration unit

C. You see small piles of sawdust that appear to have fallen from the ceiling

D. You find webs and wings in the dry-storage area

4. Why does careful cleaning help prevent insect infestations?

A. It eliminates their food supply

B. It destroys insect eggs

C. It reduces the number of places pests can take shelter

D. All of the above

5. You want to implement an IPM program in your establishment. What should you do first?

A. Create a plan that emphasizes control measures

B. Position glue boards around the establishment

C. Work with a PCO to create a pest prevention and control plan

D. Work with local government to develop a treatment plan

6. To prevent a pest infestation, you should store food at least _____ inch(es) off of the floor.

A. one

B. two

C. four

D. six

7. Your storeroom has a white line painted on the floor around the perimeter of the room. It extends six inches from the wall. What is its purpose?

A. It seals the floor against moisture seeping from the walls

B. It serves as a reminder to employees to keep stored items away from the walls

C. It works as a repellent against nesting mice

D. It works as a repellent against cockroaches

8. What is the primary reason an establishment must act quickly to control roach infestations?

    A. Roaches can disturb the foundation of the establishment

    B. Roaches can drive away customers

    C. Roaches eat large quantities of stored food

    D. Roaches carry disease-causing microorganisms that can make customers ill

9. Anytime pesticides are stored on the premises

    A. they must be transferred from the original container to smaller containers.

    B. they must be stored away from food-preparation areas.

    C. corresponding MSDS must be kept by the PCO.

    D. they must be registered with the local regulatory agency.

ADDITIONAL RESOURCES

Books and Periodicals

Buettner, R., Fan, M., and George, T. "Guess who's crawling to dinner…uninvited: Majority of eateries in city are losing the war with vermin." *New York Daily News,* June 2, 1998, p. 7.

Mixon, J. A. 1991. *Guidelines for the storage and care of food products: A technical assistance manual: Vol. V.,* 3rd ed. Washington, DC: Food Industry Services Group, USDA Food and Nutrition Service.

Pest management in restaurants: An integrated approach. 1997. Washington, DC: National Restaurant Association, Technical Services, Public Health and Safety Department.

Web Sites

ECOLAB, INC.

http://www.ecolab.com

An excellent source of information on housekeeping and sanitation supplies for the restaurant and foodservice industry from the world's leading sanitation-product supplier.

FDA Food Code

http://vm.cfsan.fda.gov/~dms/foodcode.html

As the basis for many local sanitation codes, as well as the basis for information in this textbook, the FDA Food Code, available at this Web address, is a useful resource for information relating to food safety for the restaurant and foodservice industry.

Johnson Wax Professional

http://www.jwp.com

Johnson Wax Professional is a global marketer of hygiene and appearance products, as well as solutions for most every type of sanitation, cleaning, and maintenance task—from floor care, carpet care, and restroom and laundry care to foodservice, food safety, and pest elimination. Information on setting up systems of food safety products and food safety services tailored specifically to your type of operation can be found on this Web site.

National Integrated Pest Management Network (NIPMN)

http://www.reeusda.gov/nipmn

The National IPM Network is an evolving and cooperating group of universities, government agencies, and other organizations coming together to provide up-to-date, accurate information for pest management. This Web site has a wealth of information on pest control, arranged by commodity, pest, region, and tactics.

National Pest Management Association

http://www.pestworld.org

The mission of the National Pest Management Association is to communicate their role as protectors of food, health, property, and the environment, and to affect the success of its members through education and advocacy. This Web site provides information on how to become a member, key conferences, sales leads, and a pest control operator (PCO) locator.

National Restaurant Association

http://www.restaurant.org

The National Restaurant Association is the leading business association for the restaurant industry. Together with the National Restaurant Association Educational Foundation, the Association's mission is to represent, educate, and promote the rapidly growing restaurant and foodservice industry. This Web site should be your starting place for all issues and concerns related to your restaurant. This Web site has it all, from tips for running your establishment to vital data on your customers' spending habits.

Occupational Safety and Health Administration (OSHA)

http://www.osha.gov

OSHA's mission is to prevent work-related injuries, illnesses, and deaths. This Web site provides information related to workplace-safety conferences, technical links, and regulation and compliance documents.

Orkin Commercial

http://www.orkin.com/commercial

From the leader in pest control, Orkin's Web site contains useful information on how to prevent and control pests in your foodservice establishment. It also contains a bug guide providing all the information you'll ever want to know about ants, rodents, flies, and birds.

UNIT 4

SANITATION MANAGEMENT

Chapter 13
Food Safety Regulation and Standards

Knowledge

TEST YOUR FOOD SAFETY KNOWLEDGE

1. **True or False:** The FDA writes the food regulations that must be followed by each establishment. *(See page 13-3.)*

2. **True or False:** Health inspectors are generally employees of the Centers for Disease Control and Prevention (CDC). *(See page 13-3.)*

3. **True or False:** You should ask to accompany the health inspector during the inspection of your establishment. *(See page 13-6.)*

4. **True or False:** Critical violations noted during a health inspection should be corrected within one week of the inspection. *(See page 13-8.)*

5. **True or False:** A HACCP-based inspection focuses on the flow of food in an establishment as opposed to the sanitary appearance of the facility. *(See page 13-10.)*

For answers, please turn to Appendix A.

Learning Objectives

After completing this chapter, you should be able to

○ Identify the principles and procedures needed to comply with food safety regulations.

○ Identify state and local regulatory agencies and regulations that require food safety compliance.

○ Prepare for a regulatory inspection.

○ Identify the proper procedures for guiding a health inspector through the establishment.

○ Explain how the application of HACCP and HACCP principles can help ensure food safety and regulatory compliance.

Key Terms

U.S. Department of Agriculture (USDA)
Food and Drug Administration (FDA)
U.S. Public Health Service (USPHS)
Regulations
Health inspector

U.S. Centers for Disease Control and Prevention (CDC)
U.S. Environmental Protection Agency (EPA)
U.S. National Marine Fisheries Service (NMFS)

There are several reasons why it is important to have a foodservice inspection program. Most important is that failure to ensure food safety can jeopardize the health of your customers and could cost you your business. All establishments—including quick-service and fine-dining restaurants, delis, hospitals, nursing homes, and schools—must follow standard food safety practices critical to the safety and quality of the food served. An inspection system lets the establishment know how well it is following these practices.

OBJECTIVES OF A FOODSERVICE INSPECTION PROGRAM

All establishments serving the public must provide safe food and are subject to inspection. It does not matter whether there is a charge for the food or whether the food is consumed on or off premises.

The purpose of an inspection program is to:

○ Evaluate whether the establishment is meeting minimum sanitation and food safety standards

○ Protect the public's health by requiring establishments to provide food that is safe, uncontaminated, and presented properly

○ Convey new food safety information to the establishment

○ Provide an establishment with a written report, noting deficiencies, so it can be brought into compliance with safe food practices

GOVERNMENT REGULATORY SYSTEM FOR FOOD

Every country has a history of government involvement in the development of health laws. Today, government control of food in the United States is exercised at three levels: federal, state, and local.

At the federal level, the **U.S. Department of Agriculture (USDA),** the **Food and Drug Administration (FDA),** and the **U.S. Public Health Service (USPHS)** are directly involved in the inspection process.

The USDA is responsible for inspection and quality grading of meat, meat products, poultry, dairy products, eggs and egg products, and fruit and vegetables shipped across state boundaries. The USDA provides these services through the Food Safety and Inspection Service (FSIS) agency.

The FDA is the agency that writes the Food Code. *(See page 13-3.)* In addition it inspects foodservice operations that cross state borders (interstate

Key Point

The FDA writes the Food Code, which provides recommendations for restaurant and foodservice regulations.

establishments such as those on planes and trains, as well as food manufacturers and processors), because they overlap the jurisdictions of two or more states. The FDA shares responsibility with the USDA for inspecting food-processing plants to ensure standards of purity, wholesomeness, and compliance with labeling requirements. The USPHS inspects cruise ships crossing international waters.

In the United States, most food **regulations** affecting restaurant and foodservice operations are written at the state level (except regulations for interstate or international establishments, which are determined at the federal level). Each state decides whether to adopt the Food Code or some modified form of it.

State regulations may be enforced by local (city or county) or state health departments. In a large city, the city health department will probably be responsible for enforcing health codes. In smaller cities or in rural areas, a county or state health department may be responsible for enforcement. In any case, the manager must be familiar with the local agencies and their enforcement system. City, county, or state **health inspectors** (also called sanitarians, health officials, or environmental health specialists) conduct restaurant and foodservice inspections in most states. They generally are trained in food safety, sanitation, and public health principles and methods.

THE FOOD CODE

The Food Code is written by the FDA based on input from the Conference for Food Protection (CFP). The Food Code lists the government's recommendations for foodservice regulations. Currently, these recommendations are updated every two years to reflect developments in the restaurant and foodservice industry and the field of food safety.

The Food Code is intended to assist state health departments in developing regulations for a foodservice inspection program. It is not an actual law. Although the FDA recommends adoption by the states, it cannot require it. Rather, the Food Code represents the FDA's best advice for a uniform system of regulation to ensure food safety. Some states use the Food Code as a basis for developing their own codes, instead of adopting it in its entirety. Food and sanitation codes are written very broadly and generally cover the following areas:

○ **Foodhandling and preparation:** sources, receiving, storage, display, service, transportation

○ **Personnel:** health, personal cleanliness, clothing, practices

○ **Equipment and utensils:** materials, design, installation, storage

Recommendations for restaurant and foodservice regulations are written at the federal level, regulations are made at the state level, and enforcement is carried out at the local level.

Federal Level

Regulations recommended

State Level

Regulations written

Local Level

Regulations enforced

Key Point

Managers must contact local health departments to find out which specific regulations apply to their operations.

Key Point

An establishment can provide input on the code requirements in its area.

○ **Cleaning and sanitizing:** facility, equipment
○ **Utilities and services:** water, sewage, plumbing, restrooms, waste disposal, integrated pest management (IPM)
○ **Construction and maintenance:** floors, walls, ceilings, lighting, ventilation, dressing rooms, locker areas, storage areas
○ **Foodservice units:** mobile, temporary
○ **Compliance procedures:** restaurant and foodservice inspections, enforcement actions

Adoption and interpretation of food safety standards may vary widely from one state to another or, in some cases, from one locality to another. Therefore, restaurant and foodservice managers must consult with local health departments to find out which specific regulations apply to their operations.

Currently, when a state adopts, develops, or amends its food code, it is required to provide time for public input and comment. This is the time when any interested person or association can comment on how the proposed changes will impact their establishment(s). If you are interested in participating in this process, make sure you keep in contact with your health department or State Restaurant Association.

The lack of uniformity in local food codes can frustrate efforts by the restaurant industry to establish uniform food safety standards. For example, some jurisdictions require refrigerated food temperatures to be 45°F (7°C) or lower, while others require 41°F (5°C) or lower. States may also differ in the recommended inspection frequency. Some states require inspections for restaurant and foodservice establishments at least every six months, while others schedule inspections more or less frequently. Some states may even differ as to which establishments should be inspected; for example, some states do not inspect convenience food stores. Some states might have separate regulations or agencies for different types of foodservice operations (such as vending machines or delis in grocery stores).

Although all inspectors focus on food safety practices, the areas emphasized during the inspection can vary among jurisdictions or individual inspectors. For example, some inspectors are more concerned with refrigeration temperatures while others may focus on the physical appearance of the facility. One inspector may examine the flow of food while another may focus primarily on the personal hygiene of employees. It is the responsibility of the manager to keep food safe and wholesome throughout the establishment at all times, regardless of the inspector or the inspection process.

FOODSERVICE INSPECTION PROCESS

Well-managed establishments will perform continuous self-inspections to keep food safe, in addition to the regular inspections performed by the health department. Establishments with high standards for sanitation and food safety consider health department inspections only a supplement to their self-inspection programs.

There are many benefits that come from establishing a good self-inspection program. A good program can result in safer food, improved food quality, a clean environment for employees and customers, and higher inspection scores. Strive to exceed the expectations of both your customers and the health department. The higher you set your standards, the more likely you are to do well on your health department inspection. In addition, your customers will notice your commitment to providing them a safe, sanitary dining experience.

During health department inspections, the local health code serves as the inspector's guide. You should keep a current copy of your local or state sanitation regulations and be familiar with them. Regularly compare the code to procedures at your establishment, but remember that code requirements are only *minimum* standards to keep food safe.

Keep in mind that changes continue to occur in the inspection process. Some areas use traditional, scored inspection systems with a number or letter grade, while others use HACCP-based inspections, or a combination of the two.

Traditional Inspection System

Some health departments are required to conduct inspections at least every six months. However, the frequency will vary depending on the area, type of establishment, or food served. Many use a risk-based approach to inspection frequency. Determining factors can include:

○ **Size and complexity of the operation.** Larger operations offering a considerable number of potentially hazardous food items might be inspected more frequently.

○ **Inspection history of the establishment.** Establishments with a history of low sanitation scores or consecutive violations might be inspected more frequently.

○ **Clientele's susceptibility to foodborne illness.** Nursing homes, schools, day-care centers, and hospitals might receive more frequent inspections.

○ **Thoroughness of the operation's HACCP program.** Establishments with HACCP programs might be inspected less frequently.

○ **Workload of the local health department and the number of inspectors available.**

Key Point

Health department inspections should be viewed as a supplement to a self-inspection program.

Key Point

The frequency of regulatory inspections will vary based on several factors, including complexity of the establishment's menu, the size of the operation, and its clientele.

In most cases, an inspector will arrive without prior warning and ask for the manager or the person in charge of the operation. In your absence, make sure your employees know who is in charge. Some companies have specific policies on how to handle an inspection. Make sure you are aware of those policies. Establishments should not refuse entry. In some jurisdictions, inspectors could have the authority to gain access or to revoke the establishment's permit for refusal to allow an inspection.

It is best to accompany the sanitarian during the inspection. This allows you to answer any questions and, possibly, correct deficiencies immediately. Accompanying the inspector will also give you the opportunity to learn from the inspector's comments and suggestions, and to gain sanitation advice.

After the inspection, the inspector will discuss the results and the score (if a score is given) and arrange for any follow-up, if necessary. Managers will be asked to sign the inspection report to acknowledge they have received it. Follow your company's policy regarding this issue. A copy of the report is then given to the manager or person in charge at the time of the inspection. All deficiencies noted on the report should be acted upon. Critical deficiencies should be corrected immediately or within forty-eight hours of the inspection. All other deficiencies should be corrected as soon as possible. Copies of all reports should be kept on file in the establishment and referred to when planning improvements and assessing facility goals. Copies of reports are kept on file at the health department. These are considered public documents and may be made available to the public upon request.

The following steps will help managers and operators get the most out of sanitation inspections:

1. **Ask for identification.** Many inspectors will volunteer their credentials. Do not let anyone enter the back of the facility without proper identification. Ask the purpose of the visit; make sure you know whether it is a routine inspection, the result of a customer complaint, or for some other purpose.

2. **Cooperate.** Most inspectors have learned to expect some defensiveness and resentment from restaurant and foodservice operators. This can be interpreted as having something to hide. Answer all of the inspector's questions to the best of your ability. Instruct employees to do the same. Explain to the inspector that you wish to accompany him or her during the inspection. This will encourage open communication and a good working relationship. If a deficiency can be corrected quickly, do so, or tell the inspector when it can be corrected.

3. **Take notes.** As you accompany the inspector, make a note of any problem pointed out. Make it clear you are willing to correct problems. Taking your

Key Point

Accompany the sanitarian during the inspection so you can answer any questions.

own notes will help you remember exactly what was said. If you believe the inspector is incorrect about something, note what was commented upon. Then ask the inspector's supervisor for a second opinion.

4. **Keep the relationship professional.** Don't offer food or drink before, during, or after an inspection. This could be viewed as bribery.

5. **Be prepared to provide records requested by the inspector.** You might ask the inspector why these records are needed. If a request appears inappropriate, you can check with the inspector's supervisor or with your lawyer about limits on confidential information. Records you provide to the inspector will become part of the public record. Inspectors might ask for records of purchases to verify that food is received from an approved source, records of integrated pest management (IPM) treatments, and a list of all chemicals used in the facility. HACCP records could be requested in some cases. In fact, HACCP records very well could be an important part of an inspection, because they document the establishment's efforts to ensure food safety. Having a HACCP system in place clearly shows the inspector you are committed to food safety.

6. **Discuss violations and time frames for correction with the inspector.** The inspection report should be studied closely. Deficiencies and comments should be discussed in detail with the inspector. If any deficiencies were corrected on the spot, make sure they are noted. In order to make complete and permanent corrections, you will need to know the exact nature of the violation, how it impacts food safety, how to correct it, and whether or not the inspector will do any follow up. The inspector sees many operations and may offer expert advice on how to correct deficiencies.

Exhibit 13a

Inspection Follow Up
Determine why each violation occurred by evaluating sanitation procedures, the master cleaning schedule, and employee foodhandling practices. Establish new procedures or revise existing ones.

7. **Follow up.** Take the inspection report with you through your facility and correct the problems. Determine why each problem occurred by evaluating sanitation procedures, the master cleaning schedule, and employee foodhandling practices. Establish new procedures or revise existing ones to correct the problem permanently, if necessary. *(See Exhibit 13a.)*

Many health departments use a traditional inspection system using a demerit scoring scale. *(See Exhibit 13b.)* Usually the highest possible score is one hundred points. For every violation, between one and five points are subtracted from one hundred to get the final score. Noncritical violations worth one or two points must be corrected by the time of the next routine inspection. Critical violations worth four to five points must be corrected within a time frame specified by the inspector. Other health departments use a letter to score the establishment. Whatever scoring system is used, make sure you understand it, as well as your options for improving an unsatisfactory score.

Establishments are generally given a short amount of time (forty-eight hours or less) to correct critical violations and improve their overall score. If a low score is received upon reinspection, the establishment may be fined or even closed.

In some states, if the inspector determines a facility poses an immediate and substantial health hazard to the public, he or she may ask for a voluntary closure, or issue an immediate suspension of the permit to operate. A suspension requires the approval of the local health offices and means that operations at the establishment must cease immediately. Examples of hazards calling for closure include:

○ Significant lack of refrigeration

○ Backup of sewage into the establishment or its water supply

○ Emergency, such as a building fire or flood

○ Serious infestation of insects or rodents

○ Long interruption of electrical or water service

The suspension order may be posted at a public entrance to the establishment; however, this is not required if the establishment closes voluntarily. Regulations vary from one jurisdiction to another, but to reinstate a permit to operate, the establishment must eliminate the hazard or hazards causing the suspension, then pass a reinspection.

The health department can request a hearing if it feels that an establishment must correct a hazard, but it does not wish to suspend the establishment's permit. The establishment can also request a hearing if it believes an action by the health department is unjustified. There is often a time limit to request such a hearing (usually within five to ten days after an inspection). Check local regulations to determine these limits.

One limitation of the traditional inspection system centers on the fact that multiple violations of the same type do not cause any more demerits than a single violation. For example, leaving one potentially hazardous food at room

Key Point

Establishments can be closed when the health department feels they pose an immediate and substantial health hazard to the public.

FOOD AND SANITATION INSPECTION REPORT

Address	Inspection Date
Inspector	Inspection Time

Establishment Name

Manager's Name

Based on an inspection this day, the items marked below identify the violations in operation or facilities that must be corrected by the next routine inspection or such shorter time as may be specified in writing by the regulatory. Failure to comply with any time limits for corrections specified in this notice may result in cessation of your Foodservice operations.

Item #	Wt	
		FOOD
01	5#	From approved, licensed sources, free of spoilage, no home-processed food, no dented or damaged cans.
02	1	Food containers properly labeled.
		FOOD PROTECTION
03	5*	Potentially hazardous food must be maintained at 41°F or 140°F or above; dairy products, meat, poultry, fish, cooked or broiled potatoes, beans, and rice. Cooling: 1) Shallow pans, product 3 inches in depth, refrigerated. 2) Precooled in ice bath with frequent stirring to 41°F. **NO COOLING AT ROOM TEMPERATURE.**
04	4*	Adequate equipment to maintain proper food temperature.
05	1	Accurate thermometers; conspicuous.
06	2	Potentially hazardous food properly thawed 1) in a refrigerator, 2) under cold running water, 3) by microwave, 4) as part of cooking.
07	4*	Unwrapped and potentially hazardous food not re-served.
08	2	Food protected from contamination; i.e., covered, off floors, etc.
09	2	Handling of food (ice) minimized, proper utensils provided and used.
10	1	Food-dispensing utensils properly stored when in use.
		PERSONNEL
11	5*	Personnel with infectious disease, cuts or burns restricted from handling food, clean utensils/ equipment.
12	5*	Hands washed, good hygienic practices. No smoking, eating, or drinking in kitchen.
13	1	Clothes clean, hair restrained.
		FOOD EQUIPMENT
14	2	Food-contact surfaces nontoxic, smooth, durable, nonabsorbent, and easily cleanable.
15	1	Nonfood surfaces smooth, durable, nonabsorbent, and easily cleanable.

Item #	Wt	
16	2	Dishwashing facilities properly designed, constructed, maintained, installed, located, operated.
17	1	Dishwashing facilities provided with accurate thermometers, pressure gauge, chemical test kit.
18	1	Soiled equipment, dishes, and utensils preflushed, scraped, soaked.
19	2	Wash and rinse water clean, hot.
20	4*	Washing and Sanitizing 3-compartment sink: 1) wash 3) sanitize 2) rinse 4) air dry Sanitizers: 50 ppm chlorine, 200 ppm quaternary ammonia, or 12.5 ppm iodine for 1 minute. Warewashing machine: 180°F final rinse temperature or 50 ppm chlorine at dish level.
21	1	Wiping cloths: Stored in 100 ppm chlorine or 25 ppm iodine.
22	2	Food-contact surfaces of equipment: cutting boards, meat slicers, can openers, work counters, etc., shall be washed, rinsed, and sanitized. Surfaces free of abrasives and detergents.
23	1	Nonfood-contact surfaces of equipment and utensils clean.
24	1	Clean equipment and utensils properly stored and handled to prevent contamination.
25	1	Single-service items properly stored and used to prevent contamination.
26	2	No reuse of single-service articles.
		WATER
27	5*	Safe water source. Hot and cold water provided at all times.
		SEWAGE
28	4*	Sewage and wastewater (mop) properly disposed
		PLUMBING
29	1	Plumbing properly installed and maintained
30	5*	No cross-connection between potable and wastewater. Backflow devices present.
		TOILET AND HANDWASHING FACILITIES
31	4*	Handwashing sinks accessible.

Item #	Wt	
32	2	Handsinks provided with soap and single-service towels. Toilet rooms clean, in good repair, doors self-closing, trash receptacles.
		GARBAGE AND REFUSE
33	21	Garbage containers sanitarily maintained, covered.
34	1	Outside refuse areas maintained, clean.
		INSECT AND RODENT CONTROL
35	4*	No insects, rodents, or other animals present. Outer openings protected. IPM policies in place.
		FLOORS, WALLS, AND CEILINGS
36	1	Floors clean and constructed to be smooth, durable, nonabsorbent, and properly covered.
37	1	Walls, ceilings, and attached equipment clean and constructed to be smooth, durable, and nonabsorbent.
		LIGHTING
38	1	Lighting adequately shielded.
		VENTILATION
39	1	Rooms and equipment vented as required.
		DRESSING ROOMS
40	1	Employees' personal items properly stored in locker or separate area.
		OTHER OPERATIONS
41	5*	Chemicals/cleaning agents stored away from food-preparation and utensil-washing areas; chemicals labeled.
42	1	Premises clean, maintained. No unnecessary articles. Cleaning/ maintenance equipment properly stored. Authorized personnel only.
		FIRE EXTINGUISHERS
43	1	Sufficient number, servicing performed, personnel familiar with use.
		FOOD AND SANITATION TRAINING PROGRAM
44		

Exhibit 13b Traditional Inspection Report Form

temperature incurs the same number of demerits as leaving all food at room temperature. Clearly, leaving all food at room temperature is a greater health concern.

As a result, a new inspection system is being implemented in some areas. In this type of inspection system, a narrative format is used to report violations rather than a demerit system. When inspectors see problems with a facility, they write up the problems—as well as any recommended corrections— in paragraph form. Reference numbers to the health code and repeated violations are noted, with multiple violations of the same type receiving greater emphasis. In the old scoring system, violations were considered *critical* or *noncritical.* The new system allows inspectors to use their professional judgment regarding some of the violations. It also provides establishments with more feedback. See *Exhibit 13c* for an example of the new inspection report form recommended by the FDA.

HACCP-Based Inspections

Some health departments use HACCP-based inspections, focusing on the flow of food rather than on the sanitary appearance of the facility. Inspectors can observe the way an establishment receives, stores, prepares, cooks, serves, cools, and reheats food, and the critical control points identified. Since this type of inspection could be viewed as complex and time-consuming, it might only be performed under special circumstances. A health department might perform a HACCP-based inspection to do the following:

○ Trace the source of contamination following a report of foodborne illness

○ Evaluate an establishment with significant hazards

○ Assist an establishment in developing or implementing a HACCP system

Other health agencies are using the HACCP-based approach to decide how often different establishments should be inspected, or to determine how the inspections should be conducted. With this approach, scoring is done differently. Instead of using a point system with demerits, HACCP-based inspections determine critical violations. Many of these critical violations are similar to the four- and five-point items identified in the traditional inspection system. *Exhibit 13d* on page 13-12 shows one type of form used during a HACCP-based inspection.

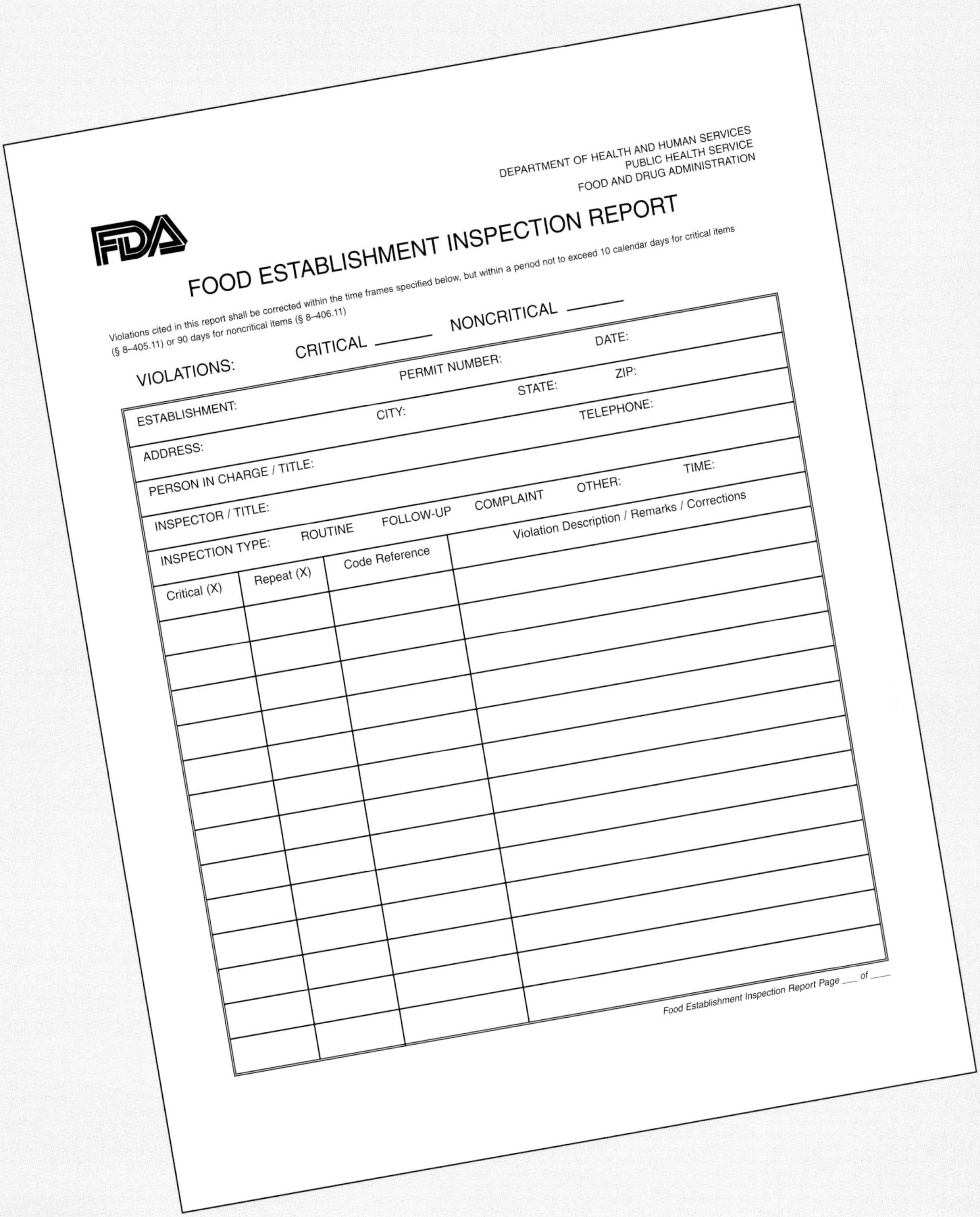

DEPARTMENT OF HEALTH AND HUMAN SERVICES
PUBLIC HEALTH SERVICE
FOOD AND DRUG ADMINISTRATION

FDA

FOOD ESTABLISHMENT INSPECTION REPORT

Violations cited in this report shall be corrected within the time frames specified below, but within a period not to exceed 10 calendar days for critical items (§ 8–405.11) or 90 days for noncritical items (§ 8–406.11)

VIOLATIONS: CRITICAL _____ NONCRITICAL _____

	PERMIT NUMBER:			DATE:		
ESTABLISHMENT:	CITY:		STATE:	ZIP:	TELEPHONE:	
ADDRESS:						
PERSON IN CHARGE / TITLE:						
INSPECTOR / TITLE:					TIME:	
INSPECTION TYPE:	ROUTINE	FOLLOW-UP	COMPLAINT	OTHER:		

Critical (X)	Repeat (X)	Code Reference	Violation Description / Remarks / Corrections

Food Establishment Inspection Report Page ____ of ____

Exhibit 13c   **New Inspection Report Form Recommended by the FDA**

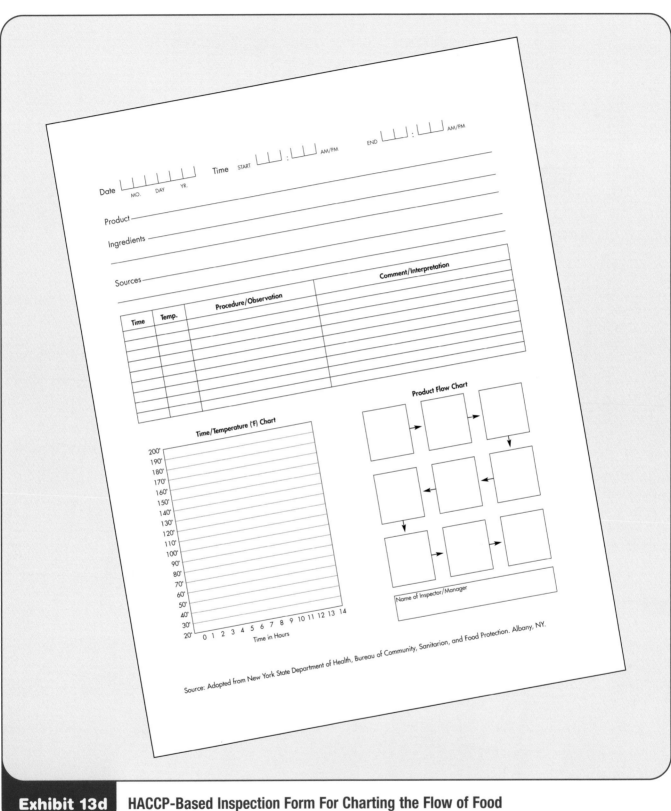

Date ⌊_⌊_⌊_⌊_⌋   MO.   DAY   YR.

Time   START ⌊_⌊_⌋ : ⌊_⌊_⌋ AM/PM   END ⌊_⌊_⌋ : ⌊_⌊_⌋ AM/PM

Product

Ingredients

Sources

Time	Temp.	Procedure/Observation	Comment/Interpretation

Time/Temperature (°F) Chart

200°
190°
180°
170°
160°
150°
140°
130°
120°
110°
100°
90°
80°
70°
60°
50°
40°
30°
20°
    0 1 2 3 4 5 6 7 8 9 10 11 12 13 14
        Time in Hours

Product Flow Chart

Name of Inspector/Manager

Source: Adopted from New York State Department of Health, Bureau of Community, Sanitarion, and Food Protection. Albany, NY.

Exhibit 13d   **HACCP-Based Inspection Form For Charting the Flow of Food**

FEDERAL REGULATORY AGENCIES

Several federal government agencies work to protect the sanitary quality of food in an establishment. Earlier in this chapter, the roles of the FDA, USDA, and USPHS were discussed. A few other agencies concerned with food safety should also be mentioned.

The **U.S. Centers for Disease Control and Prevention (CDC),** located in Atlanta, Georgia, are agencies of the USPHS. The Centers provide the following services:

○ Investigate outbreaks of foodborne illness

○ Study the causes and control of disease

○ Publish statistical data and case studies in the *Morbidity and Mortality Weekly Report (MMWR)*

○ Provide educational services in the field of sanitation

○ Conduct the Vessel Sanitation Program—an inspection program for cruise ships

The **U.S. Environmental Protection Agency (EPA)** sets air- and water-quality standards and regulates the use of pesticides (including sanitizers) and the handling of wastes.

The **U.S. National Marine Fisheries Service (NMFS)** of the U.S. Department of Commerce implements a voluntary inspection program that includes product standards and sanitary requirements for fish-processing operations.

VOLUNTARY CONTROLS WITHIN THE INDUSTRY

Few industries have devoted as much effort to regulating themselves as the restaurant and foodservice and food-processing industries. Scientific and trade associations, manufacturing firms, and foodservice corporations have vigorously pursued programs to raise the standards of the industry through research, education, and cooperation with government. Although participation in these programs is voluntary, these organizations have actively promoted professional standards, recommended legislative policy, sponsored uniform enforcement procedures, and provided educational opportunities. The overall results in food safety have included:

○ Increased understanding of foodborne illness and its prevention

○ Improvements in the sanitary design of equipment and facilities

○ Industry-wide efforts to maintain the sanitary quality of food during processing, shipment, storage, and service

○ Efforts to make foodservice laws more science-based, practical, and uniform

Key Point

The National Restaurant Association works closely with government agencies to develop and implement proper guidelines and regulations regarding the food supply.

Key Point

The National Restaurant Association Educational Foundation is dedicated to developing, promoting, and providing educational and training solutions for the restaurant and hospitality industry.

While many organizations have contributed to these endeavors, those having the most relevance for the foodservice operator and manager include:

○ **The National Restaurant Association,** founded in 1919, is the leading business association for the restaurant and foodservice industry. The Association's mission is to represent, educate, and promote the rapidly growing industry. The Association is a strong voice advocating proper food safety practices and science-based regulations. The Association's Health and Safety Regulatory Affairs Department provides information to the industry on topics such as sanitation, pest control, nutrition, ergonomics, and the Americans with Disabilities Act.

In addition, it works closely with government agencies to develop and implement proper guidelines and regulations regarding the food supply. Through its excellent Research Department, the Association generates both operator and consumer research on a variety of issues and trends, including the annual Restaurant Industry Operations Report and the Restaurant Industry Forecast. The Association's successful lobbying arm represents the interests of the industry by being in constant contact with legislators.

Each May, the National Restaurant Association holds the Restaurant, Hotel-Motel Show, one of the largest trade shows in the United States. More than one hundred thousand people come to Chicago during the show to see everything from tablecloths and plates to industrial warewashers and food safety systems.

For more information about the National Restaurant Association, its services, and member benefits, call 800.424.5156 or 202.331.5900.

○ **The National Restaurant Association Educational Foundation (NRAEF)** is a not-for-profit organization dedicated to developing, promoting, and providing educational and training solutions for the restaurant and hospitality industry. The NRAEF has developed training programs and educational curricula regarding issues such as workplace safety, responsible beverage-alcohol service, sexual-harassment prevention, and food safety and sanitation. In 1993, the NRAEF formed the International Food Safety Council as its food safety initiative to increase awareness on the importance of food safety education throughout the restaurant and foodservice industry.

As the educational resource for the industry, the NRAEF administers the largest scholarship program of its kind for the restaurant and hospitality sector. Whether students are beginning their restaurant and hospitality education or continuing a successful career, the NRAEF offers scholarships to help them. Scholarships are available for senior high school students, undergraduate and graduate students, industry professionals, and educators/administrators.

For more information about the National Restaurant Association Educational Foundation and its products and services, call 800.765.2122, ext. 701 (outside the Chicago area) or 312.715.1010, ext. 701 (within the Chicago area). For information or applications for the Scholarships Program, call 312.715.1010, ext. 733.

○ **The National Environmental Health Association (NEHA)** is an organization of environmental specialists and professionals, including those responsible for food-inspection services and environmental health programs. NEHA's Registered Sanitarian (RS) program has established national standards for education, experience, and testing for health inspectors.

○ **The International Association for Food Protection (IAFP)** was a pioneer in the highly successful United States milk-sanitation program. It publishes information on food safety and on generally accepted standard procedures for investigating a foodborne illness.

○ **The Institute of Food Technologists (IFT)** is a multidisciplinary scientific society with expertise in technology, research, education, manufacturing, and the safety of food in the foodservice industries. IFT sponsors food and restaurant industry research in a variety of areas, such as food quality, safety, and process controls.

○ **The National Society of Professional Sanitarians (NSPS)** and the **American Academy of Sanitarians (AAS)** are composed of food industry professionals and sanitarians concerned with environmental health protection policies at the national, state, and local levels.

○ **The Association of Food and Drug Officials (AFDO)** develops and publishes food sanitation codes and encourages food protection through the adoption of uniform legislation and enforcement procedures.

○ **The Council of Hotel and Restaurant Trainers (CHART)** is an association of restaurant and foodservice trainers and human resource professionals. CHART was formed in 1971 to provide members with a forum in which to grow professionally and increase their effectiveness as trainers.

○ **The National Pest Management Association (NPMA)** consists of licensed and certified Pest Control Operators (PCOs) throughout the United States. NPMA provides guidelines and training materials for IPM treatment programs, hazard communications, and other topics relating to safety and sanitation.

○ **NSF International** evaluates and lists foodservice equipment meeting its standards. It also issues the NSF International mark. NSF International develops and publishes widely accepted standards for equipment design, construction, and installation, updated every five years. A listing of equipment meeting these standards is updated every six months. NSF's approach demonstrates the kind of progress the foodservice industry can achieve through voluntary programs.

○ **Underwriters Laboratories, Inc. (UL)** performs a service similar to NSF International, listing equipment that meets NSF International standards. UL also lists products complying with their own published environmental and public health standards. In addition, UL lists electrical equipment that passes its own safety requirements.

SUMMARY

Public and private organizations and agencies can offer valuable assistance to restaurant and foodservice managers in meeting their commitment to food safety. It is up to the manager to use available help and to maintain a sanitary establishment.

Federal governmental agencies create standards affecting the establishment directly or indirectly. The Food Code, developed by the FDA, serves as a guideline for many state and local regulations. In addition, the FDA regulates the purity and safety of food in interstate commerce.

While state and local health codes vary throughout the nation, virtually all contain provisions governing food safety, personal hygiene, sanitary facilities, equipment, and utensils, safe operating practices, training, and enforcement procedures. The inspector—a representative from the state or local health department—is a professional in sanitation and public health.

To keep food safe in your establishment, implement a HACCP program and follow a daily program of self-inspection. The effectiveness of this system will be reflected in inspection reports. HACCP-based inspections—a departure from the traditional emphasis on sanitary facilities—are being adopted by some regulatory agencies. During this type of inspection, inspectors observe the way an establishment receives, stores, prepares, cooks, holds, serves, cools, and reheats food, and the critical control points identified.

Professional and trade organizations in food service and public health recommend guidelines for restaurant and foodservice sanitation, investigate sanitation problems, and promote best practices. These associations also develop educational programs and conduct research into the causes and prevention of foodborne illness.

A CASE IN POINT

Case Study

Carolyn, the inspector from the city's public health department, was standing at the door of Jerry's Diner. Jerry greeted her, and they both walked to the kitchen. Carolyn pulled out a HACCP worksheet. While Jerry explained that the cook was preparing stir-fried chicken and vegetables, Carolyn noted the ingredients and their sources. Then she observed the preparation procedures and noted times and temperatures, which she plotted on a time-temperature graph. She also filled in a product flowchart and indicated the critical control points for the stir-fried chicken. Jerry told her what corrective actions his employees were trained to carry out if critical control points were not met.

Next, Carolyn checked the concentration of the sanitizing solution in the three-compartment sink the dish washer used to clean, rinse, and sanitize equipment. She also checked the handwashing station.

Jerry and Carolyn went to Jerry's office, where they discussed the report. Jerry compared his flowchart for the stir-fried chicken and vegetables with the one Carolyn had done, determining where changes could be made to improve monitoring.

Did Jerry handle the inspection correctly? What does he need to do following this inspection?

TRAINING TIPS

1. Who Regulates What?

Purpose: *After completing this activity, trainees should be able to identify the purpose of different regulatory and food safety agencies.*

Time: 20 minutes

Directions: Create a handout with the following information on it:

List of Agencies   **Purpose of the Agency**

○ Food and Drug Administration

○ U. S. Department of Agriculture

○ Centers for Disease Control and Prevention

○ Food Safety and Inspection Service

○ National Marine Fisheries Service

○ Environmental Protection Agency

○ Local health department

   Ask trainees to identify the purpose of each agency on the handout. Allow ten minutes for the class to complete the activity, and then discuss.

2. Inspection Walkthrough

Purpose: *After completing this activity, trainees should be able to identify proper procedures for guiding a health inspector through an establishment.*

Time: 30 minutes

Directions: Conduct two role-plays.

Role-Play 1: The wrong way

   Select two trainees who have dramatic flair. Ask one to play a health department inspector conducting a health inspection, the other a restaurant manager. Ask them to refer to the proper procedures for guiding an inspector through an establishment as presented in Chapter 13, and do the *opposite* of what is suggested. Ask them to develop a dialogue that is adversarial, somewhat exaggerated, humorous, but realistic. Have them limit their role-play to five to eight minutes. After the role-play, discuss with the group what went wrong. Discuss the correct procedure for guiding an inspector through an establishment, as presented in Chapter 13.

Role-Play 2: The right way

In this role-play, demonstrate the *proper* way to guide an inspector through the establishment. Choose one trainee to act out this role-play with you. Depending on the group, you can choose to be the *manager* or the *inspector*. After the role-play, discuss with the group what went "right."

3. Inspection Report Review and Preview

Purpose: *To evaluate the effectiveness of your food safety system using the inspection requirements of your health department.*

Time: ongoing

Directions: Conduct an inspection of your establishment using the inspection form used by your health department. Don't be easy on yourself, don't make excuses, and don't make exceptions. Pretend you are conducting the inspection of a competitor's establishment!

After the inspection, determine your food safety score. You might solicit help from your health department in determining how to use and score the form. What corrective actions must be taken immediately, and which ones can be taken later?

Before you actually make any corrections, you might ask your chef or assistant manager (or both) to conduct a similar inspection. How do the inspection reports compare? Do you each see food safety from the same perspective? Chances are your inspections will differ in some areas. Discuss these differences and take corrective action.

DISCUSSION QUESTIONS

1. What is the role of the FDA regarding food protection?

2. What are the roles of federal, state, and local agencies regarding the regulation of food safety in establishments?

3. What should a manager do during and after an inspection?

4. What are some factors that determine the frequency of a health inspection in an establishment?

5. What are some of the significant industry organizations that help managers deal with employee training and sanitation regulations and standards?

MULTIPLE-CHOICE STUDY QUESTIONS

1. An establishment can be closed for all of the following reasons *except* a

   A. significant lack of refrigeration.
   B. backup of sewage.
   C. serious infestation of insects or rodents.
   D. minor violations not corrected within twenty-four hours.

2. Which of the following is a goal of the food safety inspection program?

   A. To evaluate whether an establishment is meeting minimum food safety and sanitation standards
   B. To protect the public's health
   C. To convey new food safety information to establishments
   D. All of the above

3. Which operation would most likely be subject to a food safety inspection by a federal agency?

   A. A hospital

   B. A cruise ship crossing international waters

   C. A local ice cream store with a history of safety violations

   D. A food kitchen run by church volunteers

4. A person shows up at a restaurant claiming to be a health inspector. What should the manager do?

   A. Ask to see identification.

   B. Ask to see an inspection warrant.

   C. Ask for a hearing to determine if the inspection is necessary.

   D. Ask for a twenty-four-hour postponement to prepare for the inspection.

5. How does a HACCP-based food safety inspection differ from an ordinary inspection?

   A. It focuses more on the flow of food through the establishment.

   B. It needs to be done more frequently.

   C. It focuses more on the sanitary appearance of the facility.

   D. It uses a point system with demerits.

6. Which of the following agencies enforce food safety in a restaurant?

   A. The FDA

   B. The CDC

   C. State or local health departments

   D. The USDA

7. Violations noted on the health inspection report should be

   A. discussed in detail with the inspector.

   B. corrected within forty-eight hours or less if they are critical.

   C. explored to determine why they occurred.

   D. All of the above

8. The responsibility for keeping food safe in an establishment rests with the:

   A. State health department   C. Health inspector

   B. Manager/operator   D. FDA

9. Food codes developed by state agencies are

   A. minimum standards necessary to ensure food safety.

   B. maximum standards necessary to ensure food safety.

   C. voluntary guidelines for establishments to follow.

   D. inspection practices for grading meats and meat products.

For answers, please turn to Appendix A.

ADDITIONAL RESOURCES

Books and Periodicals

Frable, F. 1997. Industry needs central resource for code, regulation information. *Nation's Restaurant News.* 31(20):44.

Hertneky, P. B. 1996. You and your health inspector. *Restaurant Hospitality.* 80(6):57.

Marriott, N. G. 1994. *Principles of food sanitation.* New York: Chapman & Hall.

Murray, J. 1994. The model food code. *FoodService Director,* 7(4):82.

Penner, K. 1992. *Exploring public policy options: Food safety in food service.* Manhattan, KS: Kansas State University Cooperative Extension Service.

Solis, O. C. 1997. Partner or adversary? *Food & Service.* 58(3):16.

Web Sites

Association for Food and Drug Officials (AFDO)

http://www.afdo.org

AFDO is a leader and trusted resource in people developing strategies to resolve and promote public health and consumer protection issues related to the regulation of food, drugs, medical devices, and consumer products.

Audits International

http://www.audits.com

Audits International is a worldwide leader in providing independent information about safety and food quality at retail. Services include restaurant food safety evaluations, product and package performance evaluations, crisis assistance or resolution, and product retrieval. From this Web site you can learn more about the services they provide, as well as access 1999 and 2000 home food safety surveys.

Centers for Disease Control and Prevention (CDC)

http://www.cdc.gov

The mission of the CDC is to promote health and quality of life by preventing and controlling disease, injury, and disability. To prevent and control foodborne illness, the CDC collects data on outbreaks. This Web site provides general information on foodborne illnesses and their prevention.

Code of Federal Regulations (CFR)

http://access.gpo.gov/nara/cfr

The CFR provides electronic access to the general and permanent rules published in the Federal Register by the executive departments and agencies of the Federal Government. The codes of interest to the restaurant and foodservice industry include Chapters 7 and 21.

Conference for Food Protection (CFP)

http://www.foodprotect.org/

The Conference for Food Protection is a nonprofit organization providing a representative and equitable partnership among regulators, industry, academia, professional organizations, and consumers to identify problems, formulate recommendations, and develop and implement practices ensuring food safety. This Web site houses previous Conference issues and proceedings, its constitution and by-laws, and current Conference activities.

Council of Hotel and Restaurant Trainers (CHART)

http://www.chart.org/

CHART is one of the oldest and largest organizations dedicated to training in the hospitality industry, with a mission to help hospitality-training professionals improve operational performance by developing people. This Web site provides information on how to become a member of CHART, information on conferences, and a sample of their newsletter.

Department of Health and Human Services (HHS)

http://www.hhs.gov.phs/

HHS is the United States government's principal agency for protecting the health of all Americans and providing essential human services. The department includes more than 300 programs, covering a wide spectrum of activities, including the Food and Drug Administration (FDA) and the Centers for Disease Control and Prevention (CDC). The various food safety programs the Department oversees can be accessed from this site.

Environmental Protection Agency (EPA)

http://www.epa.gov

The EPA endeavors to abate and control pollution systematically, by proper integration of a variety of research, monitoring, standard setting, and enforcement activities. This Web site provides information on all areas of pollution prevention and control.

Food and Drug Administration (FDA)

http://www.fda.gov

FDA's mission is to promote and protect the public's health by helping safe and effective products reach the market in a timely manner, as well as to monitor products for continued safety after they are in use. This Web site provides a wealth of information on all areas they regulate, including food and product recalls, enforcement activities, all field programs and their inspectional references, and projects to keep industry and consumers aware of health-related topics.

FDA Enforcement Report Index

http://www/fda/gpv/opacom/Enforce.html

The FDA Enforcement Report, accessible through this Web site, is published weekly by the Food and Drug Administration and contains information on actions taken in connection with agency regulatory activities

FDA Food Code

http://vm.cfsan.fda.gov/~dms/foodcode.html

As the basis for many local sanitation codes, as well as the basis for information in this textbook, the FDA Food Code, available at this Web address, is a useful resource for information relating to food safety for the restaurant and foodservice industry.

FDA-Center for Food Safety and Applied Nutrition (CFSAN)

http://www.cfsan.fda.gov

As the center within the FDA responsible for food safety and nutrition, CFSAN promotes and protects public health by researching and implementing guidelines, policies, and standards to ensure that food is safe, nutritious, wholesome, and properly labeled. This Web site provides a wealth of information on food safety and sanitation, including corresponding guidelines, policies, and standards.

Food and Drug Law Institute

http://www.fdli.org

The Food and Drug Law Institute is a nonprofit institute dedicated to advancing the public health by providing a neutral forum for critical examination of the laws, regulations, and policies related to drugs, medical devices, other healthcare technologies, and foods. This Web site contains information on pertinent publications, conferences, and sources for experts in this field.

Food Marketing Institute (FMI)

http://www.fmi.org

The FMI is a nonprofit association conducting programs in research, education, industry relations, and public affairs on behalf of its members, which include large multistore chains, small regional firms, and independent supermarkets.

foodsafety.gov

http://www.foodsafety.gov

foodsafety.gov is a gateway Web site providing links to selected information related to government food safety. This Web site is a great starting point for accessing all government-related food safety information.

Grocery Manufacturers of America (GMA)

http://www.gmabrands.com

GMA is the world's largest association of food, beverage, and consumer-product companies. With U.S. sales of more than $460 billion, GMA members employ more than 2.5 million workers in all fifty states. The organization applies legal, scientific, and political expertise from its member companies to vital food, nutrition, and public-policy issues affecting the industry. Their Web site contains information on public policy and industry affairs, products and services, and facts and figures concerning the grocery industry.

Institute of Food Technologists (IFT)

http://www.ift.org

The Institute of Food Technologists is a scientific, educational society with an interest in providing a safe and wholesome food supply. Their goal is to provide guidance on relevant issues in the field of food. The Web site provides scientific articles on food and food safety, and lists daily happenings in the food industry.

International Association for Food Protection (IAFP)

http://www.foodprotection.org

Members of the International Association for Food Protection are a diverse group, representing all areas of food protection, industry, government, and academia. Members are kept current on rapidly changing technologies, innovations, and regulations in the area of food safety through two monthly scientific journals, as well as annual meetings.

International Dairy-Deli-Bakery Association (IDDBA)

http://www.iddba.org

IDDBA is a resource for information and services across all food channels for the dairy, deli, and bakery industries. This Web site provides information relevant to all people who work in the dairy, deli, and bakery industries.

Morbidity and Mortality Weekly Report (MMWR)

http://www.cdc.gov/mmwr

The MMWR is prepared by the Centers for Disease Control and is available—free-of-charge—on this Web site. The reports contain data based on weekly accounts of reportable diseases from state health departments. Report databases can be searched for information on foodborne-illness outbreaks.

National Association of Convenience Stores (NACS)

http://www.cstorecentral.com

NACS is a proactive organization representing the convenience store segment. This informative Web site provides up-to-date industry happenings and contains reports outlining vital industry statistics.

National Automatic Merchandising Association (NAMA)

http://www.vending.org

NAMA is the national trade association of the merchandising vending and contract-foodservices management business. This Web site provides up-to-date information on all issues concerning this segment of the foodservice industry, as well as provides a resource for conferences and education for those working in this segment.

National Environmental Health Association (NEHA)

http://www.neha.org

NEHA's goals include fostering cooperation and understanding among environmental health professionals while remaining solidly committed to improving the environment in cities, towns, and rural areas throughout the world. This Web site includes many useful resources for restaurant and foodservice inspectors, sanitarians, and the industry.

National Food Processors Association (NFPA)

http://www.nfpa-food.org

The NFPA is the voice of the food processing industry on scientific and public policy issues involving food safety, nutrition, technical and regulatory matters, and consumer affairs. Their Web site offers many resources on food safety and food processing for the entire food industry.

National Frozen Food Association (NFFA)

http://www.nffa.org

NFFA's mission is to promote the sales and consumption of frozen food through education training, research, sales planning, and menu development, and to provide a forum for industry dialogue. Publications and information on frozen foods and how to market them to consumers are available at this Web site.

National Marine Fisheries Service

http://wwwnmfs.noaa.gov/

This organization provides a voluntary inspection service to the industry, assuring compliance with all applicable food regulations. In addition, product-quality evaluation, grading, and certification services on a product-lot basis are also provided. Benefits include the ability to apply official marks—such as the U.S. Grade A, Processed Under Federal Inspection (PUFI), and lot-inspection marks. This Web site provides information on National Oceanic and Atmospheric Administration (NOAA) fisheries and their initiatives, as well as information on the voluntary inspection service.

National Pest Management Association (NPMA)

http://www.pestworld.org

The mission of the National Pest Management Association is to communicate their role as protectors of food, health, property, and the environment, and to affect the success of its members through education and advocacy. This Web site provides information on how to become a member, key conferences, sales leads, and a pest control operator (PCO) locator.

National Restaurant Association

http://www.restaurant.org

The National Restaurant Association is the leading business association for the restaurant industry. Together with the National Restaurant Association Educational Foundation, the Association's mission is to represent, educate, and promote the rapidly growing restaurant and foodservice industry. This Web site should be your starting place for all issues and concerns related to your restaurant. This Web site has it all, from tips for running your establishment to vital data on your customers' spending habits.

National Restaurant Association Educational Foundation (NRAEF)

http://www.nraef.org

The NRAEF's Web site is a comprehensive overview of the initiatives, products, and services offered by the organization, and an effective tool to help you become engaged in NRAEF initiatives, as well as its education and training opportunities.

NSF International

http://www.nsf.org

Use this Web site to find a wealth of information related to food standards and resources. NSF works closely with the food-equipment industry, regulators, and foodservice organizations to develop public health safety standards used by the restaurant and foodservice industry.

Underwriters Laboratories, Inc. (UL)

http://www.ul.com/eph

UL is an independent, not-for-profit product-safety testing and certification organization offering certification programs to demonstrate compliance with public health codes to manufacturers of foodservice equipment. UL classifies products in accordance with food-related NSF International standards. This Web site contains a foodservice-equipment and drinking water product directory, information on the indoor air quality program, ISO 14001, and other information related to the foodservice industry.

U.S. Department of Agriculture (USDA)

http://www.usda.gov

The USDA touches many topic areas involved with land and its use in the U.S. These areas include ensuring a safe, affordable, nutritious, and accessible food supply; caring for agricultural, forest, and range lands; providing economic opportunities for farm and rural residents; and working to reduce hunger in America and throughout the world. This Web site provides information on all of these topics and particularly the safety of food from animal and plant farms.

USDA-FSIS

http://www/fsis.usda.gov

This branch of the U.S. Department of Agriculture focuses on the safety of food produced on farms through the regulation and inspection of meat, poultry, and egg products. This Web site provides a wealth of information on food safety practices and procedures at all levels for meat, poultry and eggs.

Chapter 14
Employee Food Safety Training

Knowledge

TEST YOUR FOOD SAFETY KNOWLEDGE

1. **True or False:** Practicing a task without feedback will not be effective. *(See page 14-3.)*

2. **True or False:** A training need is a gap between what an employee knows and what they need to know to do their job. *(See page 14-4.)*

3. **True or False:** The ideal length for a training session is forty to sixty minutes. *(See page 14-11.)*

4. **True or False:** An objective states what the learner will be able to do after training. *(See page 14-5.)*

5. **True or False:** A written test is the only way to determine if training objectives have been met. *(See page 14-15.)*

For answers, please turn to Appendix A.

Key Terms

Presentation
Feedback
Application
Training program
Training need
Objective

Demonstration
Lecture
Role-plays
Job aids
Technology-based training
Evaluation

Table of Contents

Learning Objectives

After completing this chapter, you should be able to

- Assess the training needs of employees.
- Design, implement, and reinforce a training program.
- Evaluate the success of a training program.
- Identify when and where food safety training should take place.
- Recognize the importance of food safety certification.

Training means teaching employees how to do a job properly. This sounds simple, but training is actually more involved. In your establishment, it is up to you to train all employees in a wide range of tasks and to include food safety training as a key element in your training program.

Training programs must be established for both new and current employees. For new employees, food safety training must be mandatory. It is dangerous to assume they know the safe and sanitary procedures used in your establishment. While employees already on staff might know the correct procedures, they might not always follow them since they are often in a hurry, forget, or lack motivation. For these employees, it might be important to schedule short retraining sessions, update meetings on new procedures, or motivational sessions that reinforce methods and practices.

In this chapter, the following aspects of food safety training for employees will be explored:

○ Assessing and analyzing food safety-training needs

○ Planning, executing, and reinforcing the training program

○ Evaluating the success of the training effort

PURPOSE OF FOOD SAFETY TRAINING

Food safety training provides employees with the knowledge and skills needed to handle food safely in your establishment. The final responsibility for food safety rests with the manager.

Employee training is an important factor in every operation's financial statement. At first glance, training appears costly. The manager, of course, must evaluate every facet of the business for cost and might wonder how training will affect the bottom line. It is true that food safety training might require time away from regular tasks for both employees and managers. It might also require the services of professional trainers and the selection and use of training materials to reinforce safe practices, such as videos, slides, books, and CD-ROMs. However, staff training will have a positive return on investment in the long run. Benefits from food safety training include:

○ **Avoiding the costs associated with a foodborne-illness outbreak.** These costs may include legal fees and medical bills.

○ **Preventing the loss of revenue and reputation incurred when an establishment is forced to close due to a foodborne-illness outbreak.**

○ **Improving employee morale and reducing turnover.** Most employees want to do their jobs right and expect to receive training, which helps instill employee confidence.

○ **Increasing customer satisfaction.** When customers see an establishment is committed to serving safe food, satisfaction will be higher.

A successful training program has these essential elements:

○ Clearly defined and measurable objectives

○ Training that supports the objectives

○ Evaluation to ensure the objectives have been achieved

○ A work climate that reinforces training

○ Management support

In order for the training program to be effective, employees must see that the commitment to food safety comes from the top down. Management should lead by example. If managers show a commitment to food safety by behavior and attitude, employees are likely to follow.

KEY ELEMENTS OF EFFECTIVE TRAINING

Successful training involves three key elements: *presentation, feedback,* and *application.* **Presentation** is the delivery of content to the learner, which can be accomplished through a variety of methods. Once the content is presented, the learner must have the opportunity to practice, apply, or respond to the content in order to retain it. While learners are practicing or applying the content, they must receive **feedback** (positive or negative reinforcement) on their performance. The feedback received must be specific and immediate. **Application** or practice without feedback will be ineffective.

As a general rule, one-third of the time spent training should be devoted to the presentation of content, while the remaining two-thirds should be devoted to activities that allow trainees to apply what they have learned, with feedback.

An effective trainer will have good presentation skills and be knowledgeable in the science of food safety. In addition, a good trainer recognizes that each trainee learns at a different rate and through different media.

DEVELOPING THE TRAINING PROGRAM

A **training program** is a structured sequence of events that leads to learning. To be effective, the training program must be well-organized. Some large establishments set up a training department and have a training professional on staff to develop and deliver their programs. Other establishments rely on the manager to develop and deliver training. A foodservice manager can develop an effective training program through careful planning and the use of good support tools.

Key Point

As a general rule, one-third of the time spent training should be devoted to the presentation of content, while the remaining two-thirds should be spent allowing trainees to apply the information learned, with feedback.

There are several steps in the development and delivery of an effective training program. These include:

1. Assessing training needs
2. Establishing learning objectives
3. Choosing training delivery methods
4. Selecting an instructor
5. Selecting training materials
6. Scheduling a training session
7. Selecting a training area
8. Preparing the trainer

Assessing Training Needs

The first task in developing a training program is to assess the training needs in your establishment. A **training need** is a gap between what your employees are required to know to perform their job and what they actually know. For new hires, the need might be apparent. For employees already on staff, the need is not always as obvious. Determining the gap in performance may require work. To identify food safety training gaps, a manager or trainer can do the following:

○ Observe employee job performance
○ Question or survey employees to identify areas of weakness
○ Review past health inspection reports for violations relating to employee performance
○ Test employees' food safety knowledge

Your staff needs the correct food safety knowledge and skills. From their first day on the job, new foodhandlers will require training in the following areas:

○ **Overview:** The importance of food safety
○ **Personal hygiene:** Health, personal cleanliness, proper work attire, and hygienic practices, including handwashing
○ **Food preparation:** Time-temperature control, the prevention of cross-contamination, and safe practices for preparing, cooking, holding, serving, cooling, and reheating food
○ **Cleaning and sanitizing:** Procedures for cleaning and sanitizing food-contact surfaces
○ **Chemicals:** Procedures for safely handling the chemicals used in the establishment
○ **Pests:** Pest identification and prevention measures

Key Point

A training need is a gap between what your employees know and what they *need* to know.

Training needs will differ for each job within the establishment. All employees will need general information. Other information will be job-specific. For example, all employees need to know the proper way to wash their hands, but only cooks need to know the required minimum internal cooking temperature for chicken.

Regardless of the position, employees need to be retrained periodically on the food safety practices previously listed. It is the manager's responsibility to keep employees informed of the changes in the science of food safety and best practices in the industry.

Establishing Learning Objectives

Once training needs have been identified, the next step is to define the objectives of the training program. An **objective** states what the learner will be able to do after instruction is completed. Objectives need to be stated clearly and in measurable terms. Use action verbs such as *operate, demonstrate,* or *practice,* rather than *be aware, observe, understand,* or *notice.* Examples of clearly stated objectives include:

- Demonstrate the proper procedure for calibrating a bimetallic stemmed thermometer using the ice-point method
- List the internal cooking temperature for meat, seafood, and poultry

Choosing Training Delivery Methods

There are many methods for delivering content to the learner. Some delivery methods are better suited to one-on-one learning, and others to group learning. No single delivery method is best for training at all levels. Using several delivery methods will result in more effective learning. Some methods for delivering training include:

- Demonstration
- Lecture
- Role-play
- Job aids
- One-on-one training (on-the-job training)
- Technology-based training
- Group training

Remember, the delivery method chosen should allow trainees to apply what they have learned while they receive feedback on their performance. However, the choice will also depend upon the number of people needing training, the cost, and—most importantly—how the trainees will learn best.

Key Point

An objective states what the learner will be able to do after instruction is completed.

Key Point

The delivery method chosen should allow trainees to apply what they have learned, with feedback on their performance.

Every participant learns differently. Ask yourself if the delivery methods you have chosen allow learners to do one or more of the following:

○ Talk to each other
○ Reflect on the content and determine how it applies to their job
○ See the task being performed
○ Be physically active and perform the task
○ Hear the instructions spoken
○ Read the materials and take notes
○ Reason through real-life situations

The delivery method chosen must also be based on the specific training objective being taught. For example, if a manager wants to train an employee to wash, rinse, and sanitize the salad-preparation counter properly, the best method could be to demonstrate the process, then have the employee perform the task. Demonstration allows immediate feedback. The manager can explain key points of a task while performing it, then observe progress as the employee practices the same task.

Demonstration

Demonstration is the process of illustrating a skill or task before another person or group. When conducting a demonstration, remember the following steps:

1. Preface the demonstration with an explanation of what trainees should look for when the skill or task is demonstrated.

2. Emphasize key points as you demonstrate the task or skill.

3. Explain how each step fits into the task sequence. Trainees need to know where they are, step by step, in the process of performing the task.

4. Demonstrate the skill or task slowly, so trainees can see what is happening. Then repeat the skill or task at normal speed.

5. Before giving trainees an opportunity to perform the same task, ask them to explain each of the steps in sequence.

6. Ask trainees to demonstrate the skill or task. Provide appropriate feedback, correcting any errors as they occur.

Lecture

A **lecture** is a prepared oral presentation used to deliver content to a group of participants. Lectures are most effective when mixed with other presentation methods and media, and are better received and accepted when the following techniques are used:

Key Point

The choice of delivery method must be based on the specific training objective being taught.

- Start with an interesting statement, observation, quotation, or question
- Use relevant humor where appropriate
- Use interesting and relevant examples, anecdotes, analogies, and statistics
- Ask frequent questions to solicit audience participation
- Use frequent small-group discussions and activities
- Build in a review

Only one-third of the training time should be spent lecturing, while two-thirds of the time should be spent allowing participants to apply the information learned, with feedback.

Role-Play

In **role-plays,** trainees enact a situation in order to try out new skills or apply what has been learned. Different types of role-plays are suitable for different types of learning situations. They usually are set up so a trainee is confronted by another trainee and must answer questions, handle problems, provide satisfaction, solve a complaint, etc. For example, a role-play can be used to teach a manager how to deal with a health inspector during an inspection.

When using a role-play, you should:

- Make certain new content is understood before beginning
- Provide trainees with detailed instructions
- Explain and model the situation before trainees begin
- Keep the role-play simple

Job Aids

Job aids can also be used to deliver content to employees. They may include worksheets, checklists, samples, flowcharts, procedural guides, glossaries, diagrams, decision tables, etc. Other job aids might include visual reminders, such as posters illustrating the proper steps in handwashing, manual warewashing, or other tasks. Employees can then use these job aids once they return to their jobs. Job aids are particularly useful when:

- Tasks are performed infrequently
- Tasks are complex
- The sequence of performance is critical
- The consequences of making a mistake are severe
- Safety is a concern

Key Point

Lectures are most effective when mixed with other presentation methods and media.

Key Point

Job aids should be used when tasks are performed infrequently, are complex, when the consequences of making a mistake are severe, or when safety is a concern.

One-on-One Training

One-on-one training has the following advantages:

○ Takes into consideration the special needs of individual employees

○ Can take place on the job, eliminating the need for a separate training location

○ Enables the manager to monitor employee progress

○ Allows for immediate feedback

○ Offers the opportunity to apply information that has been learned

One-on-one training does have some disadvantages, however. Most importantly, its effectiveness depends upon the ability of the person delivering the training. Therefore, the trainer must be selected very carefully. Many establishments certify trainers or validate that they have the appropriate skills before allowing them to train.

Technology-Based Training

Web-based training, interactive CD-ROM, and other **technology-based training** programs offer yet another way to deliver training. The advantages of technology-based delivery methods include:

○ **Standardized delivery.** The training is delivered the same way every time.

○ **Standardized feedback.** Each time a trainee responds to a situation, the program can provide standardized feedback.

○ **Customizable instruction.** Each trainee can choose their own learning path.

○ **Increased performance practice.** Trainees are allowed to practice a skill until they are proficient.

Group Training

If several employees require food safety training, group sessions might be more practical than one-on-one training. *(See Exhibit 14a.)* Group training can be delivered through a number of methods, including those discussed earlier in this chapter. It has several advantages:

○ It is more cost effective

○ Training is more uniform

○ You know precisely what employees have been taught

One disadvantage of group training is that it is designed to meet the overall needs of a group of employees rather than the needs of each individual. Slower learners, less-skilled employees, or those with limited proficiency in English may not understand all the material presented. A trainer must remember that it is not enough simply to cover the material. Trainees must be involved in the training process too. If they are involved, they will retain the information. Studies on training effectiveness show that trainees retain:

- Ten percent of what they read
- Twenty percent of what they hear
- Thirty percent of what they see
- Fifty percent of what they hear and see
- Seventy percent of what they say
- Ninety percent of what they say and do

Exhibit 14a **Group Training**
Group training can be cost effective and might offer more uniform content delivery.

Selecting an Instructor

The manager often is the most likely person to teach food safety to their staff. He or she knows the operation and is responsible for food safety practices within the establishment. If the manager chooses not to conduct the training personally, an instructor should be selected based on the following criteria:

- Knowledge of food safety practices
- Understanding of the operation's food safety challenges
- Demonstrated skill teaching others
- Good communication skills

Individuals who may be able to conduct food safety training include:

- **Immediate supervisors.** Often these people are a logical choice because of their working relationship with the employee.
- **Staff trainer.** In large establishments, a training professional is often on staff to deliver training and support the training needs of the operation.

Key Point

Trainees retain ninety percent of what they *say* and *do*.

Key Point

Contact
your local
health department for
help in teaching food
safety in your
establishment.

○ **Representative of the health department.** The local sanitarian may agree to teach a food safety training session. This fosters a cooperative working relationship with their agency.

○ **Representative from a professional or educational organization.** These organizations often provide trainer-preparation courses for operators and also have staff instructors who conduct courses on site or at other locations. Check with suppliers and your State Restaurant Association.

Very often, you will experience greater results when you use a combination of these resources.

Selecting Training Materials

Training materials save time, add interest, help participants learn and retain information, and make the trainer's job easier.

The manager or trainer should be guided by the "Three A's" when selecting materials. To be useful, training materials must be: *accurate, appropriate,* and *attractive.*

Accurate. Materials must be factual, up-to-date, and complete. To ensure that training materials are accurate, they should be purchased from professional or educational organizations, industry suppliers, or from an authority in foodservice training.

Key Point

To be
useful,
training materials
must be accurate,
appropriate, and
attractive.

Appropriate. Materials must be suitable for the purpose they are to serve. Written materials must be matched to the reading-comprehension levels of trainees. For employees with limited English proficiency, training materials in other languages are available. In addition, materials should suit the abilities of the trainer. Limitations imposed by the training location will influence your choice of training materials. Reliable equipment must be available in order to use videotapes, slides, CD-ROMs, Web-based programs, overhead transparencies, and other audiovisual or technological tools.

Attractive. In order to teach people, you must first get their attention. When teaching subjects that generally do not have a wide appeal, make information exciting and memorable with eye-catching materials, whenever possible.

Materials Used in a Typical Lesson Plan

The manager or trainer should start each training session by clearly stating the objectives of the session. It is a good idea to provide handouts to students with these objectives clearly stated, or to write them on a flipchart, easel, or chalkboard.

Sometimes a pretest or quiz is given to assess the trainees' prior knowledge. The new information is then presented, using audiovisual materials to supplement the presentation. Afterwards, a brief written, oral, or practical exam is given to test the trainees' knowledge.

The trainer can use written examinations prepared by an educational or training organization, or they can write examinations themselves. The advantage of using prepared exams is that they are usually validated for content and evaluated for reading-level comprehension. Any mistakes made by the trainee should be discussed and corrected immediately after the exam is scored.

To reinforce the training, use signs, posters, bulletin boards, reminders in pay envelopes, and other materials.

Scheduling Training Sessions

Developing and implementing a master training schedule at your establishment can be a useful method to determine training priorities and show your commitment to training. Both orientation and ongoing training should be included in this schedule. For legal reasons, it is important to document the training that took place, even if the training session was brief.

Training sessions should not be too long. The ideal length is probably about twenty to thirty minutes per segment. Successful training sessions, however, may be as long as an hour and a half, if well-planned. The ideal length depends on the type of learning activity used. Training sessions might need to be given more than once during the day in order to cover all employee shifts.

Training sessions should be scheduled during slow times for the establishment: before opening, after closing, or on a day when the establishment is normally closed. If there is a separate area in the facility for training purposes, scheduling can be more flexible.

Selecting a Training Area

Conduct training in a comfortable location. An on-site location allows for demonstrations and provides the opportunity to relate instruction to your establishment. Employee work stations can be used as training areas, particularly for one-on-one sessions. In most establishments, an employee break room, an executive office, or a section of the facility not in use can also be used.

Consider the following when choosing an area for group training:

○ The area should be an appropriate size; no one should feel crowded

○ Tables or desks should be available so trainees can take notes

Key Point

Training content should ideally be broken up into segments twenty to thirty minutes long.

○ Seating should be adequate, comfortable, and arranged to encourage open discussion

○ A blackboard or flip chart should be provided

○ The room should have enough electrical outlets for projectors, sound equipment, videotape players, computers, and other electronics

○ If audiovisual materials are to be used, it must be possible to darken the room so trainees can see the visuals adequately

○ The area should be free of distractions

Preparing the Trainer

Training employees requires excellent communication and organizational skills. The following suggestions will help you conduct a successful training session.

○ **Make sure you are knowledgeable in all areas of food safety.** Feeling comfortable with the subject is important, in case trainees ask questions. If you are uncertain of your subject, trainees will know.

○ **Prepare for the presentation.** Make an outline of the presentation. Practice it until it feels natural. Familiarize yourself with the training room and check all audiovisual equipment to make sure it is in working order.

○ **Maintain eye contact with trainees during your presentation.**

○ **Keep your delivery conversational and informal.** This makes the atmosphere more comfortable. Vary the tone of your voice for emphasis and do not speak too quickly. A moderate rate will allow trainees to take notes and ask questions.

○ **Use simple language.** Using unnecessary technical terms will cause trainees to lose interest or become lost. Pause often and ask questions to make sure your audience understands the material.

○ **Treat all questions and comments made by the trainees seriously.** Answer them in a straightforward way. You can give trainees positive feedback by saying, "That's a good question."

○ **Look for cues indicating employees do not understand the information or are bored and losing interest.** These cues could include fidgeting, looking at watches, doodling, and not asking or answering questions. Vary the way in which you present the information to hold the trainees' attention.

- ○ **Keep the sessions short.**
- ○ **Keep the training as practical as possible and relate information to the employee's job.** If the subject is too general, employees will tend to lose interest. Give specific examples of how the training content relates to the job they do.

CONDUCTING THE TRAINING SESSION

For training to be effective, make sure trainees know how the material will benefit them. Answer the question, "What's In It For Me (WIIFM)?" Trainees want to know what they will gain from participating in the training session. They need to know how the training will enable them to do their jobs faster, easier, or better.

It is the trainer's job to capture their attention and to help them remember the material. Here are some guidelines to help you achieve this.

Keep It Short and Simple

It is important to take a simple, straightforward approach. Because time is often limited, focus on five to nine major points in the training session.

Individualize Training

Make trainees feel you are talking to them individually and that you have a real interest in them. Encourage employee participation during sessions and show you are interested in what they have to say.

Recognize Employee Achievement in the Training Program

Wall charts tracking individual progress through the training course, as well as certificates of completion, can serve as incentives.

Be Creative

Put questions and answers in a game-show format. Create teams and have trainees compete to give correct answers for points or small prizes. *(See Exhibit 14b on the next page.)* Ask one trainee to draw a sanitation procedure on a chalkboard or flip chart while the rest of the group guesses what they are drawing. If the group is not interested in drawing, have them act out the words.

At the end of each chapter in this text there are specific training tips that can be used during training sessions in your establishment. Also, a variety of unit-level training tools and games are available through the National Restaurant Association Educational Foundation.

Key Point

Give specific examples of how the training content relates to the employee's job.

Key Point

Participants want to know how the training will enable them to do their jobs faster, easier, better.

Key Point

Present no more than five to nine major points in a short training session.

Exhibit 14b   **Training Games**
Games can be used by trainers to hold the attention of trainees and help them retain information.

Provide Feedback

Praise and positive feedback are part of all good learning experiences, and they should be part of the training process in your establishment. Praise indicates to employees that they are performing well and have been successful in the training program. Generous praise reinforces employees' positive attitudes about their jobs. When they see that management values their best effort, they will continue their good practices.

Evaluate

The training program is not complete until it has been evaluated. The **evaluation** process is important because it tells you if training has provided the participants with the knowledge and skills needed to do their job. To evaluate the training program, the manager must carefully judge the trainee's performance against the learning objectives.

When evaluating the training program's effectiveness, ask the following questions: Did the training produce results on the job? If the intended results were not produced, why not?

If training was ineffective, several factors might be involved. For example, the trainee could have acquired the knowledge, but is simply not applying it. Or, perhaps equipment used during the training session is different from the equipment used on the job. Perhaps there are negative consequences for doing the job the way the employee was trained to do it.

In some cases, an employee has learned the material, but is simply applying it incorrectly. There may be several reasons for this. Perhaps the employee has been improperly trained and his or her performance is consistent with the bad practices taught. Perhaps the employee has learned the skill, but it is not being reinforced on the job. This requires management intervention to ensure that the proper skills are being taught, practiced, and reinforced.

Evaluation should always be based on the training objectives. There are several ways in which objectives can be measured. Most commonly, objectives are measured through written or oral tests. Test results can help a manager determine if a trainee needs to review the content. Objectives can also be measured by evaluating an employee while he or she performs a task or skill required by the objective. Evaluation works best when a combination of written and performance-based tests are used.

Key Point

Evaluation should always be based on the training objectives.

FOOD SAFETY CERTIFICATION

The National Restaurant Association and federal and state regulatory officials recommend food safety training and certification, particularly for managers and supervisors. Certification demonstrates that a person comprehends basic food safety principles and recommended food safety practices that prevent foodborne illness.

Manager certification is already a requirement in some states. In other states, some cities and counties require certification although the state as a whole does not require it. The National Restaurant Association Educational Foundation Web site provides a jurisdictional summary of training and certification requirements for the entire country. *(See page 14-25.)*

Regardless of state regulations, many proactive establishments and managers have made a strong commitment to food safety by making food safety training and education an integral part of their establishments. You, too, should make a commitment to training, as well as certification, when possible, even if your state or city does not require it. As a conscientious foodservice manager, you should train your employees, monitor their practices, and make food safety a part of everyone's job description. In this way, you will be able to ensure that the food you serve is safe.

SUMMARY

Food safety training provides employees with the knowledge and skills needed to handle food safely in your establishment. Some of the benefits include preventing a foodborne-illness outbreak, the loss of revenue and reputation following an outbreak, improving employee morale, and increasing customer satisfaction.

Successful training involves presentation, application, and feedback. Once the content is presented, the learner must have the opportunity to practice and apply it, with feedback, in order to retain it. There are several steps in the development and delivery of an effective training program. First, you must assess the training need in your establishment. This is the gap between what your employees are required to know and what they actually know. To identify food safety-training needs, you can observe employee performance, review past health inspection reports, and test your employees' knowledge. Once training needs have been identified, clear and measurable objectives need to be defined stating specifically what the learner will be able to do after instruction is complete. The trainer then must determine how content will be delivered to the trainees. Content can be delivered through demonstration, lecture, role-play, job aids, and technology-based programs. Using several delivery methods will result in more effective learning. However, whatever method is chosen, it should allow trainees to apply what they have learned with feedback on their performance.

Instructors and training materials need to be selected. Instructors should be knowledgeable, must have the ability to teach others, and must have good communication skills. Training materials must be accurate, appropriate, and attractive. A master training schedule should be developed to schedule around training priorities and to show your commitment to training. Training sessions should be short—from twenty to thirty minutes—and should be scheduled when the operation is slow. Comfortable training areas, free of distractions, should be identified within the establishment. Finally, the trainer should be prepared before the training session begins. This includes making sure you are knowledgeable in the areas you are teaching and practicing your presentation.

For training to be effective, make sure you let trainees know what they will gain from participating in the training session. They should be able to see how the training will help them do their job faster, easier, or better. Keep the training short and simple—focusing on five to nine major points in the session. Encourage employee participation during sessions and recognize employee achievement during the program. Praise and positive feedback are part of all good learning experiences and should be a part of the training process in your establishment. Finally, evaluate the training program to determine if employees can do what you trained them to do.

The National Restaurant Association and federal and state regulatory officials recommend food-safety training and certification, particularly for managers. Certification shows that a person comprehends basic food safety principles and recommended food safety practices that prevent foodborne illnesses.

A CASE IN POINT

Case Study

Paul has two employees who need food safety training. One is an assistant cook, Albert, who used to be a buser and needs further training. The other, Maria, is an inexperienced server. Albert received general training in the fundamentals of food safety about three months ago, but Maria has had no training. Both Albert and Maria require specialized training. Paul maintains an ongoing food safety training program, so he has already trained the rest of his employees and evaluated the effectiveness of his original program.

Paul decides to have the kitchen supervisor train Albert using the one-on-one method, so the supervisor can demonstrate procedures and give immediate feedback to Albert as he practices them. By relying on the supervisor to train the cook, Paul can have both the cook and server trained at the same time.

Paul gives Maria a copy of the job description and learning objectives and sets up a food safety videotape in his office for her to watch. While the videotape is playing, Paul goes to the back of the facility to talk with a clerk about an expected shipment. After half an hour, Paul returns to his office to give Maria a written test. While Maria is completing the test, Paul goes to the kitchen to observe the supervisor training Albert, the cook. Noting that everything looks good in the kitchen, Paul goes to the dining room.

Do you think Paul's training program for the cook and server will be effective?

What else does Paul need to be doing?

TRAINING TIPS

1. Keys to Being a Successful Trainer

Purpose: *After completing this activity, class participants will be able to identify the traits, characteristics, and qualities of a successful trainer.*

Time: 30 minutes

Directions: After discussing Chapter 14, ask trainees to make a list of the traits, characteristics, or qualities a trainer should possess in order to be effective.

Allow five minutes for trainees to write their lists. Then, ask a trainee to explain one item on his or her list. Next, solicit thoughts, comments, or reactions from the rest of the group. Some questions to stimulate conversation might include, "Is this item on anyone else's list?" or "Do you think it's a critical success factor for a trainer?" Suggest that the group add the item to their lists, if they agree it is valid. Then, have the first trainee select another trainee to discuss one item on his or her list, and continue on until everyone in the group has had a chance to contribute at least one item.

2. Training Tools That Work

Purpose: *After completing this activity, trainees will be able to identify the advantages and disadvantages of using various training tools.*

Time: 30 minutes

Directions: Many different training tools can be used in the process of food safety training. All have advantages and disadvantages. Training tools can include the following:

- Books
- Training guides or manuals
- Training videos
- Slides
- Overheads
- Handouts
- Posters
- Computer-based presentations
- Role-plays
- Dry-erase board or chalkboard
- Interactive CD-ROMs
- Flipcharts

Make a list of training tools, such as those listed above. You can print the list as a handout, or ask trainees to generate the list while you write the items on a flipchart. Then divide the trainees into two groups. Ask one group to list the advantages of each of the training tools on the list. Ask the other group to list the disadvantages of the training tools.

After a short time (about ten minutes), review with the entire group the advantages and disadvantages of each training tool. Write these on the flipchart. Tell the group that the effectiveness of specific tools may depend on a number of factors, including size of the group, type of audience, layout of a room, time allotted, quality of the training tool, and personal preferences of the trainer.

Point out that no training tool works all the time, that every tool has some advantages and disadvantages, that every trainer has favorite tools, and that all training tools must be thoughtfully selected and properly used.

3. Conduct a Food Safety Training Materials Audit

Purpose: *To audit existing food safety training materials to determine if they are appropriate and accurate.*

Time: ongoing

Directions: The training materials your operation uses may or may not include valid and up-to-date food safety information. An audit should help you assess your materials as they relate to current food safety information, and help you determine how they can be improved.

First, collect all training materials currently in use, and highlight anything related to food safety. Then assess the materials, by asking the following questions:

○ Are they accurate and do they reflect current company policy?

○ Do they meet or exceed local health department codes?

○ Are they job-specific?

○ Are they based on clearly stated and measurable objectives?

○ Are they user-friendly?

○ Are they thorough?

○ Are they necessary?

○ Do they convey the information effectively ?

○ Are materials sufficient in size and scope?

Based on your answers, you may need to modify or enhance your training materials.

DISCUSSION QUESTIONS

1. What are the benefits of teaching food safety in an establishment?

2. What are the three key elements of successful training? How do they fit together?

3. How can an establishment determine its food safety training needs?

4. What is an objective? Give an example of a good objective.

5. What are some methods that can be used to deliver training?

MULTIPLE-CHOICE STUDY QUESTIONS

1. Mary has hired college students to run her lakeside hot-dog stand for the summer. She would like to develop a brief training program to prepare them for the job. What should Mary do first?
   A. Write training objectives
   B. Choose training delivery methods
   C. Assess her training needs
   D. Evaluate her training program

2. The purpose of implementing a food safety-training program is to
   A. avoid the costs associated with a foodborne-illness outbreak.
   B. prevent the loss of reputation associated with a foodborne-illness outbreak.
   C. improve employee morale.
   D. All of the above

3. The ideal length for a training session is
   A. five to ten minutes.   C. forty to sixty minutes.
   B. twenty to thirty minutes.   D. one to two hours.

4. Which of the following is *not* a good way for a restaurant or foodservice manager to identify food safety-training needs in their establishment?

   A. Reading the comments on the health inspection report.
   B. Testing employees to find out what they know.
   C. Observing employees while they perform their jobs.
   D. Reading a textbook on training restaurant employees.

5. As a general rule, one-third of the time in a training session should be spent presenting content while the remaining two-thirds of the time should be spent

   A. measuring knowledge with a test.
   B. allowing employees to apply what was learned, with feedback.
   C. determining the training need.
   D. determining a delivery method.

6. A WIIFM shows a trainee all of the following *except*

   A. what's in it for them.
   B. how the training will help them do their job faster, easier, or better.
   C. how the training will benefit them.
   D. how the training will benefit the establishment.

7. An instructor should choose a training delivery method that

   A. supports the training objectives.
   B. allows trainees to apply what they have learned.
   C. complements how trainees learn.
   D. All of the above

8. During the training session, a trainer should focus on five to _____ major points.

   A. nine
   B. ten
   C. twelve
   D. fifteen

9. Which of the following objectives is stated clearly and in measurable terms?

   After completing this course you will:

   A. Be able to demonstrate the proper procedure for calibrating a thermometer
   B. Understand how to calibrate a thermometer
   C. Be aware of how to calibrate a thermometer
   D. Know how to calibrate a thermometer

10. Which of the following statements about training objectives is *not* true?

    A. They should describe the end result that the training is intended to produce.
    B. They should be stated in measurable terms.
    C. They should state what the learner will be able to do prior to training.
    D. They should be phrased using action verbs.

For answers, please turn to Appendix A.

ADDITIONAL RESOURCES

Books and Periodicals

Bax, B. 1997. *Handbook for safe food service management.* Upper Saddle River, NJ: Prentice-Hall.

Craig, R. L., ed. 1987. *Training and development handbook: A guide to human resource development,* 3rd ed. New York: McGraw-Hill. (Sponsored by the American Society for Training and Development.)

National Restaurant Association Educational Foundation. 2001. *The Manager's Guide to Employee-Level Training.* Chicago, IL: National Restaurant Association Educational Foudation

National Restaurant Association Educational Foundation. 2001. *ServSafe® Train-the-Trainer Program.* Chicago, IL: National Restaurant Association Educational Foudation

Pike, R. W. 1994. *Creative Training Techniques Handbook,* 2nd ed. Minneapolis MN: Lakewood Books.

Rothwell, W. J., and H. C. Kazanas 1998. *Mastering the instructional design process: A systematic approach,* 2nd ed. San Francisco: Jossey-Bass.

Web Sites

Council of Hotel and Restaurant Trainers (CHART)

http://www.chart.org/

CHART is one of the oldest and largest organizations dedicated to training in the hospitality industry, with a mission to help hospitality-training professionals improve operational performance by developing people. This Web site provides information on how to become a member of CHART, information on conferences, and a sample of their newsletter.

FDA Food Code

http://vm.cfsan.fda.gov/~dms/foodcode.html

As the basis for many local sanitation codes, as well as the basis for information in this textbook, the FDA Food Code, available at this Web address, is a useful resource for information relating to food safety for the restaurant and foodservice industry.

Foodservice Consultants Society International (FCSI)

http://www.fcsi.org

The FCSI links many professionals working in the foodservice industry, helping consultants advance their careers, providing all-important professional recognition, and supplying numerous networking opportunities with colleagues. This Web site allows you to become a member, take continuing education classes, and even find a job.

The International Council on Hotel, Restaurant, and Institutional Education

http://chrie.org/

The International Council on Hotel, Restaurant, and Institutional Education is the global advocate of hospitality and tourism education for schools, colleges, and universities, offering programs in hotel and restaurant management, foodservice management, and culinary arts. The Web site provides a place for exchanging information, ideas, research, and products and services related to education, training, and resource development for the hospitality and tourism industry.

International Food Safety Council

http://www.foodsafetycouncil.org/

In 1993, the National Restaurant Association Educational Foundation recognized the need for food safety awareness and created the International Food Safety Council as its food safety awareness initiative. The Council's mission is to heighten the awareness of the importance of food safety education throughout the restaurant and foodservice industry. This Web site is an excellent resource center on food safety issues from the restaurant and foodservice industry.

National Agricultural Library (NAL)

http://www.nalusda.gov

NAL has a mission to increase the availability and utilization of agricultural information for researchers, educators, policymakers, consumers of agricultural products, and the public. Use this Web site to locate valuable food safety and HACCP information.

National Association of College and University Food Services (NACUFS)

http://www.nacufs.org/

NACUFS is the trade association for foodservice professionals at over 650 institutions of higher education in the United States, Canada, and abroad. This Web site provides a full range of educational programs, industry-specific publications, management services, and networking opportunities.

National Restaurant Association

http://www.restaurant.org

The National Restaurant Association is the leading business association for the restaurant industry. Together with the National Restaurant Association Educational Foundation, the Association's mission is to represent, educate, and promote the rapidly growing restaurant and foodservice industry. This Web site should be your starting place for all issues and concerns related to your restaurant. This Web site has it all, from tips for running your establishment to vital data on your customers' spending habits.

National Restaurant Association Educational Foundation (NRAEF)

http://www.nraef.org

The NRAEF's Web site is a comprehensive overview of the initiatives, products, and services offered by the organization, and an effective tool to help you become engaged in NRAEF initiatives, as well as its education and training opportunities.

Appendixes

APPENDIX A
ANSWER KEY

Chapter 1
Providing Safe Food

Test Your Food Safety Knowledge (page 1-1)

1. T
2. T
3. T
4. T
5. F

Multiple-Choice Study Questions (page 1-20)

1. B   6. B
2. B   7. C
3. C   8. A
4. B   9. D
5. D

Chapter 2
The Microworld

Test Your Food Safety Knowledge (page 2-1)

1. F
2. F
3. T
4. F
5. T

Multiple-Choice Study Questions (page 2-28)

1. B   6. D
2. C   7. A
3. D   8. A
4. B   9. A
5. D   10. A

Chapter 3

Contamination, Food Allergies, and Foodborne Illness

Test Your Food Safety Knowledge (page 3-1)

1. F
2. F
3. T
4. F
5. T

Multiple-Choice Study Questions (page 3-14)

1. C	6. C
2. C	7. B
3. D	8. B
4. B	9. C
5. A	10. B

Chapter 4

The Safe Foodhandler

Test Your Food Safety Knowledge (page 4-1)

1. T
2. F
3. T
4. T
5. F

Multiple-Choice Study Questions (page 4-17)

1. D	6. B	11. B
2. B	7. A	12. D
3. D	8. A	13. C
4. A	9. C	14. D
5. B	10. C	

Chapter 5

Purchasing and Receiving Safe Food

Test Your Food Safety Knowledge (page 5-1)

1. T
2. T
3. T
4. T
5. T

Multiple-Choice Study Questions (page 5-29)

1. A	6. B	11. C
2. B	7. B	12. B
3. C	8. D	13. C
4. D	9. A	14. D
5. C	10. B	15. D

Chapter 6

Keeping Food Safe in Storage

Test Your Food Safety Knowledge (page 6-1)

1. F
2. T
3. F
4. T
5. F

Multiple-Choice Study Questions (page 6-19)

1. D	6. D
2. C	7. B
3. A	8. A
4. B	9. C
5. D	10. B

Chapter 7

Protecting Food During Preparation

Test Your Food Safety Knowledge (page 7-1)

1. F
2. T
3. T
4. F
5. F

Multiple-Choice Study Questions (page 7-27)

1. A		6. D
2. C		7. C
3. D		8. D
4. D		9. B
5. C		10. C

Chapter 8

Protecting Food During Service

Test Your Food Safety Knowledge (page 8-1)

1. F
2. T
3. T
4. T
5. F

Multiple-Choice Study Questions (page 8-19)

1. A		6. A
2. B		7. A
3. C		8. A
4. D		
5. C		

Chapter 9

Principles of a HACCP System

Test Your Food Safety Knowledge (page 9-1)

1. T
2. T
3. F
4. T
5. F

Multiple-Choice Study Questions (page 9-19)

1. B	6. D	11. D
2. B	7. D	12. B
3. C	8. A	13. D
4. A	9. D	14. A
5. C	10. A	15. C

Chapter 10

Sanitary Facilities and Equipment

Test Your Food Safety Knowledge (page 10-1)

1. T
2. F
3. F
4. F
5. T

Multiple-Choice Study Questions (page 10-31)

1. D	6. D	11. A
2. D	7. C	12. B
3. D	8. D	
4. C	9. B	
5. C	10. D	

Chapter 11

Cleaning and Sanitizing

Test Your Food Safety Knowledge (page 11-1)

1. F
2. F
3. F
4. F
5. F

Multiple-Choice Study Questions (page 11-29)

1.	D	6.	D
2.	C	7.	B
3.	B	8.	C
4.	C	9.	C
5.	B		

Chapter 12

Integrated Pest Management

Test Your Food Safety Knowledge (page 12-1)

1. T
2. T
3. F
4. T
5. F

Multiple-Choice Study Questions (page 12-21)

1.	B	6.	D
2.	C	7.	B
3.	B	8.	D
4.	D	9.	B
5.	C		

Chapter 13

Food Safety Regulations and Standards

Test Your Food Safety Knowledge (page 13-1)

1. F
2. F
3. T
4. F
5. T

Multiple-Choice Study Questions (page 13-20)

1. D   6. C
2. D   7. D
3. B   8. B
4. A   9. A
5. A

Chapter 14

Employee Food Safety Training

Test Your Food Safety Knowledge (page 14-1)

1. T
2. T
3. F
4. T
5. F

Multiple-Choice Study Questions (page 14-20)

1. C   6. D
2. D   7. D
3. B   8. A
4. D   9. A
5. B   10. C

APPENDIX B
STORAGE TEMPERATURES FOR FRESH FRUIT

Source: Produce Marketing Association

Fruit	Storage Temperatures (°F/°C)	Relative Humidity (rh) (%)	Comments
Apples	32°F to 35°F (0°C to 2°C)	90–95%	Will ripen at room temperature.
Avocados	CA: 40°F to 42°F (4°C to 6°C), FL: 55°F (13°C)	85–90% 85–95%	Will ripen at 65°F to 70°F (18°C to 21°C). FL: 55°F (13°C). Will ripen at room temp at 85 to 95% rh.
Bananas	60°F	85–90%	Will ripen at 62°F to 64°F (17°C to 18°C) at 85–95% rh.
Blueberries	31°F to 34°F (-1°C to 1°C)	90–95%	Should not be washed before storage.
Cantaloupe	38°F to 40°F (3°C to 4°C)	85–90%	Will ripen at room temperature.
Cherries	32°F to 34°F (0°C to 1°C)	90–95%	Will not ripen after picking; keep dry; store away from other food with strong odors.
Cranberries	38°F to 40°F (3°C to 4°C)	85–90%	
Grapefruit	CA/AZ: 55°F to 58°F (13°C to 14°C), FL/TX: 50°F to 55°F (10°C to 13°C)	85–90%	Will not ripen after picking; chill damage occurs at temps below 40°F (4°C).
Grapes	32°F to 35°F (0°C to 2°C)	90–95%	Will not ripen in storage; do not wash before storage; do not handle excessively.
Kiwifruit	32°F to 35°F (0°C to 2°C)	90–95%	Can be stored up to 12 weeks at 32°F (0°C); will ripen quickly if stored near ethylene-producing fruit.
Lemons	50°F to 55°F (10°C to 13°C)	85–90%	Long refrigerated shelf life; store away from food with strong odors.
Limes	50°F to 55°F (10°C to 13°C)	85–90%	Do not hold as long as lemons.
Mangos	50°F to 55°F (10°C to 13°C)	90–95%	Will ripen at room temperature; refrigerate to retard ripening.
Melons	45°F to 50°F (7°C to 10°C)	85–90%	Will ripen at room temperature; refrigerate to retard ripening.
Nectarines	32°F (0°C)	95%	Immature fruit will not ripen further.
Oranges	CA: 45°F to 48°F (7°C to 9°C), FL: 34°F to 40°F (1°C to 4°C), TX: 32°F to 35°F (0°C to 2°C)	85–90% 85–90% 85–95%	
Peaches	32°F (0°C)	95%	Will not ripen properly after picking.
Pears	32°F to 35°F (0°C to 2°C)	90–95%	
Pineapple	45°F (7°C)	85–90%	Prone to chill damage; store as close to 45°F (7°C) as possible.
Plums	32°F (0°C)	95%	Does not have a long shelf life, even under refrigeration.
Raspberries	32°F to 35°F (0°C to 2°C)	90–95%	Do not sprinkle before storage; refrigerate immediately after receiving; use quickly.
Strawberries	32°F to 35°F (0°C to 2°C)	90–95%	Will not ripen after being picked; do not wash until just before use; very perishable.
Tangerines	40°F (4°C)	85–90%	
Tomatoes	55°F to 60°F (13°C to 16°C)	85–95%	Do not refrigerate.
Watermelon	55°F to 70°F (13°C to 21°C)	85–90%	Holding at room temperature improves flavor; quality deteriorates at temperatures of 50°F (10°C) or below.

This table is a general guideline for best product quality and overall safety. Where applicable, always use any product by its use-by date marked on package. If purchase date is unknown, or if quality or safety is compromised in any way, discard product.

ADDITIONAL RESOURCES

Resources

Books and Periodicals

Beck, B. 1984. *Produce: A fruit and vegetable lovers' guide.* New York: Friendly Press.

The import issue: How safe is U.S. produce? 1997. *FoodService Director.* 10 (6):72.

National Restaurant Association. 1999. *Fruits and Vegetables: How to Choose Produce.* Washington, DC: National Restaurant Association.

The Produce Marketing Association. 1998. *Food Safety for Produce Distribution: Getting Fresh Produce Safely from the Shipper's Dock to the Consumer.* Newark, DE: Produce Marketing Association.

The Produce Marketing Association. 1996. *Fresh Produce Manual.* Newark, DE: Produce Marketing Association.

The Produce Marketing Association. 1997. *Produce Identification Guide.* Newark, DE: Produce Marketing Association.

Schneider, E. 1986. *Uncommon fruits and vegetables: A commonsense guide.* New York: Harper & Row.

Web Sites

International Fresh-cut Produce Association (IFPA)

http://www.fresh-cuts.org

IFPA serves commercial producers, suppliers, and distributors of fresh-cut produce. It provides members with a forum for representation, education, technical support, networking, information, and resources. This site contains information about recent industry happenings and regulations, and provides guidance documents for purchasing, receiving, and storing fresh-cut produce.

Produce Marketing Association

http://www.pma.com

The Produce Marketing Association is a not-for-profit trade association serving members who market fresh fruit, vegetables, and floral products worldwide. Its members are involved in the production, distribution, retail, and foodservice sectors of the industry. This Web site provides promotional ideas and information about hot topics in the produce industry.

United Fresh Fruit & Vegetable Association

http://www.uffva.org

United members provide the leadership to shape business, trade, and public policies that drive the fruit and vegetable industry. Working with thousands of industry members, United provides a fair and balanced forum to promote business solutions, helps build strong partnerships among all segments of the industry, promotes increased produce consumption, and provides scientific and technical expertise essential to competing effectively in today's marketplace.

USDA Agricultural Marketing Service

http://www.ams.usda.gov/standards/frutmrkt.htm

Comprehensive site featuring a list of standards for fresh fruit from the USDA's Agricultural Marketing Service.

APPENDIX C
STORAGE TEMPERATURES FOR FRESH VEGETABLES

Source: Produce Marketing Association

Vegetables	Storage Temperatures (°F/°C)	Relative Humidity (%)	Comments
Artichokes	32°F to 34°F (0° to 1°C)	90–95%	Do not dampen when unrefrigerated, may cause mold.
Asparagus	32°F to 35°F (0° to 2°C)	90–95%	Shelf life is improved when stood on end in an inch of water.
Beans	45°F to 50°F (7°C to 10°C)	90–95%	Do not rinse until before use; allow adequate air flow around containers.
Broccoli	32°F to 35°F (0° to 2°C)	85–90%	
Cabbage	32°F (0°C)	90–95%	Will lose moisture at room temperatures; refrigerate whole heads with leaves intact.
Carrots	32°F to 34°F (0° to 1°C)	90–95%	Extremely long shelf life; remove tops to prevent moisture loss.
Cauliflower	32°F to 35°F (0° to 2°C)	85–90%	Susceptible to damage in handling and storage; store boxes upside down; dry atmosphere will cause curd to dry out.
Celery	32°F to 35°F (0° to 2°C)	90–95%	Highly perishable; wilts in high humidity.
Corn	32°F to 34°F (0° to 1°C)	85–90%	Refrigerate to slow conversion of sugar to starch.
Cucumbers	45°F to 50°F (7°C to 10°C)	85–95%	Most cucumbers are shipped waxed to prevent moisture loss.
Eggplant	45°F to 50°F (7°C to 10°C)	85–90%	Easily damaged; will decay if bruised.
Lettuce	32°F to 35°F (0° to 2°C)	90–95%	Rotate FIFO; revive wilted lettuce by plunging in cold water; store away from ethylene-producing produce.
Mushrooms	32°F to 35°F (0° to 2°C)	85–90%	Do not wash or cut until just before use.
Onions (green)	32°F (0°C)	90–95%	Very perishable; trimming end and soaking in water will revive wilting.
Onions (white)	45°F to 50°F (7°C to 10°C)	65–70%	Can withstand long-term storage; air circulation is beneficial to prevent other produce from absorbing odors.
Peas	32°F to 35°F (0° to 2°C)	50%	Very perishable; refrigerate and rotate supplies.
Peppers	45°F to 50°F (7°C to 10°C)	85–90%	Very perishable at room temperature.
Potatoes	45°F to 50°F (7°C to 10°C)	85–90%	Store in cool, dark area; should not be exposed to light or freezing temperatures.
Spinach	32°F (0°C)	90–95%	Wilts quickly at room temperature; adding ice to containers prolongs life.
Sprouts	36°F to 40°F (2°C to 4°C)	90–95%	Cover during storage to prevent moisture loss.

This table is a general guideline for best product quality and overall safety. Where applicable, always use any product by its use-by date marked on package. If purchase date is unknown, or if quality or safety is compromised in any way, discard product.

ADDITIONAL RESOURCES

Books and Periodicals

Beck, B. 1984. *Produce: A fruit and vegetable lovers' guide.* New York: Friendly Press.

The import issue: How safe is U.S. produce? 1997. *FoodService Director.* 10 (6):72.

National Restaurant Association. 1999. *Fruits and Vegetables: How to Choose Produce.* Washington, DC: National Restaurant Association.

The Produce Marketing Association. 1998. *Food Safety for Produce Distribution: Getting Fresh Produce Safely from the Shipper's Dock to the Consumer.* Newark, DE: Produce Marketing Association.

The Produce Marketing Association. 1996. *Fresh Produce Manual.* Newark, DE: Produce Marketing Association.

The Produce Marketing Association. 1997. *Produce Identification Guide.* Newark, DE: Produce Marketing Association.

Schneider, E. 1986. *Uncommon Fruits and Vegetables: A commonsense guide.* New York: Harper & Row.

Web Sites

International Fresh-cut Produce Association (IFPA)

http://www.fresh-cuts.org

IFPA serves commercial producers, suppliers, and distributors of fresh-cut produce. It provides members with a forum for representation, education, technical support, networking, information, and resources. This site contains information about recent industry happenings and regulations, and provides guidance documents for purchasing, receiving, and storing fresh-cut produce.

Produce Marketing Association

http://www.pma.com

The Produce Marketing Association is a not-for-profit trade association serving members who market fresh fruit, vegetables, and floral products worldwide. Its members are involved in the production, distribution, retail, and foodservice sectors of the industry. This Web site provides promotional ideas and information about hot topics in the produce industry.

United Fresh Fruit & Vegetable Association

http://www.uffva.org/

United members provide the leadership to shape business, trade, and public policies that drive the fruit and vegetable industry. Working with thousands of industry members, United provides a fair and balanced forum to promote business solutions, helps build strong partnerships among all segments of the industry, promotes increased produce consumption, and provides scientific and technical expertise essential to competing effectively in today's marketplace.

USDA Agricultural Marketing Service

www.ams.usda.gov/standards

Comprehensive site featuring the USDA Agricultural Marketing Service's list of standards for various fresh foods, including poultry and vegetables.

APPENDIX D
REFRIGERATED STORAGE OF FOOD

Sources: Tyson Foods, Inc., American Egg Board, and *Safe Food Storage Times and Temperatures* by Mark L. Tamplin, Ph.D.

Food	Recommended Product Temperatures (°F/°C)	Maximum Storage Periods
Meat		
Roasts, steaks, chops	35°F to 41°F (2°C to 5°C)	2 to 5 days
Steaks	35°F to 41°F (2°C to 5°C)	2 to 5 days
Chops	35°F to 41°F (2°C to 5°C)	3 to 4 days
Ground and stewing	35°F to 41°F (2°C to 5°C)	1 to 2 days
Variety meats	35°F to 41°F (2°C to 5°C)	1 to 2 days
Whole ham	35°F to 41°F (2°C to 5°C)	7 days
Half ham	35°F to 41°F (2°C to 5°C)	3 to 5 days
Ham slices	35°F to 41°F (2°C to 5°C)	3 to 5 days
Canned ham	35°F to 41°F (2°C to 5°C)	9 months to 1 year
Frankfurters	35°F to 41°F (2°C to 5°C)	1 week
Bacon	35°F to 41°F (2°C to 5°C)	5 to 7 days unopened
Luncheon meats	35°F to 41°F (2°C to 5°C)	3 to 5 days
Leftover cooked meats	35°F to 41°F (2°C to 5°C)	1 to 2 days
Gravy, broth	35°F to 41°F (2°C to 5°C)	1 to 2 days
Poultry		
Whole chicken, turkey, duck, goose	32°F to 36°F (0°C to 2°C)	1 to 2 days
Giblets	32°F to 36°F (0°C to 2°C)	1 to 2 days
Stuffing	32°F to 36°F (0°C to 2°C)	1 day
Cut-up, cooked poultry	32°F to 36°F (0°C to 2°C)	1 to 2 days
Fish		
Fresh fish	32°F to 36°F (0°C to 2°C)	1 to 2 days
Fish (smoked)	30°F to 41°F (-1°C to 5°C)	1 to 2 days
Clams, crab, lobster (in shell)	30°F to 41°F (-1°C to 5°C)	2 days
Scallops, oysters, shrimp	30°F to 41°F (-1°C to 5°C)	1 day
Eggs		
Eggs in shell	45°F (7°C)	*4 to 5 weeks beyond pack date
Leftover yolks	40°F to 45°F (4°C to 7°C)	1 to 2 days
Leftover whites	40°F to 45°F (4°C to 7°C)	4 days
Dried eggs (whole eggs and yolks)	40°F to 45°F (4°C to 7°C)	Up to 1 year (unreconstituted)
Reconstituted dried eggs		Use immediately
Cooked Dishes with eggs, meat, milk, fish, poultry	32°F to 36°F (0°C to 2°C)	Serve day prepared
Dairy Products		
Fluid milk	35°F to 41°F (2°C to 5°C)	5 to 7 days after date on container
Butter	35°F to 41°F (2°C to 5°C)	2 weeks
Hard cheese (cheddar, parmesan, romano)	35°F to 41°F (2°C to 5°C)	1 month
Soft cheese	35°F to 41°F (2°C to 5°C)	1 week
Dry milk (nonfat)	35°F to 41°F (2°C to 5°C)	1 year unopened
Reconstituted dry milk	35°F to 41°F (2°C to 5°C)	1 week

This table is a general guideline for best product quality and overall safety. Where applicable, always use any product by its use-by date marked on package. If purchase date is unknown, or if quality or safety is compromised in any way, discard product.
*Recommended by the American Egg Board. Most eggs arrive at a distribution site within a few days of being packed.

ADDITIONAL RESOURCES

Books and Periodicals

Bertagnoli, L. 1995. Enter the new ice age. *Restaurants & Institutions.* 105 (8):124.

Bertagnoli, L. 1995. Rethinking refrigeration. *Restaurants & Institutions.* 105 (23):120.

Castagna, N. G. 1997. Know your cook-chill. *Restaurants & Institutions.* 107 (16):84.

Durocher, J. 1997. Cool breeze: Specialty refrigeration for restaurants. *Restaurant Business.* 96 (21):161.

Durocher, J. 1998. The big chill. *Restaurant Business.* 97 (8):125.

Hunt-Ashby, B. 1995. *Protecting perishable foods during transport by truck* (Handbook 669). Washington, DC: USDA Agricultural Marketing Service Transportation and Marketing Division.

National Restaurant Association. *Sanitation survival kit.* 1998. Washington, DC: National Restaurant Association.

Slomon, E. 1997. Refrigeration: The cold, hard facts. *Pizza Today.* 15 (8):60.

Tamplin, M. L. 1994. *Safe food storage times and temperatures* (Fact Sheet HE 8490). Gainesville: University of Florida. A series of the Home Economics Department, Florida Cooperative Extension Service, Institute of Food and Agricultural Sciences.

Web Sites

American Egg Board (AEB)

http://www.aeb.org

As the egg industry's promotional arm, the AEB is the U.S. egg producer's link to the consumer for communicating the value of the egg. This Web site offers a wealth of information about eggs and egg safety.

FDA Seafood Information and Resources

http://vm.cfsan.fda.gov/seafood1.html

The FDA operates an oversight compliance program for fishery products under which responsibility for product safety, wholesomeness, identity, and economic integrity rests with the processor or importer, who must comply with regulations under the Federal Food, Drug and Cosmetic (FD&C) Act. This Web site houses information on the seafood program, foodborne pathogens and contaminants associated with seafood, and HACCP compliance.

Food Marketing Institute (FMI)

http://www.fmi.org

The FMI is a nonprofit association conducting programs in research, education, industry relations, and public affairs on behalf of its members, which include large multistore chains, small regional firms, and independent supermarkets.

International Dairy-Deli-Bakery Association (IDDBA)

http://www.iddba.org

IDDBA is a resource for information and services across all food channels for the dairy, deli, and bakery industry. This Web site provides information relevant to all people who work in the dairy, deli, and bakery industry.

International Fresh-cut Produce Association (IFPA)

http://www.fresh-cuts.org

IFPA serves commercial producers, suppliers, and distributors of fresh-cut produce. It provides members with a forum for representation, education, technical support, networking, information, and resources. This site contains information about recent industry happenings and regulations, and provides guidance documents for purchasing, receiving, and storing fresh-cut produce.

The Mushroom Council

http://www.mushroomcouncil.com

This interesting and useful Web site from the Mushroom Council contains recipes and information about how to handle and store mushrooms to ensure safety and quality.

The National Chicken Council

http://www.eatchicken.com

This Web site is brought to you by the National Chicken Council and the U.S. Poultry & Egg Association, which represent a wide spectrum of companies and individuals in the chicken business. This fun and interesting Web site contains chicken safety tips and nutritional and statistical information.

National Food Processors Association (NFPA)

http://www.nfpa-food.org

The NFPA is the voice of the food processing industry on scientific and public policy issues involving food safety, nutrition, technical and regulatory matters, and consumer affairs. Its Web site offers many resources on food safety and food processing for the entire food industry.

National Frozen Food Association (NFFA)

http://www.nffa.org

NFFA's mission is to promote the sales and consumption of frozen food through education training, research, sales planning, and menu development, and to provide a forum for industry dialogue. Publications and information on frozen food and how to market it to consumers are available at this Web site.

National Pork Producers Council

http://www.nppc.org

Billing itself as the place for pork lovers, pork producers, and anyone who wants to learn more about pigs and pork, the National Pork Producers Council's Web site provides information on pork safety and quality, nutrition and industry information, and publications.

National Turkey Federation

http://www.turkeyfed.org

The National Turkey Federation advocates for all segments of the turkey industry, providing services and conducting activities which increase demand for its members' products by protecting and enhancing their ability to profitably provide wholesome, high-quality, nutritious products. Its Web site contains a special area specifically for the restaurant and foodservice industry, including recipe and menu ideas, cooking demonstrations, and a reference on purchasing, preparing, and promoting turkey.

Produce Marketing Association

http://www.pma.com

The Produce Marketing Association is a not-for-profit trade association serving members who market fresh fruit, vegetables, and floral products worldwide. Its members are involved in the production, distribution, retail, and foodservice sectors of the industry. This Web site provides promotional ideas and information about hot topics in the produce industry.

Tyson Foods, Inc.

http://www.tyson.com

Tyson Foods is the world's largest poultry producer. This Web site is a great resource about chicken products for the restaurant and foodservice industry. It includes recipes and menu ideas, information on new products, and information about Tyson Foods' efficient distribution and transportation system.

APPENDIX E
STORAGE OF FROZEN FOOD

Sources: Tyson Foods, Inc. and *Safe Food Storage Times and Temperatures* by Mark L. Tamplin, Ph.D.

Food	Maximum Storage Period at 0°F to 10°F (-12°C to -18°C)
Meat	
Beef, roasts, and steaks	6 to 9 months
Beef, ground and stewing	3 to 4 months
Pork, roasts and chops	4 to 8 months
Pork, ground	2 months
Lamb, roasts and chops	6 to 9 months
Lamb, ground	3 to 5 months
Veal	8 to 12 months
Variety meats	3 to 4 months
Ham, frankfurters, bacon, luncheon meats	2 weeks
Leftover cooked meats	2 to 3 months
Gravy, broth	2 to 3 months
Sandwiches with meat filling	1 to 2 months
Poultry	
Whole chicken, turkey, duck, goose	12 months
Giblets	3 months
Cut-up, cooked poultry	4 to 6 months
Fish	
Fresh fish	2 to 3 months
Frozen fish	3 to 6 months
Clams, lobster	3 months
Scallops, shrimp	3 months
Ice Cream	3 months; original container; quality maintained better at 10°F (-12°C)

This table is a general guideline for best product quality and overall safety. Where applicable, always use any product by use-by date marked on package. If purchase date is unknown, or if quality or safety is compromised in any way, discard product.

ADDITIONAL RESOURCES

Books and Periodicals

Tamplin, M. L. (1994). *Safe food storage times and temperatures* (Fact Sheet HE 8490). Gainesville: University of Florida. A series of the Home Economics Department, Florida Cooperative Extension Service, Institute of Food and Agricultural Sciences.

Web Site

Tyson Foods, Inc.

http://www.tyson.com

Tyson Foods is the world's largest poultry producer. This Web site is a great resource about chicken products for the restaurant and foodservice industry. It includes recipes and menu ideas, information on new products, and information about Tyson Foods' efficient distribution and transportation system.

APPENDIX F
SHELF LIFE OF DRIED GOODS

Source: *Food Storage Times and Temperatures* by Marl L. Tamplin, Ph.D.

Food	Recommended Maximum Storage Period if Unopened
Baking Material	
Baking powder	8 to 12 months
Baking soda	2 years
Chocolate, baking	6 to 12 months
Chocolate, sweetened	2 years
Cornstarch	2 to 3 years
Dried bread crumbs	6 months
Flour	6 to 8 months
Honey	12 months
Rice, white	2 years
Yeast, dry	18 months
Beverages	
Coffee, cans	2 years
Coffee, ground, not vacuum packed	2 weeks
Coffee, instant	8 to 12 months
Tea, bags	18 months
Tea, loose	2 years
Tea, instant	3 years
Canned Goods	
Fruit (in general)	1 year
Fruit, acidic (citrus, berries, sour cherries)	6 to 12 months
Fruit juices	9 months
Seafood (in general)	1 year
Pickled fish	4 months
Soups	1 year
Vegetables (in general)	1 year
Vegetables, acidic (tomatoes, sauerkraut)	7 to 12 months
Dairy Food	
Cheese, Parmesan (grated)	10 months
Milk, condensed	1 year
Milk, evaporated	1 year
Nondairy creamer	9 months
Fats and Oils	
Mayonnaise	2 months
Salad dressings	10 to 12 months
Salad oil	6 to 9 months
Shortening, solid	8 months

Food	Recommended Maximum Storage Period if Unopened
Grains and Grain Products	
Cereal grains for cooked cereal	6 months
Cereals, ready-to-eat	6 to 12 months
Flour, bleached	9 to 8 months
Macaroni, spaghetti, and other dry pasta	2 years
Rice, white	2 years
Rice, flavored or herb	6 months
Seasonings	
Flavoring extracts	2 years
Monosodium glutamate	Indefinite
Mustard, prepared	2 to 6 months
Salt	Indefinite
Sauces (steak, soy, etc.)	2 years
Spices and herbs (whole)	2 years to indefinite
Paprika, chili powder, cayenne	1 year
Seasoning salts	1 year
Vinegar	2 years
Sweeteners	
Sugar, granulated	2 years
Sugar, confectioners	18 months
Sugar, brown	4 months
Syrups, corn, honey, molasses, sugar	1 year
Miscellaneous	
Dried beans	1 to 2 years
Cookies, crackers	1 to 6 months
Dried fruits	6 to 8 months
Dried prunes	6 months
Gelatin	2 to 3 years
Ketchup	1 month
Jams, jellies	1 year
Nuts (whole or packaged meats)	6 months
Pickles, relishes	1 year
Potato chips	1 month

This table is a general guideline for best product quality and overall safety. Where applicable, always use any product by use-by date marked on package. If purchase date is unknown, or if quality or safety is compromised in any way, discard product.

ADDITIONAL RESOURCES

Books and Periodicals

Bendall, D. 1998. Making the most of shelving. *Restaurant Hospitality.* 82(11):146.

Food Marketing Institute. 1996. *The food keeper.* Washington, DC: Food Marketing Institute.

Lorenzini, B. 1994. Storing supplies safe and sound. *Restaurants & Institutions.* 104(14):106.

Mixon, J. A. 1991. *Guidelines for the storage and care of food products: A technical assistance manual: Vol. V.,* 3rd ed. Washington, DC: Food Industry Services Group, USDA Food and Nutrition Service.

National Restaurant Association. 1998. *Sanitation survival kit* Washington, DC: National Restaurant Association.

Tamplin, M. L. 1994. *Safe food storage times and temperatures* (Fact Sheet HE 8490). Gainesville: University of Florida. A series of the Home Economics Department, Florida Cooperative Extension Service, Institute of Food and Agricultural Sciences.

Web Site

United States Department of Agriculture (USDA)

http://www.usda.gov

The USDA touches many topic areas involved with land and its use in the United States. These areas include ensuring a safe, affordable, nutritious, and accessible food supply; caring for agriculture, forest, and range lands; providing economic opportunities for farm and rural residents; and working to reduce hunger in America and throughout the world. This Web site provides information on all of these topics and, particularly, on the safety of food from animal and plant farms.

APPENDIX G: RESPONDING TO A FOODBORNE-ILLNESS OUTBREAK

The *ServSafe Coursebook* is designed to help you take steps to ensure that the food you serve is safe. Despite your best efforts, however, a foodborne-illness outbreak can occur in your establishment at any time. How you respond when that happens can determine whether or not you end up in the middle of a crisis.

Some of the most respected companies in the industry have suffered from crises involving foodborne illness. Most companies survive, but not without a tremendous loss of both credibility and business. In some cases, crises have been severe enough to shut down businesses altogether.

Public relations experts say there are two ways to determine if you're in a crisis. First, an event occurs that could potentially threaten your business and your reputation. Next, the media hear about it.

The media's job is to look for and report news. A problem or incident in your establishment can escalate into a crisis fueled by media attention. Often, the size and duration of the crisis can be judged by the number of reporters responding to the event.

Dealing with the media may be the most difficult facet of any crisis. At the same time you are trying to determine what went wrong and how to fix the problem, the media want to know what happened, who was responsible, and what you're going to do about it.

In many cases, operators don't have answers to these questions right away, which is why crises often play out in phases, including:

○ Surprise
○ Lack of reliable information
○ Escalating flow of events
○ Loss of control
○ Intense outside scrutiny from customers and the media
○ Short-term focus
○ Panic

Averting a Crisis

You may be able to avert a crisis by responding quickly when you do receive customer complaints. Take all customer complaints seriously. Express your concern and be sincere, but do not admit responsibility or accept liability. Listen carefully and promise to investigate and respond.

Write down all facts about the incident. Record the exact food and beverages in question and the time when the customer became ill. These facts can help you identify the illness and determine whether your food is the cause. Consider developing an incident report to guide you through the process. Questions to ask include:

○ What did you eat and drink at our establishment and when?

○ When did you become ill? What were the symptoms, and how long did you experience them?

○ Did you eat anything else before or after eating at our establishment? What and where? Who else ate the same food, and did they become ill?

○ Did you seek medical attention? Where and how soon after becoming ill? What diagnosis and treatment did you receive?

Evaluate the complaint. If it is isolated, use it as an opportunity to review food safety procedures with employees. Remind them of the importance of personal hygiene. If you have a HACCP system in place, review all documentation on the date in question and verify that the system is working. Pay close attention to temperature and equipment issues.

If more than one person has complained, you have the potential for a crisis on your hands. Take steps to control the situation and reassure customers that you are doing everything you can to identify and fix the problem. If you have a crisis-management plan in place, call your crisis team together.

First, contact your local health department. Health department officials often can help you determine the cause and source of the foodborne illness. They also can act as an information resource for the media. It's important that local health officials be partners, not adversaries, while you investigate the possible cause of a foodborne illness. Therefore, make sure you develop a working relationship with them. Don't wait for a crisis to call them.

You may also want to call in outside experts. You may need help isolating the cause of the illness or dealing with the situation. Resources you can call on include food-testing labs, public relations firms or issues experts, and even your own management personnel.

Managing a Crisis

If you have received two or more complaints, it's usually only a matter of time before the media hear about it and you do have a crisis on your hands. Every crisis evolves differently, but in general, you can minimize potential damage by taking certain steps.

○ Put a team together to gather information, plan courses of action, and manage events as they unfold. In large, multiunit operations, the team may include the president and senior managers from finance, operations, marketing, franchising, human resources, public relations, and training departments. Actual crisis-management teams are usually much smaller and may vary in makeup depending on the establishment and the situation. An independent establishment's team might include the owner, general manager, and chef.

○ Appoint a single spokesperson to handle all media queries and communications. Designating a point person usually results in more consistent messages and allows you to control media access to your staff. Train the spokesperson in media relations and interview skills so he or she knows what to expect and how to respond. Crisis situations can be very stressful. Training can help your spokesperson handle these situations more easily. Make sure all of your staff knows who the spokesperson is, and instruct them to direct any and all questions to that person.

○ Work with, not against, the media. Be as proactive as you can, as early as you can. Contact the media and arrange a press conference to communicate what you know before they contact you. That way you can control what the media reports. Stick to the facts, and be as honest as possible. If you don't have all the facts, say so, and let the media know that you'll communicate them as soon as you do know. Keep a cool head and don't be defensive. The easiest way to magnify or prolong a crisis is to deny, lie, or change your story.

○ Show concern and be sincere. If health officials have confirmed that your establishment is the source of the illness, accept responsibility. Accepting responsibility is not the same as admitting liability. While customers may have become ill from eating food in your operation, the cause may have been beyond your control and not your fault. If you don't express your concern, and mean it, you'll lose credibility with the public, not just customers.

○ Develop a communication system to get information directly to all of your key audiences. Don't depend on the media to relay all the facts. Tell your side of the story to employees, customers, stockholders, and the community. Use any means possible such as a newsletter, a Web site, flyers, and newspaper or radio advertising.

○ Fix the problem and communicate what you've done both to the media and to your customers. Each time you take a step to resolve the problem, let the media know. Hold news briefings when you have news, and go into each briefing or press conference with an agenda. Take control. Don't simply respond to questions.

Crisis Preparedness

The best way to manage a crisis is to avoid having one in the first place. Just as a HACCP plan lets you monitor and correct potential problems to prevent foodborne illnesses, a crisis-management plan can avert much of the negative publicity associated with an outbreak of foodborne illness.

The process of preparing for a crisis usually decreases the chance of a crisis actually occurring. There are a number of steps you can take to prepare for the possibility of a crisis. First, assemble a team from different areas of your operation. Using this appendix as a guide to some of the issues the team should address, assign tasks and responsibilities.

○ Certify all your managers in food safety. Thoroughly train employees in good food safety practices. Implement a HACCP plan to give you greater control over food safety and create a paper trail in case something ever goes wrong.

○ Cultivate a good relationship with your local health department. Don't wait for local health officials to inspect your establishment as required by law. Work with them to develop a food safety program and invite them to monitor your progress. Hire an outside firm to inspect your establishment, using higher standards than your health department requires. If something does go wrong, you want the health department to be able to come to your defense.

○ Identify and assess all potential risks. While the greatest threat to customers may be from foodborne illnesses, don't forget that other crises can include robberies, severe weather, fire, or some other trauma.

○ Develop simple instructions on what to do in each type of crisis. In a foodborne-illness outbreak, steps include isolating the suspect food, obtaining samples of the suspect food, preventing further sale of the food, excluding suspect employees from handling food, contacting the local health department, and so forth. In each case, decide:

- Who will manage the situation
- Who will act as spokesperson
- Who you want to communicate to
- What should be said

○ Assemble a contact list of names and numbers and post it by the phones. The list should include all crisis-management team members and outside resources, such as police, fire and health departments, testing labs, issues experts, and management or headquarters personnel.

○ Train your spokesperson to be professional, truthful, understanding, and concerned when dealing with the media and others. Public relations firms or companies specializing in media training can help.

○ Develop a list of media responses or a Q&A sheet suggesting what to say in the event of each type of crisis, including foodborne illness, robbery, fire, or other disaster you have identified. Create sample press releases that can be tailored quickly to each incident.

○ Put together a list of media contacts to call for press conferences or news briefings. Include a media relations plan with do's and don'ts of dealing with the media.

○ Include instructions on how to communicate with employees, whether through shift meetings, email, a telephone tree, or some other vehicle.

○ Assemble everything in a crisis kit for the establishment. The kit can be in the form of a three-ring notebook or binder enclosing these materials. Keep the kit in an accessible place, such as the manager's or chef's office.

○ Test the plan by running a simulation. Hire a public relations or consulting firm with crisis-management experience to enact a crisis and test your team's readiness. In most cases, the firm will design a simulated crisis that will be as close as possible to what could happen in a real-life situation.

A foodborne-illness outbreak has the potential to damage your business beyond repair. Investing time and resources in a crisis management plan can insure against that. Remember these three key rules of crisis management.

1. **Take steps to prevent a crisis from occurring by practicing good food safety habits.**

2. **Prepare for the possibility of a crisis by developing contingency plans.**

3. **If a crisis does occur, take control of the situation. Use your plan to manage the crisis thoughtfully, honestly, and as quickly as possible.**

ADDITIONAL RESOURCES

Books and Periodicals

Augustine, N. R. 1995. Managing the crisis you tried to prevent. *Harvard Business Review.* 73 (6):147.

Barton, L. 1994. Crisis management: Preparing for and managing disasters. *Cornell Quarterly.* 35 (2):59.

Confronting a crisis. 1998. *Best Practices.* 2 (3):6-10.

Doeg, C. 1995. *Crisis management in the food and drinks industry: A practical approach.* In *Practical Approaches to Food Control and Food Quality: Vol. 2.* New York: Chapman & Hall.

Fitzpatrick, K. R. 1995. Ten guidelines for reducing legal risks in crisis management. *Public Relations Quarterly.* 40 (2):33.

In the event of a crisis, the crisis management team would be notified, day or night. 1990. *NCA/Club Director.* 1 (9):10.

National Restaurant Association. 1997. *Effective media training.* Washington, DC: National Restaurant Association.

Glossary

Note: The number in parentheses at the end of each entry refers to the chapter in which the term is discussed in detail.

Abrasive cleaners

Cleaners containing a scouring agent that helps scrub off hard-to-remove soils. They may scratch some surfaces. *(11)*

Acid cleaners

Acid cleaners (pH below 7.0) are used on mineral deposits and other soils that alkaline cleaners can't remove, such as scale, rust, and tarnish. *(11)*

Acidity

Level of acid in a food. An acidic substance has a pH below 7.0. Foodborne microorganisms typically do not grow in highly acidic food, while they grow best in food with a neutral to slightly acidic pH. *(2)*

Aerobic microorganisms

Microorganisms that require oxygen to grow. *(2)*

Air curtains

Devices installed above or alongside doors that blow a steady stream of air across an entryway, creating an air shield around open doors. Insects avoid them. Also called air doors or fly fans. *(12)*

Air gap

Air space used to separate a water-supply outlet from any potentially contaminated source. A properly designed and installed sink has air gaps to prevent backflow. The air space between the floor drain and drain pipe of a sink is an example. An air gap is the only completely reliable method for preventing backflow. *(10)*

Alkaline cleaners

Alkaline cleaners (pH above 7.0) are used to clean fresh soil from floors, walls, ceilings, prep surfaces, and most equipment and utensils. *(11)*

Alkalinity

Level of alkali in food. An alkaline substance has a pH above 7.0. Most food is not alkaline. *(2)*

Americans with Disabilities Act (ADA)

Federal law requiring reasonable accommodation for access to a facility by patrons and employees with disabilities. *(10)*

Anaerobic microorganisms

Microorganisms that grow only when oxygen is absent. *(2)*

Application

Applying what was learned, with feedback from the instructor/trainer. *(14)*

Backflow

Unwanted reverse flow of contaminants through a cross-connection into a potable water system. It occurs when the pressure in the potable water supply drops below the pressure of the contaminated supply. *(10)*

Bacteria

Living, single-celled microorganisms that can cause food spoilage and illness. Some form spores that can survive freezing and very high temperatures. *(2)*

Bacterial growth

Reproduction of bacteria by splitting in two. When conditions are favorable, bacterial growth can be rapid—doubling the population as often as every twenty minutes. Their growth can be broken down into four phases: lag phase, log phase, stationary phase, and death phase. *(2)*

Bimetallic stemmed thermometer

The most common and versatile type of thermometer, measuring temperature through a metal probe with a sensor in the end. Most can measure temperatures from 0°F to 220°F (-18°C to 104°C) and are accurate to within ±2°F (±1°C). They are easily calibrated. *(5)*

Biological contaminant

Microbial contaminant that can cause foodborne illness. These contaminants include bacteria, viruses, parasites, fungi, and biological toxins. Many are destroyed by cooking. *(2)*

Biological toxins

Poisons produced by pathogens, plants, or animals. They can also occur in animals as a result of their diet. *(3)*

Blast chiller

Equipment designed to cool food quickly. Many are able to cool food from 140°F to 37°F (60°C to 3°C) within ninety minutes. *(10)*

Boiling-point method

Method of calibrating a thermometer based on the boiling point of water. *(5)*

Booster heater

Water heater attached to hot-water lines leading to warewashing machines or sinks. Raises water to temperatures required for heat sanitizing of tableware and utensils (180°F [82°C]). *(11)*

Calibration

Process of ensuring that a thermometer gives accurate readings by adjusting it to a known standard, such as the freezing point or boiling point of water. *(5)*

Cantilever-mounted

Equipment that is attached to a wall with a bracket, allowing easier cleaning behind and underneath it. *(10)*

Carrier

Person who carries pathogens and infects others, yet never becomes ill himself. *(4)*

Centers for Disease Control and Prevention (CDC)

Agency of the U.S. Public Health Service that investigates foodborne-illness outbreaks, studies the causes and control of disease, publishes statistical data, and conducts the Vessel Sanitation Program. *(13)*

Ceramic tile

Hard, nonresilient, nonporous tile; commonly used in restrooms and high soil areas. *(10)*

Chemical contaminants

Chemical substances that can cause foodborne illness, including toxic metals, pesticides, and improperly used cleaning products, sanitizers, and lubricants. *(3)*

Chemical sanitizing

Using a chemical solution to reduce the number of microorganisms on a clean surface to safe levels. Items can be sanitized by immersing them in a specific concentration of sanitizing solution for a required period of time, or by rinsing, swabbing, or spraying the items with a specific concentration of sanitizing solution. *(11)*

Chemical storage

Cleaning tools and chemicals should be kept in their own storage area. Store these supplies away from food in clean, dry rooms or in cabinets that can be locked. *(6)*

Chlorine

A commonly used chemical sanitizer, due to its low cost and effectiveness. It kills a wide range of microorganisms. *(11)*

Ciguatera poisoning

Illness that occurs when a person eats fish that has consumed the ciguatera toxin. This toxin occurs in certain predatory tropical reef fish, such as amberjack, barracuda, groupers, and snapper. *(3)*

Clean

Free of visible soil. It refers only to the appearance of a surface. *(11)*

Cleaning

Process of removing food and other types of soil from a surface. *(11)*

Cleaning agents

Chemical compounds that remove food, soil, rust stains, minerals, or other deposits from surfaces. *(11)*

Cold-holding equipment

Equipment specifically designed to keep cold food at an internal temperature of 41°F (5°C) or lower. *(8)*

Cold paddle

Plastic paddle that can be filled with water and frozen. When used to stir hot food, it cools the food quickly. *(7)*

Contact spray

Spray used to kill insects on contact. They are usually used on groups of insects, such as clusters of roaches and a nest of ants. *(12)*

Contamination

Presence of harmful substances in food. Some food safety hazards occur naturally, while others are introduced by humans or the environment. *(1)*

Control point (CP)

Any step in a food's flow where a physical, chemical, or biological hazard can be controlled. *(9)*

Corrective action

Predetermined step taken when food doesn't meet a critical limit. *(9)*

Coving

Curved, sealed edge placed between the floor and wall to eliminate sharp corners or gaps that would be impossible to clean. Coving also eliminates hiding places for pests and prevents moisture from deteriorating walls. *(10)*

Critical control point (CCP)

In a HACCP system, the last step where you can intervene to prevent, eliminate, or reduce the growth of microorganisms before food is served to customers. *(9)*

Critical limit

In a HACCP system, the minimum and maximum limit a critical control point (CCP) must meet in order to prevent, eliminate, or reduce a hazard to an acceptable level. *(9)*

Cross-connection

A physical link through which contaminants from drains, sewers, or other wastewater sources can enter a potable water supply. A hose connected to a faucet and submerged in a mop bucket is an example. *(10)*

Cross-contamination

The transfer of microorganisms from one surface or food to another. *(1)*

Death phase

The phase in bacterial growth in which the number of bacteria dying exceeds the number growing, resulting in a population decline. *(2)*

Demonstration

Process of illustrating a skill or task before another person or a group. *(14)*

Detergent

Cleaning agent designed to penetrate and soften soil to help remove it from a surface. *(11)*

Dry storage

Storage used to hold dry and canned food at temperatures between 50°F and 70°F (10°C and 21°C) and at a relative humidity of 50 to 60 percent. *(6)*

Electronic insect eliminator ("zapper")

Mechanical device that uses light to attract flying insects to an an electrically charged grid which kills them. *(12)*

Environmental Protection Agency (EPA)

Federal agency that sets standards for environmental quality—including air and water quality—and regulates pesticide use and waste handling. *(10)*

Evaluation

Judging the performance of training participants against learning objectives. *(14)*

Facultative microorganisms

Microorganisms that grow with or without the presence of oxygen. Most bacteria that cause foodborne illness are facultative. *(2)*

FAT TOM

Acronym for the conditions needed by most foodborne microorganisms to grow: Food, Acidity, Time, Temperature, Oxygen, Moisture. *(2)*

FDA Food Code

Science-based reference for retail food establishments on how to prevent foodborne illness. These recommendations are written by the FDA to assist state health departments in developing regulations for a foodservice inspection program. *(13)*

Feedback

Evaluation given to employees about their performance, including constructive criticism given to correct a mistake, or praise to reinforce proper performance of a skill or procedure. *(14)*

Finger cot

Protective covering used to cover a properly bandaged cut or wound on the finger. *(4)*

First in, first out (FIFO)

Method of stock rotation in which products are shelved based on their use-by or expiration dates, so oldest products are used first. Products with the earliest use-by or expiration dates are stored in front of products with later dates. *(6)*

Flood rim

Spillover point on a sink. *(10)*

Flow of food

Path food takes through an establishment, from receiving and storage, through preparation and cooking, holding, serving, cooling, and reheating. *(1)*

Food allergy

The body's negative reaction to a particular food protein. *(3)*

Food and Drug Administration (FDA)

The federal agency that writes the Food Code. The FDA also inspects operations that cross state borders (interstate establishments such as food manufacturers and processors, and planes and trains). In addition, the FDA shares responsibility with the USDA for inspecting food-processing plants. *(13)*

Food bar

Self-service buffet at which patrons can choose what they want to eat as they serve themselves. *(8)*

Food-contact surface

Surfaces that come into direct contact with food, such as utensils. *(1)*

Food-grade sealant

Nontoxic sealant used to seal equipment to a countertop or a masonry base. *(10)*

Food irradiation

Process of exposing food to an electron beam or gamma rays to reduce pathogenic and spoilage microorganisms. Also known as cold pasteurization. *(2)*

Food quality

Proper appearance, flavor, texture, consistency, and nutritional value in food. Food that is shipped, received, stored, prepared, and served properly is more likely to have high quality. *(1)*

Foodborne illness

Disease carried or transmitted to people by food. *(1)*

Foodborne-illness outbreak

According to the Centers for Disease Control and Prevention (CDC), an incident in which two or more people experience the same illness after eating the same food. *(1)*

Foodborne infection

Result of a person eating food containing pathogens, which then grow in the intestines and cause illness. Typically, symptoms of a foodborne infection do not appear immediately. *(2)*

Foodborne intoxication

Result of a person eating food containing toxins that cause an illness. The toxins may have been produced by pathogens found on the food or may be the result of a chemical contamination. The toxins could also come from a plant or animal that was eaten. Typically, symptoms of foodborne intoxication appear quickly, within a few hours. *(2)*

Foodborne toxin-mediated infection

Result of a person eating food containing pathogens, which then produce illness-causing toxins in the intestines. *(2)*

Food Safety and Inspection Service (FSIS)
Agency of the USDA that inspects and grades meat, meat products, poultry, dairy products, egg and egg products, and fruit and vegetables shipped across state boundaries. *(13)*

Foot-candle
Unit of lighting equal to the illumination one foot from a uniform light source. *(10)*

Four-hour rule
Potentially hazardous food must not be exposed to the temperature danger zone for more than four hours. The exposure time accumulates during each stage of handling, from the time food arrives at the receiving dock to the time it is cooked. Exposure time begins again when the food is held for service, cooled, and reheated. *(7)*

Frozen storage
Storage typically designed to hold food at 0°F (-18°C) or lower. Some food requires a different temperature. *(6)*

Fungi
Ranging in size from microscopic, single-celled organisms to very large, multicellular organisms, fungi most often cause food spoilage. Molds, yeasts, and mushrooms are examples of fungi. *(2)*

Garbage
Wet waste matter, usually containing food, that cannot be recycled. It attracts pests and has the potential to contaminate food, equipment, and utensils. *(10)*

Gastrointestinal illness
Illness relating to the stomach or intestine. *(4)*

Glue board
Pest-control device in which mice are trapped by glue and then die from exhaustion or lack of water or air. They are also used to identify the type of cockroaches that might be present. *(12)*

HACCP plan
Written document based on HACCP principles describing procedures a particular establishment will follow to ensure the safety of food served. *(9)*

Hair restraint
Device used to keep a foodhandler's hair away from food and to keep them from touching it. *(4)*

Hand sanitizer
A liquid used to lower the number of microorganisms on the skin surface. Hand sanitizers should be used after proper handwashing, not in place of it. *(4)*

Handwashing station
Sink designated for handwashing only. Handwashing stations must be conveniently located in restrooms, food-preparation areas, service areas, and warewashing areas. *(10)*

Harborage
A shelter for pests. *(12)*

Hard water
Water containing minerals such as calcium and iron in concentrations higher than 120 parts per million (ppm). *(11)*

Hazard analysis
The process of identifying and evaluating potential hazards associated with food in order to determine what must be addressed in the HACCP plan. *(9)*

Hazard Analysis Critical Control Point (HACCP)

Food safety system designed to keep food safe throughout its flow through an establishment. HACCP is based on the idea that if hazards are identified at specific points in a food's flow, the hazards can be prevented, eliminated, or reduced to safe levels. *(9)*

Hazard Communication Standard (HCS)

OSHA standard also known as Right-to-Know or HAZCOM requiring employers to tell their employees about chemical hazards to which they may be exposed at the establishment. It also requires employers to train employees on how to use the chemicals they work with safely. (*11*)

Hazards, biological

Pathogenic microorganisms that can contaminate food, such as certain bacteria, viruses, parasites, and fungi; also exist in certain plants, mushrooms, and fish in the form of harmful toxins. *(1)*

Hazards, chemical

Chemical substances that can contaminate food, such as pesticides, food additives, preservatives, cleaning supplies, and toxic metals that are leached from cookware and equipment. *(1)*

Hazards, physical

Foreign objects that can accidentally get into food and contaminate it, such as dirt, metal staples, and broken glass. *(1)*

Health inspectors

City, county, or state employees who conduct foodservice inspections in most states. They generally are trained in food safety, sanitation, and public health principles and methods. Also called sanitarians, health officials, or environmental health specialists. *(13)*

Heat sanitizing

Using heat to reduce the number of microorganisms on a clean surface to safe levels. The most common way to heat sanitize tableware, utensils, or equipment is to submerge them in or spray them with hot water. *(11)*

Hepatitis A

Disease causing inflammation of the liver. It is transmitted to food by poor personal hygiene or contact with contaminated water. *(4)*

High-risk population

People susceptible to foodborne illness due to age or health status, such as infants and preschool-age children, pregnant women, older people, people taking certain medications, and those with certain diseases or weakened immune systems. *(1)*

Histamine

Biological toxin associated with temperature-abused scombroid fish that causes scombroid poisoning. *(3)*

Host

Person, animal, or plant on or in which another organism lives and takes nourishment. *(2)*

Hot-holding equipment

Equipment such as chafing dishes, steam tables, and heated cabinets specifically designed to hold potentially hazardous food at 140°F (60°C) or higher. *(8)*

Hygrometer

Instrument used to measure relative humidity in storage areas. *(6)*

Ice-point method

Method of calibrating thermometers based on the freezing-point of water. *(5)*

Ice-water bath

A method of cooling food in which a container holding hot food is placed into a larger container of ice water. The ice water surrounding the hot food container disperses the heat quickly. *(7)*

Immune system

The body's defense system against illness. People with compromised immune systems are more susceptible to foodborne illness. *(1)*

Infected lesion

Wound or injury contaminated with a pathogen. *(4)*

Infestation

Situation that exists when pests overrun or inhabit an establishment in large numbers. *(12)*

Inspection

Process of making sure food deliveries meet your food safety standards, including proper temperature, appearance, and packaging. *(5)*

Integrated pest management (IPM)

Program using prevention measures to keep pests from entering an establishment and control measures to eliminate any pests that do get inside. *(12)*

Interstate establishments

Establishments that operate across state borders. Examples include foodservice operations on trains, planes, and ships. *(13)*

Iodine

Sanitizer effective at low concentrations and not as quickly inactivated by soil as chlorine. It might stain surfaces and is less effective than chlorine. *(11)*

Jaundice

A yellowing of the skin and eyes that could indicate a person is ill with hepatitis A. *(4)*

Job aids

Materials or visual reminders used to deliver training content to employees. *(14)*

Lag phase

The phase in bacterial growth in which bacteria are first introduced to a new environment. In this phase, bacteria go through an adjustment period in which their numbers are stable as they prepare to grow. To control the growth of bacteria, prolong the lag phase as long as possible. *(2)*

Lecture

Prepared oral presentation used to deliver content to a group. *(14)*

Log phase

The phase in bacterial growth in which conditions are favorable for bacteria to multiply very rapidly. Food quickly becomes unsafe during this phase. *(2)*

Master cleaning schedule
Detailed schedule that lists all cleaning tasks in an establishment, when and how they are to be performed, and who will do the cleaning. *(11)*

Material Safety Data Sheets (MSDS)
Sheets supplied by the chemical manufacturer listing the chemical and its common names, its potential physical and health hazards, information about using and handling it safely, and other important information. OSHA requires employers to store these sheets so they are accessible to employees. *(11)*

Microorganisms
Small, living organisms that can be seen only with the aid of a microscope. Four types of microorganisms with the potential to contaminate food and cause foodborne illness are bacteria, viruses, parasites, and fungi. *(2)*

Minimum internal cooking temperature
Required cooking temperature the internal portion of food must reach—specific to the type of food being cooked—in order to sufficiently reduce the number of illness-causing microorganisms that might be present. *(7)*

Mobile units
Portable foodservice facilities, ranging from concession vans to full field kitchens capable of preparing and cooking elaborate meals. *(8)*

Modified atmosphere packaging food (MAP)
Packaging process by which air is removed from a food package and replaced with gases, such as carbon dioxide and nitrogen, to help extend the product's shelf life. *(5)*

Mold
Type of fungus that causes food spoilage. Some produce toxins that can cause foodborne illness. *(2)*

Monitoring
In a HACCP system, the process of analyzing whether critical limits are being met and things are being done right. *(9)*

National Marine Fisheries Service (NMFS)
Agency of the U.S. Department of Commerce that provides a voluntary inspection program that includes product standards and sanitary requirements for fish processing operations. *(13)*

NSF International
Organization that develops and publishes standards for sanitary equipment design. They also assess and certify that equipment has met these standards. Restaurant and foodservice managers should look for an NSF International mark (or UL EPH product mark) on commercial foodservice equipment. *(10)*

Objective
Statement of what a trainee will be able to do after training or instruction is completed. *(14)*

Off-site service
Service of food to someplace other than where it is prepared or cooked, including catering and vending. *(8)*

Occupational Safety and Health Administration (OSHA)

Federal agency that regulates and monitors workplace safety. *(11)*

Parasite

Organism that needs to live in or on a host organism to survive. Parasites can live in many animals that humans use for food, including cows, chickens, pigs, and fish. *(2)*

Paralytic shellfish poisoning (PSP)

Illness caused by shellfish containing toxins that occur because of algae upon which they feed. Generally associated with mussels, clams, cockles, and scallops. *(3)*

Pathogens

Disease-causing microorganisms. *(1)*

Personal hygiene

Sanitary health habits that include keeping body, hair, and teeth clean, maintaining good health, wearing clean clothes, and washing hands regularly, especially when handling food and beverages. *(4)*

Pest control operator (PCO)

Licensed professional who uses safe, up-to-date methods to prevent and control pests. *(12)*

Pesticide

Chemical used to control pests, usually insects. *(12)*

pH

Measure of a food's acidity or alkalinity. The pH scale ranges from 0 to 14.0. A pH above 7.0 is alkaline, while a pH below 7.0 is acidic. A pH of 7.0 is neutral. Pathogenic bacteria grow well in food with a pH between 4.6 and 7.5 (slightly acidic to neutral). *(2)*

Physical contaminants

Foreign objects that accidentally find their way into food, such as dirt, glass, fingernails, or metal. Some physical hazards occur naturally, such as bones in fish or chicken. *(3)*

Plant toxins

Poisons found naturally in some plants. *(3)*

Pooled eggs

Eggs that have been cracked open and combined in a common container. *(7)*

Porosity

Extent to which water and other liquids are absorbed by a substance. Term usually used in relation to flooring material. *(10)*

Potable water

Water that is safe to drink or use as an ingredient in food. *(10)*

Potentially hazardous food

Food in which microorganisms can grow rapidly. Potentially hazardous food has a history of being involved in foodborne-illness outbreaks, has potential for contamination due to production and processing methods, and has characteristics that generally allow microorganisms to grow rapidly. Potentially hazardous food is often moist, high in protein, and has a neutral or slightly acidic pH. *(1)*

Prerequisite programs

The programs or standard operating procedures in an establishment designed to protect food from contamination, minimize microbial growth, and ensure the proper functioning of equipment. *(9)*

Pulper

Device used to grind food and other waste into small parts that are flushed with water, which is then removed. The processed solid wastes weigh less and are more compact for easier disposal. *(10)*

Quarry tile

Type of stone tile, generally reddish-brown in color, often used in public restrooms or high-soil areas. *(10)*

Quaternary ammonium compounds (quats)

Group of sanitizers all having the same basic chemical structure. They work in most temperature and pH ranges, are noncorrosive, and remain active for short periods of time after they have dried. However, quats may not kill certain types of microorganisms, and they leave a film on surfaces. *(11)*

Ready-to-eat food

Properly cooked food, as well as raw, washed whole or cut fruit and vegetables (including those that have had their rinds, peels, husks, or shells removed). *(1)*

Reasonable care defense

Defense against a food-related lawsuit stating that an establishment did everything that could be reasonably expected to ensure that the food served was safe. *(1)*

Receiving

Process of taking delivered food into your operation. Receiving includes unloading the supplier's truck, inspecting products, accepting or rejecting items, labeling, and storing items in a timely manner. *(5)*

Record keeping

In a HACCP system, the process of collecting documents that allow you to document that you are continuously preparing and serving safe food. *(9)*

Refrigerated storage

Storage used for holding potentially hazardous food at an internal temperature of 41°F (5°C) or lower. Some jurisdictions allow food in refrigerators to be held at an internal temperature of 45°F (7°C) or lower. Check with the local regulatory agency for specific regulations. *(6)*

Residual spray

Type of pesticide spray that leaves behind a film that insects absorb as they crawl across it. Used in cracks and crevices like those along baseboards, these sprays can can be liquid or a dust, such as boric acid. *(12)*

Resiliency

Ability of a surface to react to a shock without breaking or cracking, usually used in relation to a flooring material. *(10)*

Role-play

Training method in which trainees enact a situation in order to try out new skills or apply new knowledge. *(14)*

Sanitary

State that exists when the number of pathogens on a clean surface have been reduced to safe levels. *(11)*

Sanitizer

Compound used to reduce the number of pathogens on a clean surface to safe levels. *(11)*

Sanitizing

Process of reducing the number of microorganisms on a clean surface to safe levels. *(11)*

Scombroid poisoning

Illness that occurs when a person eats a scombroid fish that has been time-temperature abused. Scombroid fish include tuna, mackerel, bluefish, skipjack, and bonito. *(3)*

Service sink

Sink used exclusively for cleaning mops and disposing of waste water. At least one service sink or one curbed drain area is required in an establishment. *(10)*

Shelf life

Recommended period of time food may be stored and remain suitable for use. *(6)*

Shellstock tags

Identification tag that accompanies each container of live molluscan shellfish, on which the delivery date must be written. Tags are to be kept on file for ninety days after the last shellfish is used. *(5)*

Single-use gloves

Disposable gloves designed for one-time use which provide a barrier between hands and the food they come in contact with. *(4)*

Single-use item

Disposable tableware or packaged food designed to be used once only, including plastic flatware, paper or plastic cups, plates, and bowls, as well as single-serve food and beverages. *(8)*

Single-use paper towel

Paper towels designed to be used once, then discarded. *(4)*

Slacking

Process of gradually thawing frozen food in preparation for deep frying. *(7)*

Sneeze guard

Food shield placed in a direct line between food on display and the mouth and nose of a person of average height, usually placed fourteen to forty-eight inches above the food. *(8)*

Solvent cleaners

Alkaline detergents, often called degreasers, that contain a grease-dissolving agent. *(11)*

***Sous vide* food**

Food vacuum-packed in individual pouches, partially or fully cooked, and then chilled. This food is often heated for service in the establishment. *(5)*

Spoilage microorganism

Foodborne microorganisms that cause food to spoil, but typically do not cause foodborne illness. *(2)*

Spore

Alternative form for some bacteria, with a thick wall to protect it from adverse conditions, such as high and low temperatures, low moisture, and high acidity. Capable of turning back into a vegetative microorganism when conditions again become favorable. *(2)*

Stationary phase

The phase of bacterial growth in which just as many bacteria are growing as are dying. Follows the log phase of bacterial growth. *(2)*

Surfactants

Agents in all detergents that reduce surface tension between the soil and the surface of the item being cleaned. Surfactants make it easier for the detergent to penetrate and soften the soil. *(11)*

Technology-based training

Training programs delivered via a computer or other technology. *(14)*

Temperature danger zone

Temperature range between 41°F and 140°F (5°C to 60°C) within which most foodborne microorganisms rapidly grow and reproduce. *(2)*

Temporary unit

Establishment operating in one location for no more than fourteen consecutive days in conjunction with a special event or celebration. They usually serve prepackaged food or food requiring only limited preparation. *(8)*

Terrazzo

Mixture of marble chips and portland cement used for flooring. An attractive, nonporous, nonresilient surface good for use in the back of the house, dining rooms, and restrooms. *(10)*

Thermometer

Device for accurately measuring the internal temperature of food, the air temperature inside a freezer or cooler, or the temperature of equipment. *(5)*

Time-temperature abuse

Allowing food to remain too long at temperatures favorable to the growth of foodborne microorganisms. *(1)*

Time-temperature indicator (TTI)

Time and temperature monitoring device attached to a food shipment to determine if the product's temperature has exceeded safe limits during shipment or subsequent storage. *(5)*

Toxic metal poisoning

Illness caused when toxic metals are leached from utensils or equipment containing them. *(3)*

Toxins

Poisons produced by pathogens, plants, or animals. Most occur naturally and are not caused by the presence of microorganisms. Some occur in animals as a result of their diet. Many chemicals are also toxic. *(3)*

Training need

Gap between what employees are required to know to perform their jobs and what they actually know. *(14)*

Training program

Structured sequence of events that leads to learning. *(14)*

Tumble chiller

Equipment designed to cool food quickly. Prepackaged hot food is placed into a drum rotating inside a reservoir of chilled water. The tumbling action increases the effectiveness of the chilled water in cooling the food. *(10)*

Two-stage cooling

Criteria by which cooked food is cooled from 140°F (60°C) to 70°F (21°C) within two hours and from 70°F (21°C) to 41°F (5°C) or lower in an additional four hours, for a total cooling time of six hours. *(7)*

Ultra-high temperature (UHT) food

Food that is heat-treated at a very high temperature for a short time to kill microorganisms (pasteurized), then packaged under sterile conditions. *(5)*

Underwriters Laboratories (UL)

Provides sanitation classification listings for equipment found in compliance with NSF International standards. Also lists products complying with their own published environmental and public health standards. *(10)*

U.S. Department of Agriculture (USDA)

Federal agency responsible for the inspection and quality grading of meat, meat products, poultry, dairy products, eggs and egg products, and fruit and vegetables shipped across state lines. *(13)*

U.S. Public Health Service (USPHS)

Federal agency that inspects international cruise ships. *(13)*

Vacuum breaker

Device used for preventing the backflow of contaminants into a potable water system. *(10)*

Vacuum-packed food

The process of removing air from around a food product sealed in a package. This process increases the products shelf life. *(5)*

Vegetative microorganisms

Bacteria in the process of reproducing (growing) by splitting in two. *(2)*

Vending machine

Machines dispensing hot and cold food, beverages, and snacks. *(8)*

Verification

In a HACCP system, the process of confirming that critical control points and critical limits are appropriate, that monitoring is alerting you to hazards, that corrective actions are adequate to prevent foodborne illness from occurring, and that employees are following established procedures. *(9)*

Virus

Smallest of the microbial food contaminants, viruses rely on a living host to reproduce. Some survive freezing and cooking temperatures. They usually contaminate food through a foodhandler's improper personal hygiene. *(2)*

Warranty of sale

Rules stating how food must be handled in an establishment. *(1)*

Water activity (a$_w$)

Amount of moisture available in food for microorganisms to grow. Potentially hazardous food items have water-activity values of 0.85 or above. *(2)*

Yeast

Type of fungus that causes food spoilage. *(2)*

Index

abrasive cleaner, 11-4
acid cleaner, 11-4
acidity, 2-3, 2-5
aerobic microorganism, 2-6
Agriculture, U.S. Department of: *see* U.S. Department of Agriculture
air curtain, 12-3
air gap, 10-19, 10-20
air probe, 5-6
alkalinity, 11-4, 11-7
allergies: *see* food allergy
American Academy of Sanitarians (AAS), 13-15
Americans with Disabilities Act (ADA), 4-3, 10-2
anaerobic microorganism, 2-6, 5-20
anisakiasis, 2-17
ants, 12-8, 12-13
aseptically packaged food, 5-22, 6-11
audiovisual equipment, 14-12

backflow, 3-8, 10-19
Bacillus cereus gastroenteritis, 2-11
bacteria
    characteristics of, 2-2

growth and reproduction, 2-3
illnesses caused by, 2-10, 2-11, 2-12, 2-13
spore formation, 2-4
vegetative stages, 2-4
bandages, 4-11
bats, 12-10
batters, 7-9, 7-10
beef: *see* meat
bees, 12-8
beetles, 12-8
behavioral objectives, 14-5, 14-6, 14-14, 14-15
bimetallic stemmed thermometer, 5-5, 5-6
biological contaminants, 1-8, 3-2, 3-3, 3-4, 3-6
biological toxins, 3-2, 3-3, 3-4, 3-6
birds, 12-10, 12-15
blast chiller, 10-13
boiling-point method, 5-9
booster heater, 10-18
botulism, 2-12, 5-22
breading, 7-9, 7-10
brooms, 11-15
brushes, 11-15
buffets, 8-7, 10-21
butter, 5-18

calibration, 5-5, 5-6, 5-8, 5-9
campylobacteriosis, 2-12
canned products, 5-21, 5-22, 6-11
cantilever-mounted equipment, 10-16
carbonated-beverage dispensers, 3-5
carousel machine, 10-10
carpenter ants, 12-8
carpeting, 10-4, 10-5
carrier, 4-2
casseroles, 7-13
catering, 8-10
CCP: *see* critical control point
CDC: *see* Centers for Disease Control and Prevention
ceilings, 10-6, 11-13
Centers for Disease Control and Prevention (CDC), 13-13
ceramic tile, 10-4, 10-6
certification, food safety, 14-15
cheese, 2-19, 5-18
chemical bait, 12-13, 12-14
chemical contaminant, 1-8
    chemicals, 3-7, 3-8
    pesticides, 3-7, 3-8, 12-13
    toxic metals, 3-5, 3-8
chemical sanitizing, 11-4, 11-8, 11-9, 11-10, 11-18
chemical storage, 6-4, 6-12
chemical toxins, 3-5, 3-7, 3-8
chicken: *see* poultry

T

More ServSafe® tools to help you train

SERVSAFE® MANAGER TOOLKIT

This training aid is designed to teach managers how to train employees, introduce various energizing training techniques, walk managers through a step-by-step training plan and simplify the training process, by offering flexibility and variety.

Complete ServSafe Manager Toolkit

English	SSMT
Spanish	SSMTS

Toolkit includes a convenient storage case filled with:

The Manager's Guide to Employee-Level Training

Handy resource that takes managers step-by-step through the training process—from developing a training plan, training methods, follow-up tips, to a special section on how to develop a successful peer-training program.

To order additional quantities, use the product codes below.

English	MGET
Spanish	MGETS

ServSafe Employee Guide, 10-Pack

Newly enhanced guide that focuses on the basics of food safety. Includes activities, exercises, quizzes and a certificate of completion.

To order additional quantities, use the product codes below.

English	10-pack	SSEGR10
Spanish	10-pack	SSEGRS10

ServSafe Poster Package

Three helpful tools to reinforce learning through manager and employee interaction:

Posters–set of 12, each featuring a different food safety topic

Quiz Sheets–10 copies, including set of masters, of 12 different quizzes that relate to the poster topics

Activity Booklet–additional learning activities and exercises to facilitate with employees

To order additional quantities, use the product codes below.

English	SSPP
Spanish	SSPPS

ServSafe Manager Toolkit Implementation Sheet

Key information that addresses how to use the kit components together for effective training.

your employees in food safety.

SERVSAFE STEPS TO FOOD SAFETY SIX-VIDEO SET

Walk your staff through the steps to food safety with realistic scenarios in multiple industry segments, then challenge them to apply what they've learned. English and Spanish are on the same tape, with accompanying video guide.

STEPSET

SERVSAFE STEPS TO FOOD SAFETY SIX-IN-ONE VIDEO

All six Steps to Food Safety videos on one tape. Includes set of two—English and Spanish on separate tapes due to running times.

SSONE

SERVSAFE INTRODUCTION TO FOOD SAFETY CD-ROM

Provide both new and experienced employees with essential food safety information through a consistent and convenient interactive format. Includes printable certificate of completion and English and Spanish on the same disc.

CDR60

Visit our Web site for more information at www.nraef.org.